The Elements
of
Chemical Kinetics
and
Reactor Calculations
(A Self-Paced Approach)

PRENTICE-HALL INTERNATIONAL SERIES
IN THE PHYSICAL AND CHEMICAL ENGINEERING SCIENCES

NEAL R. AMUNDSON, EDITOR, *University of Minnesota*

ADVISORY EDITORS

ANDREAS ACRIVOS, *Stanford University*
JOHN DAHLER, *University of Minnesota*
THOMAS J. HANRATTY, *University of Illinois*
JOHN M. PRAUSNITZ, *University of California*
L. E. SCRIVEN, *University of Minnesota*

AMUNDSON *Mathematical Methods in Chemical Engineering: Matrices and Their Application*
AMUNDSON AND ARIS *Mathematical Methods in Chemical Engineering: Vol. II, First Order Partial Differential Equations with Applications*
ARIS *Elementary Chemical Reactor Analysis*
ARIS *Introduction to the Analysis of Chemical Reactors*
ARIS *Vectors, Tensors, and the Basic Equations of Fluid Mechanics*
BALZHISER, SAMUELS, AND ELIASSEN *Chemical Engineering Thermodynamics*
BOUDART *Kinetics of Chemical Processes*
BRIAN *Staged Cascades in Chemical Processing*
CROWE et al. *Chemical Plant Simulation*
DOUGLAS *Process Dynamics and Control: Vol. I, Analysis of Dynamic Systems*
DOUGLAS *Process Dynamics and Control: Vol. II, Control System Synthesis*
FOGLER *The Elements of Chemical Kinetics and Reactor Calculations*
FREDRICKSON *Principles and Applications of Rheology*
FRIEDLY *Dynamic Behavior of Processes*
HAPPEL AND BRENNER *Low Reynolds Number Hydrodynamics with Special Applications to Particulate Media*
HIMMELBLAU *Basic Principles and Calculations in Chemical Engineering*, 3rd edition
HOLLAND *Fundamentals and Modeling of Separation Processes*
HOLLAND *Multicomponent Distillation*
HOLLAND *Unsteady State Processes with Applications in Multicomponent Distillation*
KOPPEL *Introduction to Control Theory with Applications to Process Control*
LEVICH *Physicochemical Hydrodynamics*
MEISSNER *Processes and Systems in Industrial Chemistry*
MODELL AND REID *Thermodynamics and Its Applications in Chemical Engineering*
NEWMAN *Electrochemical Systems*
OHARA AND REID *Modeling Crystal Growth Rates from Solution*
PERLMUTTER *Stability of Chemical Reactors*
PETERSEN *Chemical Reaction Analysis*
PRAUSNITZ *Molecular Thermodynamics of Fluid-Phase Equilibria*
PRAUSNITZ AND CHUEH *Computer Calculations for High-Pressure Vapor-Liquid Equilibria*
PRAUSNITZ, ECKERT, ORYE, O'CONNELL *Computer Calculations for Multicomponent Vapor-Liquid Equilibria*
RUDD, et al. *Process Synthesis*
SCHULTZ *Polymer Materials Science*
SEINFELD AND LAPIDUS *Mathematical Methods in Chemical Engineering: Vol. III, Process Modeling, Estimation, and Identification*
WILDE *Optimum Seeking Methods*
WHITAKER *Introduction to Fluid Mechanics*
WILLIAMS *Polymer Science and Engineering*

PRENTICE-HALL, INC.
PRENTICE-HALL INTERNATIONAL, INC.,
UNITED KINGDOM AND EIRE
PRENTICE-HALL OF CANADA, LTD., CANADA

The Elements of Chemical Kinetics and Reactor Calculations
(A Self-Paced Approach)

H. SCOTT FOGLER

Department of Chemical Engineering
The University of Michigan

PRENTICE-HALL INC., Englewood Cliffs, New Jersey

Library of Congress Cataloging in Publication Data

FOGLER, H. SCOTT.
   The elements of chemical kinetics and reactor calculations.

   (Prentice-Hall international series in the physical and chemical engineering sciences)
   Includes bibliographical references.
   1. Chemical engineering—Programmed instruction.
   2. Chemical reaction, Rate of—Programmed instruction.
   I. Title.
TP155.F658   660.2'8'0077   73-14611
ISBN 0-13-263442-2

© 1974 by
PRENTICE-HALL, Inc.
Englewood Cliffs, New Jersey

All rights reserved. No part of this book may be reproduced in any form or by any means without permission in writing from the publisher.

10 9 8 7 6 5 4 3 2 1

Printed in the United States of America

PRENTICE-HALL INTERNATIONAL, INC., London
PRENTICE-HALL OF AUSTRALIA, PTY. LTD., Sydney
PRENTICE-HALL OF CANADA, LTD., Toronto
PRENTICE-HALL OF INDIA PRIVATE LTD., New Delhi
PRENTICE-HALL OF JAPAN, INC., Tokyo

To Janet

Contents

Preface — xiii

Introduction — 1

1

Mass and Mole Balances — 7

1.1 DEFINITION OF THE RATE OF REACTION, $-r_A$ 9
1.2 THE GENERAL BALANCE EQUATION   13
1.3 BATCH REACTORS   17
1.4 CONTINUOUS FLOW REACTORS   21
     1.4A Continuous Stirred Tank Reactor (CSTR), 21
     1.4B Tubular Reactor, 23
1.5 INDUSTRIAL REACTORS   25

2

Stoichiometric Relationships and Reactor Staging — 35

2.1 STOICHIOMETRIC TABLE   35
     2.1A Batch Systems, 37
     2.1B Flow Systems, 41
2.2 DESIGN EQUATIONS (MOLE BALANCES)   45
     2.2A Batch, 45
     2.2B CSTR (Backmix Reactor), 47
     2.2C Tubular Flow Reactor, 49
2.3 PROBLEMS   49
     2.3A Stoichiometric Table, 49
     2.3B Applications of the Design Equations, 51
     2.3C Reactor Staging, 55

3

Elementary Forms of the Rate Law 73

- 3.1 BASIC DEFINITIONS 75
- 3.2 VOLUME CHANGE WITH REACTION 81
- 3.3 EXPRESSING $-r_A$ AS A FUNCTION X 85
  - 3.3A Irreversible Reactions, 85
  - 3.3B Reversible Reactions, 91
  - 3.3C Nonelementary Reactions, 93
  - 3.3D Reactions with Phase Change, 95

4

Elementary Reactor Design 103

- 4.1 TOTAL PRESSURE VARIATIONS IN BATCH REACTORS 103
- 4.2 DECOMPOSITION OF 1, 1 DCE EXAMPLE PROBLEM 107
  - 4.2A Batch Data, 109
  - 4.2B CSTR Design, 111
- 4.3 DECOMPOSITION OF DIHYDROFURAN EXAMPLE PROBLEM 113
  - 4.3A Batch Data, 113
  - 4.3B Tubular Flow Design, 115
- 4.4 NITRIC OXIDE EXAMPLE PROBLEM 117
  - 4.4A Role of Nitric Oxide in Smog Formation, 117
  - 4.4B Batch Reactor, 121
  - 4.4C Tubular Reactor, 121
  - 4.4D CSTR, 125

5

Analysis of Rate Data 133

- 5.1 DIFFERENTIAL METHOD 135
- 5.2 INTEGRAL METHOD 141
- 5.3 GRAPHICAL TECHNIQUES IN THE PLOTTING OF REACTION RATE DATA 145
- 5.4 METHOD OF INITIAL RATES 155
- 5.5 DIFFERENTIAL REACTOR 157
- 5.6 METHOD OF HALF-LIVES 165
- 5.7 EXPERIMENTAL DESIGN 167
  - 5.7A Finding the Rate Law, 167
  - 5.7B Experimental Planning, 171
- 5.8 METHODS OF CHEMICAL ANALYSIS OF REACTING SPECIES 173

6

Heterogeneous Reactions   183

6.1 FUNDAMENTALS   183
     6.1A Definitions, 185
     6.1B Steps in a Catalytic Reaction, 189
     6.1C Adsorption, 191
     6.1D Surface Reaction, 193

6.2 CUMENE DECOMPOSITION EXAMPLE   195
     6.2A Is the Adsorption of Cumene Rate Limiting?, 199
     6.2B Is the Surface Reaction Rate Limiting?, 205
     6.2C Is the Desorption of Benzene Rate Limiting?, 207

6.3 HYDROMETHYLATION OF TOLUENE EXAMPLE   209
     6.3A Searching for the Mechanism, 209
     6.3B Evaluation of the Rate Law Parameters, 217
     6.3C Design of Gas-Solid Catalytic Reactors, 219

6.4 HOUDRY HYDROGENATION REACTOR   225

6.5 CLASSIFICATION OF CATALYSTS   227
     6.5A Alkylation and Dealkylation Reactions, 227
     6.5B Isomerization Reactions, 229
     6.5C Hydrogenation and Dehydrogenation Reactions, 229
     6.5D Oxidation Reactions, 231
     6.5E Hydration and Dehydration Reactions, 231
     6.5F Halogenation and Dehalogenation Reactions, 233

7

Non-Elementary Homogeneous Reactions   243

7.1 FUNDAMENTALS   245
     7.1A Active Intermediates, 245
     7.1B Pseudo-Steady State Hypothesis (PSSH), 249

7.2 SEARCHING FOR A MECHANISM   251
     7.2A General Considerations, 253
     7.2B Hydrogen Bromide Illustrative Example, 255

7.3 ENZYMATIC REACTION FUNDAMENTALS   265
     7.3A Definitions and Mechanisms, 265
     7.3B Michaelis-Menten Equation, 269
     7.3C Batch Reactor Calculations, 273

7.4 INHIBITION OF ENZYME REACTIONS   275
     7.4A Competitive Inhibition, 277
     7.4B Uncompetitive Inhibition, 277
     7.4C Noncompetitive Inhibition, 279

7.5 MULTIPLE ENZYME AND SUBSTRATE SYSTEMS   285
    7.5A Enzyme Regeneration, 285
    7.5B Enzyme Cofactors, 289
    7.5C Multiple Substrate Systems, 293
    7.5D Multiple Enzymes Systems, 297

8

Chemical Reactor Energy Balances   307

8.1 ENERGY BALANCES   307
8.2 THE ADIABATIC OPERATION OF A CSTR   327
8.3 ADIABATIC OPERATION OF A PLUG FLOW REACTOR   329
    8.3A Rate Laws, 331
    8.3B Numerical and Graphical Techniques, 333
8.4 OXIDATION OF SULFUR DIOXIDE EXAMPLE   335
8.5 ADIABATIC OPERATION OF A BATCH REACTOR   349

9

Competing Reactions   361

9.1 SCHEMES FOR MINIMIZING THE UNDESIRED PRODUCT   363
    9.1A Maximizing S for One Reactant, 365
    9.1B Maximizing S for Two Reactants, 369
9.2 THE STOICHIOMETRIC TABLE FOR MULTIPLE REACTIONS   375
9.3 APPLICATIONS OF THE DESIGN EQUATIONS TO MULTIPLE REACTIONS   379
9.4 HYDRODEALKYLATION OF MESITYLENE   385
    9.4A Packed Bed Reactor Calculations, 387
    9.4B Optimization of Xylene Production in a Packed Bed Reactor, 391
    9.4C Fluidized Bed Reactor Calculations, 395
    9.4D Optimization of Xylene Production in a Fluidized Bed Reactor, 397

10

Diffusion Limitations in Heterogeneous Reactions   409

10.1 MASS TRANSFER FUNDAMENTALS   411
    10.1A The Molar Flux, 411
    10.1B Fick's First Law, 413
    10.1C Evaluating the Bulk Flow Term, 415
    10.1D Diffusion Through a Stagnant Film, 419
    10.1E Modelling Diffusion with Chemical Reaction, 421
    10.1F Temperature and Pressure Dependence of D_{AB}, 423

10.2 Diffusion with First Order Homogeneous Reaction 425
10.3 Internal Diffusion and Reaction in Catalyst Pores 427
    10.3A First Order Reaction in Spherical Catalyst Pellets, 427
    10.3B Evaluation of the Internal Effectiveness Factor and the Thiele Modulus, 437
10.4 External Resistance to Mass Transfer 443
    10.4A The Mass Transfer Coefficient, 443
    10.4B Mass Transfer from a Single Particle, 447
    10.4C The Overall Effectiveness Factor, 447
10.5 Mass Transfer and Reaction in a Packed Bed 451
10.6 Other Forms of the Mass Transfer Coefficient 455

Appendices 465

A Graphical and Numerical Techniques 465
    A.1 Integration—The Graphical Form of Simpson's Rule, 465
    A.2 Useful Integrals in Reactor Design, 467
    A.3 Equal Area Graphical Differentiation, 468
    A.4 Solutions to Differential Equations, 471

B Ideal Gas Constant and Conversion Factors 472

C Thermodynamic Relationship Involving the Equilibrium Constant 474

D Prediction of Binary Gas Diffusivities 475

E Measurement of Slopes 477
    E.1 Linear Plots (Zero Order Reactions), 477
    E.2 The Use of Log-Log Plots in Reaction Order Analysis, 477
    E.3 Use of Semi-Log Plots in Rate Data Analysis, 481

F Guided Designs 487
    F.1 Solid Waste Disposal to Produce Crude Oil, 487
    F.2 Automobile Catalytic Afterburners, 488
    F.3 Hollow Fiber Artificial Kidney, 489
    F.4 Experimental Study of Fermentation Kinetics, 490

Index 492

Preface

This book is an introduction to chemical reaction engineering. While it is primarily aimed at the junior-senior level student in chemical engineering, certain sections of the text are more advanced and could be used at the graduate level. In addition, the text has been used successfully in a junior level course for non-chemical engineers enrolled in the Engineering Science and Environmental Engineering programs. In this course the fundamental principles of chemical kinetics are presented, but the applications, example problems, and home problems are drawn not only from chemical engineering but also from medicine and from environmental engineering.

While a number of universities who class-tested this text used it just as one would use a non-programmed text, I should like to briefly discuss how the text is complemented at The University of Michigan. The concepts are disseminated entirely by the programmed text which helps the student develop both *inductive* and *deductive* reasoning. The students utilize inductive reasoning in developing and deriving the general balance and design equations from first principles. Deductive reasoning is applied by working through specific examples in the various frames.

A one hour recitation-problem session and a two-hour "Guided Design Session" are held each week. In the guided design sessions open-ended problems are studied which examine current literature on kinetics as well as industrial processes that emphasize chemical reactor design. Occasional home problems and conferences with the instructor round out the instructional program. During these conferences, which are held approximately every four weeks, one, two, or three students at a time meet with the instructor for thirty to forty-five minutes to discuss the material covered. Hopefully, all intensive learning tends often to have enthusiasm. Variation in class techniques offers rewarding situations.

It may not be possible to cover all ten chapters of the book in a three hour undergraduate course. Although the first four chapters should be completed in sequence, the remaining chapters may be studied in any order. However, it may be preferable, but not necessary, to complete chapter six before beginning chapter ten.

Many of the problems at the end of the various chapters were selected from those which have appeared in *California Board of Registration for Professional Engineers—Chemical Engineering Examinations* (PECEE) over the last five years. The permission for use of these problems is gratefully acknowledged. [Note: These problems have been copyrighted by the California Board of Registration and may not be reproduced without their permission.] While the majority of the remaining problems were hour examinations and home problems given at the University of Michigan, a few of the problems at the end of some of the chapters were suggested by Professor L. F. Brown of the University of Colorado, and are designated by [CU]. For the most part the problems at the end of each chapter are arranged in order of increasing difficulty.

Many contributions to the material presented here have been made by a number of my colleagues (students and faculty) over the past few years through their development of and participation in the chemical reaction engineering course at The University of Michigan. I would like to sincerely acknowledge Professors R. L. Curl, G. Parravano and J. S. Schultz for their contributions and *in particular*, Professor J. D. Goddard for the many helpful discussions and suggestions on the text and on teaching kinetics. I would also like to acknowledge M. S. Peters and Octave Levenspiel who were the first to stimulate my interest in kinetics.

The comments received from class-testing the preliminary edition of the text by the following faculty and their respective undergraduate classes are greatly appreciated: Professor L. F. Brown, University of Colorado; Professor L. C. Eagleton, Pennsylvania State University; Professor O. M. Fuller, McGill University; Professor T. Owens, University of North Dakota; Professor R. A. Schmitz, University of Illinois; and Professor J. T. Sears, University of West Virginia. I would also like to acknowledge a number of students at The University of Michigan for their detailed comments on the text, especially M. L. Cadwell, K. Lund, and D. Fink. In addition, M. L. Cadwell took major responsibility for preparation of the solutions manual while G. Wong helped with proofreading the final manuscript. The continuing encouragement provided by Professors K. D. Timmerhaus, J. D. Goddard, R. E. Balzhiser, and M. S. Peters, added to that of my family and my parents, is greatly appreciated. Certainly this undertaking would not have been possible without the understanding and help of my wife, Janet, who aided in so many ways in preparing this manuscript.

I am also indebted to Mrs. J. F. May for the fine and careful typing of the preliminary edition of the text and of the final manuscript.

Ann Arbor, Michigan   H.S.F.

Introduction

A–The Approach—How To Use The Text

The success of disseminating information through programmed learning depends to some extent upon its implementation in the classroom, but primarily upon the manner in which the reader utilizes the programmed text. If the student were merely to read the programmed text as a standard textbook, the unique benefits of this type of learning would be diminished. One of the major advantages of programmed learning over lectures or textbooks is that it encourages the reader to exercise a higher level of thinking than these latter two methods of information dissemination. The thinking process can be conceived as consisting of the following intellectual abilities: recollection, manipulation, translation, interpretation, prediction, and evaluation; the lowest level of thinking being that of recall and the highest level being evaluation. The majority of the "frames" presented here will require the reader to exercise recall, manipulation, translation, and interpretation. However, a number of frames have been designed to require the highest levels of thinking. In any event, it is hoped that the frames which follow *will allow the reader to exercise and develop his thinking patterns*. Inductive reasoning will be enhanced through developing and deriving the general balance and design equations from first principles, while deductive reasoning will be increased by working through specific examples in the various frames.

It is *imperative* that the reader try diligently to answer the questions in each frame before looking at the answer on the back of the page and proceeding further in the programmed text. To answer in detail a number of the frames will probably require *re-reading* of some of the material preceding that frame. One should also read carefully the *entire solution* to each frame, as additional information is sometimes presented and important principles emphasized. I believe it is worthwhile to mention a technique developed and adopted by a number of students for using the text that appears quite reasonable and successful. The first time through the text, the students write out the solutions to the frames on a separate piece of paper and do not write in the book. When reviewing for an hour exam, they go through and fill in each frame completely. After completing the frames in this fashion, it is possible to review for the final exam by reading the text as a regular text. Owing to the book's makeup, it will not be possible to write out the entire solution in the space provided. In these instances, I would suggest writing out the complete solution on a separate sheet of paper and then outlining the salient features of the solution in the space provided. The reader should not be discouraged if he feels that it is taking slightly longer to cover the material than with a conventional text because it has been documented that their efforts, *if* done correctly, will be rewarded by a significantly greater mastering of the material.[1] Although the text is designed so that the reader may proceed at his own rate, he should try to set weekly goals for himself in order to achieve a uniform rate of progression in learning the material.

B–THE CONTENT

Chemical kinetics and reactor design are the heart of the production of almost all industrial chemicals. *It is primarily a knowledge of chemical kinetics and reactor design that distinguishes the chemical engineer from other engineers*. The selection of a reaction system which operates in the most efficient manner can be the key to the economic success or failure of the chemical plant. For example, if a reaction system produced a large amount of unwanted products, subsequent purification and separation of the desired product could prove the entire process to be economically unfeasible. In addition to industrial applications, the chemical kinetic principles learned here will have applications to the study of living systems, waste treatment, and air and water pollution. Some of the examples and problems used in the text to illustrate the principles of chemical reaction engineering are: the kinetics of nitric oxide formation and its relation to *smog formation*, the application of enzyme kinetics to improve an *artificial kidney*, the deoxygenation of *hemoglobin*, and the formation of benzene from *petroleum* feed stocks. In addition to these systems, reactor schemes which minimize unwanted products are discussed.

The manner in which the material on chemical kinetics and reactor design is to be developed is shown schematically in Figure I–4. We begin our study by performing mass balances on each chemical species in the system. Here, the term *chemical species* refers to any chemical compound or element with a given identity. For example the chemical species called nicotine (a tobacco alkaloid) contains a number of atoms of different elements arranged in a definite "stereo" configuration (see Figure I–1).

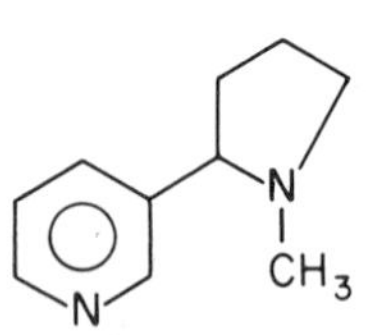

Nicotine

Figure I–1

[1] *Educational Research and Methods* (*ERM*). Vol. 5, No. 4, p. 110 (1973)

Even though two chemical compounds have exactly the same number of atoms of each element, they could still be different species, because of configurations. For example, 2-butene has four carbon atoms and eight hydrogen atoms; however, the atoms in this compound can form two different arrangements (see Figure I–2).

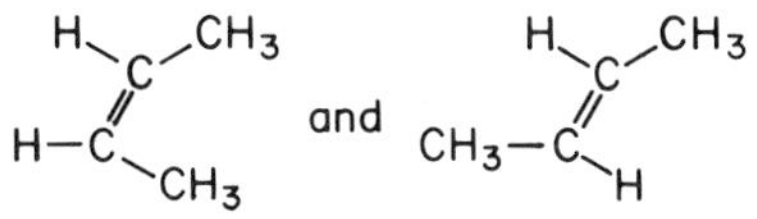

cis-2-Butene   trans-2-Butene

Figure I–2

Consequently we can have two different species even though each has the same number of atoms of each element. We say a *chemical reaction* has taken place when a detectable number of molecules of one or more species have *lost their identity* and assumed a new form by a change in the number of atoms in the compound and/or by a change in structure or configuration of these atoms.

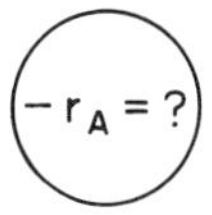

Figure I–3

After learning how to write *mole balances* for each species, we develop the general balance relationships that exist among reacting species in a given system (*stoichiometry*). We then define fractional (or percent) *conversion* and develop the reactor design equations which specify either the reactor volume (flow system) or time (batch system) necessary to achieve a given conversion. In order to numerically evaluate the design equations, with a species "A" as our basis, we must determine the relationship between the concentration or conversion of A and the *reaction rate* of A; that is, we need to know the algebraic rate law for $-r_A$ (Figure I–3).

Referring to Figure I–4, we see that there are three ways to obtain this information: (1) from previous experiments reported in the literature (e.g., technical journals, books), (2) from our own experimental design and observation, and (3) from theoretical knowledge of the reaction mechanism and rate constants. These methods of obtaining information on the rate law are shown in parallel branches in the figure to signify that the methods may be studied concurrently, or in any order desired.

Once we know the reaction rate law and the specific reaction rate(s) for a particular reaction and at a particular temperature, we shall be able to design a number of isothermal reaction systems for that temperature. One may then wish to study the material on catalysis, either homogeneous or heterogeneous, or to proceed directly to Chapter 5 to learn how to analyze reaction rate data.

Before proceeding to the design of *adiabatic* reactors with temperature changes due to heat effects, the concepts related to energy balances on reacting systems are first developed along with the temperature dependence of the reaction rate law. Once these concepts are mastered, the evaluation of the design equations for adiabatic and other types of nonisothermal conditions becomes apparent.

When carrying out industrial reactions, one usually finds various side reactions occurring, in addition to the commercially desired reaction. Therefore, in studying *multiple reactions*, this text will address itself to the problem of minimizing the unwanted products from side or parallel reactions. In this context we shall consider how the minimization may sometimes be accomplished by control of temperature and mixing, through proper choice of reactor.

The last chapter of the text concerns diffusion limitations in heterogeneous reactions. After presenting the fundamentals of binary diffusion, we develop the concepts of accessibility of the catalytic surfaces. The chapter comes to an end with a discussion of mass transfer coefficients and the factors influencing external diffusion limitations.

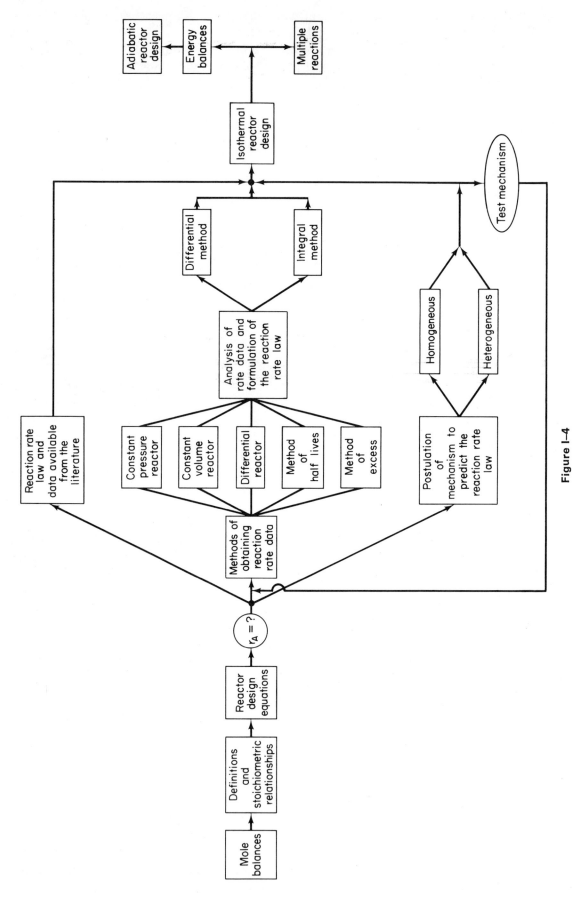

Figure I-4

An example of a typical sequencing of the material in a 3-hour course might be:

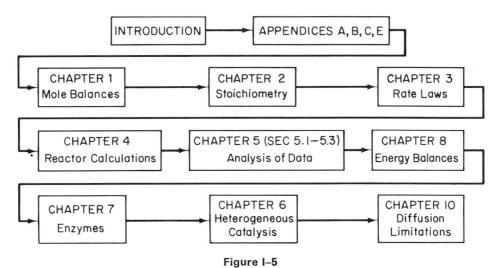

Figure I–5

Chapter 9 could just as easily be included, say, at the expense of the last sections of Chapters 5, 6, 7, and 10.

After completion of each chapter, it would be quite beneficial for the reader to refer back to Figure I–4 so as to review how the newly completed chapter fits into his overall study of chemical reaction engineering. In addition, the reasons for studying the subsequent sections should also become clearer from this periodic consultation of Figure I–4.

At the end of a number of the chapters, problems from the California Board of Registration for Professional Engineers—Chemical Engineering Examinations (PECEE) are given. For the most part, these problems are reasonably difficult, and they are presented to give the reader an idea of the level of competence that he is encouraged to achieve. If one closely follows the instructions on pages 1 and 2 to acquire the fundamentals and to develop a solution methodology by working diligently through the "frames" and review problems, he should be able to rise to the level of competence necessary to pass the PECEE problems and to carry through reactor calculations encountered in Engineering practice.

1
Mass and Mole Balances

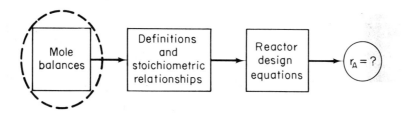

Before entering into discussions on the conditions that affect chemical reaction rates and reactor design, it is first necessary to account for the various chemical species entering and leaving the reaction system. This accounting process is achieved through mass and mole balances on the reacting system. In this chapter we shall develop a general mole balance that can be applied to any species (i.e., compound) entering, leaving, and/or remaining within the reaction system volume. After defining the rate of reaction, $-r_A$, and discussing how the reaction rate has previously been incorrectly defined, we shall then show how the general balance equation may be used to develop a preliminary form of the design equations of the most common industrial reactors: batch, continuous stirred tank (CSTR), and tubular. In developing these equations, the assumptions pertaining to the modeling of each type of reactor are delineated. Finally, a brief summary and a series of short review questions are given at the end of this chapter.

1.1 – Definition of the Rate of Reaction, $-r_A$

In the classic approach to chemical change it is assumed that mass is neither created nor destroyed when a chemical reaction takes place. The mass referred to is the total collective mass of all the species in the system. However, when considering the individual species involved in a particular reaction we do speak of the rate of disappearance of mass of a particular species. This rate of disappearance is the rate at which the species molecule loses its original chemical identity through the breaking and subsequent reforming of chemical bonds during the course of the chemical reaction. In the study of chemical kinetics we shall be concerned with the rate at which chemical reactions take place, along with the rate-controlling mechanisms that describe the reaction process. In the space provided below, state where, in a discussion of chemical reactions, the phrase "a rate of formation of mass" might be applicable.

(A1) _____

_____

For a particular species to suddenly appear in the system, some prescribed fraction of another species must lose its identity. For example, if an oxygen atom and two hydrogen atoms are suddenly "formed" from a water molecule,

$$H_2O \longrightarrow 2H + O$$

the H_2O molecule has ruptured its bonds to form these atoms and has consequently lost its identity (i.e., disappeared). In the example above we have seen how a molecule may lose its chemical identity by being broken down into atoms or into smaller molecules. State two ways by which a molecule can lose its identity, other than by being broken into smaller molecules, atoms, or atom fragments.

(A2) _____

_____

To summarize this point, we say that a molecule or number of molecules (e.g., mole) of a particular chemical species have reacted or disappeared when they lose their chemical identity.

The *rate* of reaction, i.e., the *rate* of disappearance, is the number of moles reacting per unit time per unit volume. For example, the insecticide DDT (dichlorodiphenyltrichloroethane) is produced from chlorobenzene and chloral in the presence of fuming sulfuric acid, i.e.,

$$2C_6H_5Cl + CCl_3CHO \longrightarrow (C_6H_4Cl)_2CHCCl_3 + H_2O$$

Letting A represent chloral, the numerical value of the rate of reaction, $-r_A$, tells us how many moles of chloral are reacting per unit time per unit volume. The conversion X tells us how many moles of chloral have reacted per mole of chloral fed. In the next chapter we shall delineate the prescribed relationship between the rate of formation of one species (e.g., DDT) and the rate of disappearance of another species (e.g., chloral) in a chemical reaction. In heterogeneous reaction systems the rate of reaction is usually expressed in measures other than volume, such

(A1) In the classic approach we can only consider a rate of formation of mass when we are referring to a particular chemical species.

(A2)
1. A molecule may lose its identity by chemically combining with other molecules. In the example above the oxygen atom would lose its species identity if it combined with two hydrogen atoms to form a water molecule.

2. A molecule that neither adds other molecules to itself nor breaks into smaller molecules may lose its identity by isomerization. For example,

$$CH_3\overset{\overset{\displaystyle CH_2}{\|}}{C}-CH_2CH_3 \longrightarrow CH_3\overset{\overset{\displaystyle CH_3}{|}}{C}=CHCH_3$$

as reaction surface area or catalyst weight. Thus, for a gas-solid catalytic reaction the dimensions of this rate, $-r'_A$, may be moles of A reacted per unit time per unit weight of catalyst, with typical units being gmoles of A per second per gram of catalyst. Most of this introductory discussion on chemical reaction engineering will concentrate on homogeneous systems.

The mathematical definition of a chemical reaction rate has been a source of confusion in the chemical and chemical engineering literature for many years. The origin of this confusion stems from the laboratory bench-scale experiments that were carried out to obtain chemical reaction rate data. These early experiments were batch-type experiments in which the reactants were mixed together at time $t = 0$ and the concentration of one of the reactants, C_A, was measured at various times t. The reaction vessel was closed and rigid; consequently the ensuing reaction took place at constant volume. These investigators then defined and reported the chemical reaction rate as

$$r_A = \frac{dC_A}{dt} \tag{1-1}$$

where r_A is the rate of formation of A per unit volume (e.g., gmoles/sec-cu cm). Select the conditions or restrictions under which Equation (1-1) was developed.

(A3)

1. Unsteady-state operation
2. Steady-state operation
3. Constant volume
4. Constant pressure
5. No inflow or outflow of reactants or products

As a result of the limitations and restrictions given in Frame (A3), Equation (1-1) is a rather limited and confusing definition of the chemical reaction rate. For further amplification on this point, consider the following steady flow system in which the saponification of ethyl acetate is carried out. Sodium hydroxide and ethyl acetate are continuously fed to a rapidly stirred tank in which they react to form sodium acetate and ethanol. The product stream containing sodium acetate and ethanol, along with the unreacted sodium hydroxide and ethyl acetate, is continuously withdrawn from the tank at a rate equal to the total feed rate. The contents of the tank in which this reaction,

$$NaOH + CH_3COOC_2H_5 \longrightarrow CH_3COONa + C_2H_5OH$$

is taking place may be considered to be *perfectly mixed*, denoting that there are no *spatial variations in concentration* within the tank. Because the system is operated at steady state, if we were to withdraw liquid samples at some location in the tank at various times and chemically analyze them, we would find that the concentrations of the individual species in the different samples were identical. In Frame (A4), determine whether Equation (1-1) can be modified to define a chemical reaction rate for the stirred-tank flow system in which the concentration, C_{NaOH}, at any specified point or position in the reactor does not change with time. (*Hint:* To what is the derivative of C_{NaOH} equal?)

(A3)
1. Unsteady-state operation.
3. Constant volume, closed container.
5. No inflow or outflow of reactants or products since the vessel was closed.

(A4)

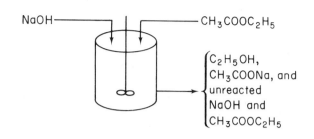

By now you should be convinced that Equation (1-1) is not the definition of the chemical reaction rate. We shall simply say that r_j is *the rate of formation of species j per unit volume*. It is the number of moles of *j* generated per unit volume per unit time. *The rate equation for* r_j *is solely a function of the properties of the reacting materials* (e.g., species concentration, temperature, pressure, or type of catalyst, if any) *at a point in the system and is independent of the type of system* (i.e., batch or continuous flow) *in which the reaction is carried out*. However, r_j can be a function of position and can vary from point to point in the system. The chemical reaction rate is an intensive quantity and the reaction rate equation (i.e. the *rate law*) is essentially an algebraic equation, *not* a differential equation.† For example, the algebraic form of the rate law, $-r_A$, for the reaction

$$A \longrightarrow \text{products}$$

may be of the form

$$-r_A = kC_A^2 \qquad (1\text{-}2)$$

or it may be some other algebraic function of concentration, such as

$$-r_A = kC_A$$

or

$$-r_A = \frac{k_1 C_A}{1 + k_2 C_A}$$

Equation (1-2) states that the rate of disappearance of A is equal to a rate constant k times the square of the concentration of A. By convention, r_A is the rate of formation of A; consequently $-r_A$ is the rate of disappearance of A.

1.2–The General Balance Equation

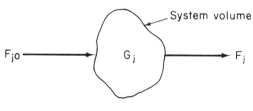

Figure 1–1

To perform a mass balance on any system, the system boundaries must first be specified. The volume enclosed by these boundaries will be referred to as the *system volume*. (In fluid mechanics texts, this volume is usually denoted as the control volume.) We shall perform a mole balance on species *j* in a system volume where species *j* represents the particular chemical species of interest, such as water, NaOH, etc. (Figure 1-1).

†For further elaboration on this point, see *Chem. Engr. Sci.* 25 (1970), 337.

(A4) It does not make any sense to define a chemical reaction rate for this system utilizing Equation (1-1). Since the reactor is operated at steady state, the species concentrations do not change with time. Where A = NaOH,

$$\frac{dC_A}{dt} = 0 \tag{A4-1}$$

The substitution of Equation (A4-1) into Equation (1-1) leads one to the conclusion that $r_A = 0$, which is incorrect since C_2H_5OH and CH_3COONa are being formed from NaOH and $CH_3COOC_2H_5$ at a finite rate. Consequently Equation (1-1) cannot be modified to define the rate of reaction for a flow system.

A mole balance on species j at any instant in time, t, yields the following equation

$$\begin{bmatrix} \text{Rate of flow} \\ \text{of } j \text{ into the} \\ \text{system} \\ \text{(moles/time)} \end{bmatrix} + \begin{bmatrix} \text{Rate of generation} \\ \text{of } j \text{ by chemical} \\ \text{reaction within the} \\ \text{system (moles/time)} \end{bmatrix} - \begin{bmatrix} \text{Rate of flow} \\ \text{of } j \text{ out of} \\ \text{the system} \\ \text{(moles/time)} \end{bmatrix} = \begin{bmatrix} \text{Rate of} \\ \text{accumulation} \\ \text{of } j \text{ within} \\ \text{the system} \end{bmatrix}$$

$$\begin{bmatrix} F_{j0} \end{bmatrix} + \begin{bmatrix} G_j \end{bmatrix} - \begin{bmatrix} F_j \end{bmatrix} = \begin{bmatrix} \dfrac{dN_j}{dt} \end{bmatrix} \tag{1-3}$$

The inlet and outlet molar flow rates may be functions of time. What does N_j represent and what are the dimensions of N_j?

(A5) ─────────────────────────────────

─────────────────────────────────

If the temperature and the concentrations of the chemical species are spatially uniform throughout the system volume, write an equation relating the rate of generation of species j, G_j, to the rate of formation, r_j, and to the system volume, V.

(A6) ─────────────────────────────────

─────────────────────────────────

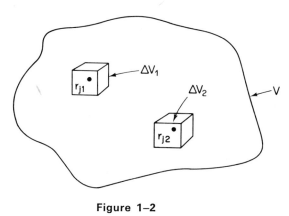

Figure 1-2

Suppose now that the rate of formation of species j for a homogeneous reaction varies with the position in the system volume. That is, it has a value r_{j1} at point 1, which is surrounded by a small volume, ΔV_1, within which the rate is uniform; similarly, the reaction rate has a value r_{j2} at point 2 and an associated volume ΔV_2. (See Figure 1-2). What is the rate of generation, ΔG_j, in terms of r_j and subvolume ΔV_1? in terms of r_j and subvolume ΔV_2?

(A7) ─────────────────────────────────

─────────────────────────────────

(A5)

N_j is the number of moles of species j that is present in the system volume at any time t. The dimensions of N_j are moles and therefore the dimensions of the rate of change of N_j with respect to time, dN_j/dt, are moles/time. These dimensions are of course consistent with the dimensions of the other terms in Equation (1-3). Typical units of dN_j/dt might be gmoles per second.

(A6)

The rate per unit volume times the volume will give the total rate of generation. Since temperature and concentrations are spatially uniform within V, we may assume r_j to be spatially uniform within V.

$$\frac{\text{Moles}}{\text{time}} = \frac{\text{Moles}}{\text{time} \cdot \text{volume}} \cdot \text{Volume}$$

$$G_j = r_j \cdot V$$

(A7)

Since the rate r_j is constant within the particular subvolume being considered, we can write

$$\Delta G_{j1} = r_{j1} \Delta V_1 \tag{A7-1}$$

and

$$\Delta G_{j2} = r_{j2} \Delta V_2 \tag{A7-2}$$

What is the total rate of generation of species j from all the subvolumes within the system volume V?

(A8) _____

_____

Rewrite the general balance Equation (1-3), replacing G_j with the appropriate expression involving r_j.

(A9) _____

_____

Equation (A9-1), which is the general mole balance equation for any chemical species j entering, leaving, reacting, and/or accumulating within any system volume V, is repeated below as Equation (1-4):

$$F_{j0} - F_j + \int^V r_j\, dV = \frac{dN_j}{dt} \tag{1-4}$$

From this general mole balance equation we can develop the design equations for the various types of industrial reactors: batch, semibatch, and continuous flow. Upon evaluation of these equations we can determine the time (batch) or reactor volume (continuous flow) necessary to convert a specified amount of the reactants to products.

1.3–BATCH REACTORS

In a batch reactor there is neither inflow nor outflow of reactants or products while the reaction is being carried out. Figure 1-3 shows two different types of batch reactors used for gas phase reactions. Reactor A is a constant-volume (variable pressure) reactor, and reactor B is a constant-pressure (variable volume) reactor.

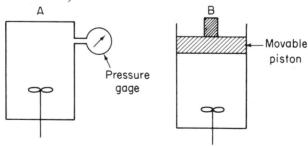

Figure 1–3

At time $t = 0$, the reactants are injected into the reactors and the reaction is initiated. Consider the decomposition of dimethylether that is taking place in these reactors to form methane,

(A8) The total rate of generation within the system volume would be the sum of all the rates of each of the subvolumes. If the total system is divided into M subvolumes,

$$G_j = \sum_{i=1}^{M} \Delta G_{ji} = \sum_{i=1}^{M} r_{ji} \Delta V_i \tag{A8-1}$$

By taking the appropriate limits (i.e., let $M \to \infty$ and $\Delta V \to 0$) we can rewrite Equation (A8-1) in the form

$$G_j = \int^V r_j \, dV \tag{A8-2}$$

From Equation (A8-2) we see that r_j will be an indirect function of position, since the properties of the reacting materials (e.g., concentration, temperature) can have different values at different points.

(A9)
$$F_{j0} - F_j + G_j = \frac{dN_j}{dt} \tag{1-3}$$

Using Equation (A8-2) to substitute for G_j, we find

$$F_{j0} - F_j + \int^V r_j \, dV = \frac{dN_j}{dt} \tag{A9-1}$$

hydrogen, and carbon monoxide:

$$(CH_3)_2O \longrightarrow CH_4 + H_2 + CO$$

Symbolically, we could write

$$[A \longrightarrow M + H + C]$$

where A = dimethyl ether, M = methane, H = hydrogen, C = carbon monoxide.

Let species j represent carbon monoxide, and apply the mole balance equation, i.e., Equation (1-4), to reactor A, eliminating the terms that do not apply.

(A10)

Repeat for reactor B.

(A11)

If the reaction mixture is *well mixed* (i.e., perfectly mixed) so that the temperature and concentration are spatially uniform throughout the reactor volume, what further simplification can be made to the mole balance equation for reactors A and B?

(A12)

Determine whether the carbon monoxide mole balance equation, $dN_{CO}/dt = r_{CO}V$, for reactor A (constant volume) can be put into a form similar or identical to Equation (1-1).

(A13)

(A10)

For the above reaction, species j in the mass balance refers to carbon monoxide ($j = CO$). Since there is no flow of carbon monoxide, CO, into or out of the system volume we have

$$F_{j0} = F_j = 0 \quad \text{(i.e., } F_{CO,0} = F_{CO} = 0) \tag{A10-1}$$

(We could also use C to represent CO, thereby eliminating the need to write a double subscript.) Then

$$\frac{dN_{CO}}{dt} = G_{CO} \tag{A10-2}$$

$$\frac{dN_{CO}}{dt} = \int^V r_{CO}\, dV \tag{A10-3}$$

(A11)

The same as (A10):

$$\frac{dN_{CO}}{dt} = \int^V r_{CO}\, dV$$

(A12)

Even though the total volume of one reactor is changing with time, the contents of the reactor are *well mixed*. Consequently the reaction rate is independent of position, i.e., it is spatially uniform, and Equation (A10-3) can be readily integrated to give

$$G_{CO} = \int^V r_{CO}\, dV = r_{CO}\, V \tag{A12-1}$$

The reaction rate, r_{CO}, will be an indirect function of time since it is a function of the species concentrations which will vary with time in batch systems. For reactor A the volume is constant, so that

$$\frac{dN_{CO}}{dt} = G_{CO}(t) = r_{CO}(t) \cdot V \tag{A12-2}$$

For reactor B the volume varies with time, so that

$$\frac{dN_{CO}}{dt} = G_{CO}(t) = r_{CO}(t) \cdot V(t) \tag{A12-3}$$

[Note: $r_{CO} \equiv r_{CO}(t)$]

(A13)

In Equation (1-1) the rate is given in terms of the derivative of concentration for a nonflow system; the concentration is defined by

$$C_{CO} = \frac{N_{CO}}{V} \tag{A13-1}$$

Dividing both sides of Equation (A12-2) by V we obtain for Reactor A

$$\frac{1}{V}\frac{dN_{CO}}{dt} = r_{CO} \tag{A13-2}$$

Since the volume, V, is constant for reactor A, we can take V inside the differential, and the rate of the formation of CO, r_{CO}, can be equated to the time rate of change of concentration of CO:

$$\frac{1}{V}\frac{dN_{CO}}{dt} = \frac{d(N_{CO}/V)}{dt} = \frac{dC_{CO}}{dt} = r_{CO} \tag{A13-3}$$

Equation (A12-2) has been arranged in a form similar to Equation (1-1).

Determine if the mole balance equation for reactor B can be put in a form similar to or identical to Equation (1-1).

(A14) _____

1.4–Continuous Flow Reactors

1.4A Continuous Stirred Tank Reactor (CSTR)

Another type of reactor used in industrial processing is the backmix reactor, which is also referred to as the Continuous Stirred Tank Reactor (CSTR). See Figure 1-4. The backmix reactor is normally operated at steady state and is normally considered to be well mixed. Using this latter concept, the backmix reactor is modeled to have no spatial variations in concentration, temperature, and reaction rate throughout the reaction vessel. Thus the temperature and concentrations of the exit stream are modeled as being identical to those inside the reactor. In systems where mixing is highly nonideal, the well-mixed model is inadequate and we must resort to other modeling techniques such as residence time distributions in order to obtain meaningful results.† However, in this text we shall restrict our discussion to perfectly mixed tank reactors.

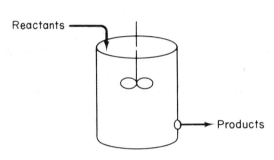

Figure 1–4. Continuous Stirred Tank Reactor

Apply the mole balance equation

$$F_{j0} - F_j + \int^V r_j \, dV = \frac{dN_j}{dt} \tag{1-4}$$

to dimethyl ether in the decomposition reaction just discussed in Section 1.3, eliminating those terms that do not apply to the CSTR system.

(A15) _____

_____

Rearrange this expression, i.e., Equation (A15-3), to determine how the reactor volume V varies with (1) the rate of reaction and (2) the difference between the inlet and outlet molar flow rates $(F_{j0} - F_j)$.

(A16) _____

_____

†For a thorough discussion on residence times, see Chapter 9 of *Chemical Reaction Engineering*, 2nd ed., by O. Levenspiel (New York: John Wiley and Sons, Inc., 1972).

(A14)

Dividing both sides of Equation (A12-3) by V we obtain for reactor B

$$\frac{1}{V(t)} \frac{dN_{co}}{dt} = r_{co} \tag{A14-1}$$

Since the concentration is spatially uniform, the number of moles in volume V at any time t is

$$N_{co}(t) = C_{co}(t) \cdot V(t) \tag{A14-2}$$

However, since V is a function of time for reactor B, we find

$$\frac{1}{V} \cdot \frac{dN_{co}}{dt} = \frac{1}{V} \frac{d(C_{co} \cdot V)}{dt} = \frac{dC_{co}}{dt} + \frac{C_{co}}{V} \frac{dV}{dt} \tag{A14-3}$$

$$\frac{1}{V} \cdot \frac{dN_{co}}{dt} = \frac{dC_{co}}{dt} + C_{co} \frac{d \ln V}{dt} \tag{A14-4}$$

Combining Equations (A14-1) and (A14-4) we obtain

$$r_{co} = \frac{dC_{co}}{dt} + C_{co} \frac{d \ln V}{dt} \tag{A14-5}$$

It can be seen by comparing Equation (A14-5) with Equation (1-1) that Equation (A12-3) *cannot* be reduced to Equation (1-1) for systems of variable volume.

(A15)

Since the system is at steady state,

$$\frac{dN_j}{dt} = 0 \tag{A15-1}$$

In addition, since the reactor is modeled as having no spatial variation in the reaction rate within the vessel,

$$G_j = \int r_j \, dV = r_j V \tag{A15-2}$$

Equation (1-4) reduces to

$$(F_{j0} - F_j) = -r_j V \tag{A15-3}$$

$j =$ dimethyl ether $=$ A:

$$F_{A0} - F_A = -r_A V \tag{A15-4}$$

(A16)

$$V = \frac{(F_{j0} - F_j)}{-r_j} = \frac{F_{A0} - F_A}{-r_A} \tag{A16-1}$$

The reactor volume is inversely proportional to the rate of reaction $-r_A$, and directly proportional to the difference of the inlet and outlet flow rates $(F_{A0} - F_A)$. For a fixed or specified difference in inlet and outlet molar flow rates $(F_{A0} - F_A)$, under what conditions should the reaction be carried out so that the smallest possible container can be used?

1.4B Tubular Reactor

In addition to the batch and backmix (CSTR) reactors, there is one other type of reactor that is commonly used in industry, the tubular reactor. This reactor is normally operated at steady state, as is the backmix reactor. For the purposes of the material presented here, we shall consider systems in which the flow is highly turbulent and the flow field may be modeled by that of plug flow. That is, there is no radial variation in concentration. In modeling the tubular reactor, we assume that the concentration varies continuously in the axial direction through the reactor. Consequently, the reaction rate, which is a function of concentration for all but zero order reactions, will also vary axially. To develop the design equation, we shall divide (conceptually) the reactor into a number of subvolumes so that within each subvolume ΔV, the reaction rate may be considered spatially uniform. We now focus our attention on the subvolume that is located a distance y from the entrance of the reactor. For this subvolume write the equation relating ΔG_j to the reaction rate and volume ΔV.

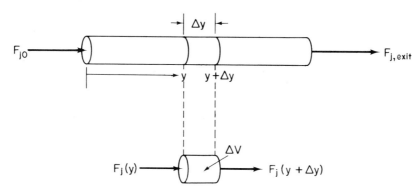

Figure 1-5. Tubular Reactor

(A17)

If $F_j(y)$ is the molar flow rate of j into volume ΔV at y, and $F_j(y + \Delta y)$ is the molar flow rate of j out of the volume at the point $(y + \Delta y)$, write the mole balance Equation (1-4) utilizing Equation (A17-1) and the simplifying assumptions that have been applied to the tubular reactor.

(A18)

The volume ΔV is the product of the cross-sectional area A of the reactor and the element of reactor length, Δy. After substituting for ΔV in the mass balance equation in Frame (A18), divide through by Δy, rearrange the terms, and then take the limit as Δy goes to zero to obtain a differential equation.

(A17)

Since the reaction rate is spatially uniform within the subvolume

$$\Delta G_j = \int^{\Delta V} r_j \, dV = r_j \, \Delta V \tag{A17-1}$$

the rate of generation of compound j, ΔG_j, within subvolume ΔV is $(r_j \, \Delta V)$. We also note that Equation (A17-1) could follow directly from either Equations (A7-1) or (A7-2) in Frame (A7).

(A18)

$$F_{j0} - F_j + \int^V r_j \, dV = \frac{dN_j}{dt} \tag{1-4}$$

$$F_{j0} \longrightarrow F_j(y)$$
$$F_j \longrightarrow F_j(y + \Delta y) \qquad V \longrightarrow \Delta V$$

For a tubular reactor operated at steady state

$$\frac{dN_j}{dt} = 0 \tag{A18-1}$$

Hence

$$F_j(y) - F_j(y + \Delta y) = -r_j \, \Delta V \tag{A18-2}$$

In this expression, Equation (A18-2), $-r_j$ is an indirect function of y.

(A19)

It is usually most convenient to have the volume V as the independent variable rather than the reactor length y. Accordingly, we shall change variables in Frame (A19), using the relation $dV = A\,dy$ to obtain one form of the design equation for a tubular reactor.

$$\frac{dF_j}{dV} = r_j \tag{1-5}$$

We also note that, for a reactor in which the cross-sectional area A varies along the length of the reactor, the design equation remains unchanged. This equation can be generalized for the reactor shown in Figure 1-6, Frame (A20), in a manner similar to that presented above, by utilizing a volume coordinate, V, system rather than a linear coordinate y. After passing through volume V, species j enters subvolume ΔV at V at a molar flow rate $F_j(V)$. Species j leaves subvolume ΔV at volume $(V + \Delta V)$ at a molar flow rate $F_j(V + \Delta V)$. Using this information in conjunction with Equation (1-4), rederive Equation (1-5) for the variable cross-sectional area reactor system shown in Frame (A20), stating all assumptions.

(A20)

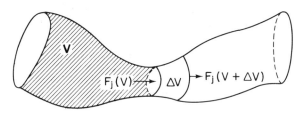

Figure 1-6

Consequently, we see that Equation (1-5) applies equally well to our model of tubular reactors of variable and constant cross-sectional area.

Although it is doubtful that one would find a reactor of the shape shown in Frame (A20), the conclusion drawn from the application of the design equation is an important one. In the remainder of this chapter we shall look at slightly more detailed drawings of some typical industrial reactors and point out a few of the advantages and disadvantages of each.[†]

1.5–Industrial Reactors

A *batch reactor* is used for small scale operation, for testing new processes that have not been fully developed, and for the manufacture of expensive products.

The reactor is charged (i.e., filled) through the two holes shown at the top. (See Fig. 1-7.) The batch reactor has the advantage of high conversions that can be obtained by leaving the reactant in the reactor for long periods of time, but it also has the disadvantages of high labor costs and small scale production.

Although a *semibatch reactor* (Fig. 1-8) has essentially the same disadvantages as the batch

[†]*Chem. Engr. 63, No. 10*, Oct. 1956, p. 211.

(A19)

$$\Delta V = A \, \Delta y \tag{A19-1}$$

We now substitute in Equation (A18-2) for ΔV and then divide by Δy to obtain

$$-\left[\frac{F_j(y + \Delta y) - F_j(y)}{\Delta y}\right] = -Ar_j \tag{A19-2}$$

The term in brackets is of the form

$$\lim_{\Delta x \to 0}\left[\frac{f(x + \Delta x) - f(x)}{\Delta x}\right] = \frac{df}{dx} \tag{A19-3}$$

Taking the limit as Δy goes to zero, we obtain

$$-\frac{dF_j}{dy} = -Ar_j \tag{A19-4}$$

$$\frac{dF_j}{dy} = Ar_j \tag{A19-5}$$

(A20)

As before, we choose ΔV small enough so that there is no spatial variation in reaction rate within the subvolume. After accounting for steady-state operation in Equation (1-4), it is combined with Equation (A17-1) to yield

$$F_j(V) - F_j(V + \Delta V) + r_j \Delta V = 0 \tag{A20-1}$$

Rearranging,

$$\frac{F_j(V + \Delta V) - F_j(V)}{\Delta V} = r_j \tag{A20-2}$$

and taking the limit as $\Delta V \to 0$, we obtain Equation (1-5).

$$\frac{dF_j}{dV} = r_j \tag{1-5}$$

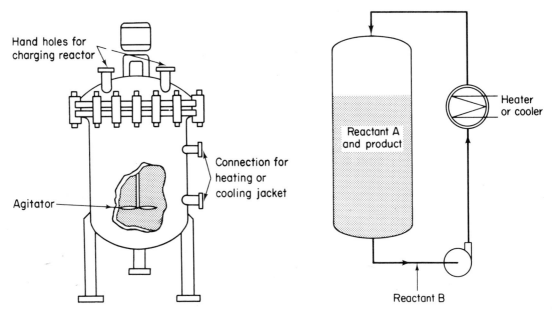

Figure 1-7. Simple Batch Homogeneous Reactor

Figure 1-8. Semibatch Reactor

reactor, it has the advantages of good temperature control and the capability of minimizing unwanted side reactions through the maintenance of a low concentration of one of the reactants.

The semibatch reactor is also used for two-phase reactions in which a gas is usually continuously bubbled through the liquid.

A *continuous stirred tank reactor* (CSTR) is used when intense agitation is required. The CSTR can either be used by itself or, in the manner shown in Fig. 1-9, as part of a series or battery of CSTR's.

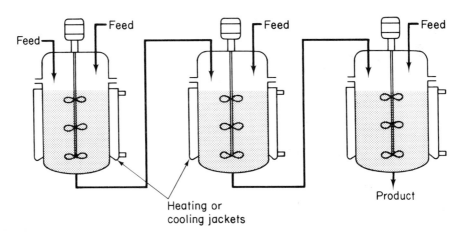

Figure 1-9. Battery of Stirred Tanks

It is relatively easy to maintain good temperature control with a CSTR; however, there is the disadvantage that the conversion of reactant per volume of reactor is the smallest of the flow reactors. Consequently, very large reactors are necessary to obtain high conversions.

The *tubular reactor* (i.e., plug flow reactor) is used for homogeneous reaction systems. In addition to being relatively easy to maintain (no moving parts), it usually produces the highest conversion per volume of reactor of any of the flow reactors. The disadvantage of the tubular

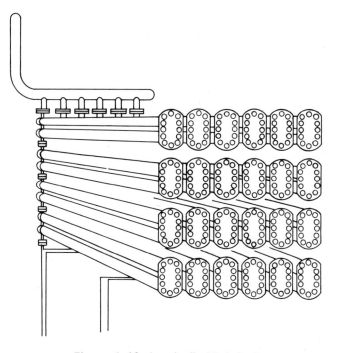

Figure 1–10. Longitudinal Tubular Reactor

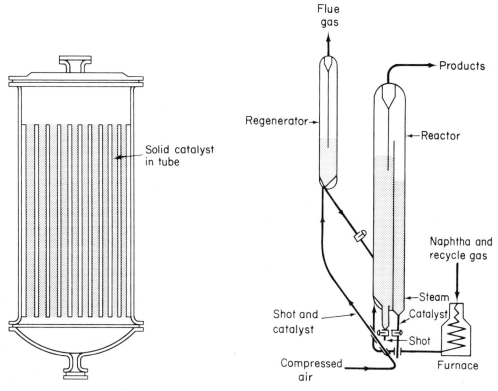

Figure 1–11. Longitudinal Catalytic Fixed-Bed Reactor

Figure 1–12. Fluidized-Bed Catalytic Reactor

29

reactor is that it is difficult to control temperature within the reactor, and hot spots can occur when the reaction is exothermic.

The tubular reactor is commonly found either in the form of one long tube or as one of a number of shorter reactors arranged in a tube bank as shown in Fig. 1-10. *Most homogeneous liquid-phase flow reactors are CSTR's while most homogeneous gas-phase flow reactors are tubular.*

A *fixed-bed* (packed bed) *reactor* is essentially a tubular reactor, which is packed with solid catalyst particles (See Fig. 1-11). This heterogeneous reaction system is mostly used to catalyze gas reactions.

Some of the limitations of this reactor are that hot spots often occur in the tubes; the catalyst is difficult to replace; and channeling of the gas flow often occurs, resulting in ineffective use of the entire catalyst bed. The advantage of the fixed-bed reactor is that for most reactions it gives the highest conversion per weight of catalyst of any catalytic reactor.

Another type of catalytic reactor in common use is the *fluidized-bed* (Fig. 1-12), which is discussed in Chapter 6 along with the fixed-bed reactor. The fluidized-bed reactor is analogous to the CSTR in that its contents, though heterogeneous, are well mixed, resulting in an even temperature distribution throughout the bed. The advantages of the ease of catalyst replacement or regeneration are sometimes offset by the high cost of the reactor and catalyst regeneration equipment.

The aim of the preceding discussion on commercial reactors is to give a more detailed picture of each of the major types of industrial reactors: batch, semibatch, CSTR, tubular, fixed-bed (packed-bed), and fluidized-bed. Many variations and modifications of these commercial reactors are in current use; for further elaboration, we refer the reader to the detailed discussion of industrial reactors given by Walas.[†]

SUMMARY

1. A mole balance on species j, which enters, leaves, reacts, and accumulates in a system volume V, is

$$F_{j0} - F_j + \int^V r_j \, dV = \frac{dN_j}{dt} \tag{S1-1}$$

2. The kinetic rate law for r_j is solely a function of the properties of the reacting materials; it is an intensive quantity. The rate law is essentially an algebraic equation and not a differential equation. For homogeneous systems, typical units of $-r_A$ may be gmoles per second per liter, and for heterogeneous systems, typical units of $-r'_A$ may be gmoles per second per gram of catalyst.

3. The differential mole balance for a tubular reactor operated at steady state is

$$\frac{dF_j}{dV} = r_j \tag{S1-2}$$

4. The mole balance on a CSTR (backmix reactor) can be expressed by the equation

$$V = \frac{F_{j0} - F_j}{-r_j} \tag{S1-3}$$

5. A mole balance on a batch reactor gives

$$\int_0^V r_j \, dV = \frac{dN_j}{dt} \tag{S1-4}$$

6. When there are no spatial variations in conditions and material properties within the reactor, Equation (S1-4) can be written

$$\frac{1}{V}\frac{dN_j}{dt} = r_j \tag{S1-5}$$

[†]S.M., Walas, *Reaction Kinetics for Chemical Engineers* (New York: McGraw-Hill, Inc., 1969) Chapter 11.

QUESTIONS AND PROBLEMS

Pl-1. What assumptions were made in the derivation of the design equation for the tubular reactor?

Pl-2. What assumptions were made in the derivation of the design equation for the batch reactor?

Pl-3. State in words the meaning of $-r'_A$ and r'_A.

Pl-4. Write the design equation for a backmix reactor that is not well mixed and consequently has a spatial variation in the reaction rate throughout the reactor volume.

Pl-5. Is the reaction rate $-r_A$ an extensive quantity? Explain.

Pl-6. What is the difference between the rate of reaction for a homogeneous system $-r_A$ and the rate of reaction for a heterogeneous system, $-r'_A$? Upon what properties does each depend?

Pl-7. How can you convert the general mole balance equation for a given species, Equation (1-4), to a general mass balance equation for that species?

Pl-8. Starting with Equation (S1-5), show that

$$\sum M_j r_j = 0 \qquad (Pl\text{-}8\text{-}1)$$

for the general reaction

$$v_3 A_3 + v_4 A_4 - v_1 A_1 - v_2 A_2 = 0 \qquad (Pl\text{-}8\text{-}2)$$

A_j represents the species, A_1 and A_2 are reactants, and A_3 and A_4 are products, M_j is the molecular weight of species j, and v_j is the stoichiometric coefficient.

Pl-9. Referring to the text material along with the additional references on commercial reactors given at the end of the chapter, fill in the following table.

Type of Reactor	Usage	Advantages	Disadvantages

Pl-10. A novel reactor used in special processing operations is the foam (liquid + gas) reactor (see Figure Pl-10). Assuming that the reaction

$$A \longrightarrow B$$

takes place only in the liquid phase, derive the differential general mole balance Equation (1-5) in terms of

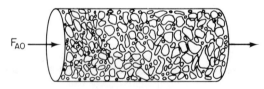

Figure P1-10 Foam Reactor.

$-r_A$ = rate of reaction (gmoles A per cu cm of liquid per sec)

e = volume fraction of gas

F_A = molar flow rate of A (gmole/sec)

V = volume of reactor

Hint: Start from a differential mole balance.

P1-11. Schematic diagrams of the Los Angeles Basin are shown below. The basin floor covers approximately 700 square miles (2×10^{10} ft^2) and is almost completely surrounded by mountain ranges. If one assumes an inversion height in the basin of 2000 feet, the corresponding volume of air in the basin is 4×10^{13} cubic feet. We shall use this system volume to model the accumulation and depletion of air pollutants. As a very rough first approximation, we shall treat the L. A. Basin as a *well-mixed* container (analogous to a CSTR) in which there are no spatial variations in pollutant concentrations.

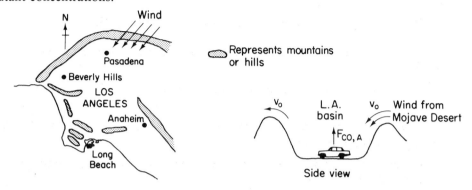

Figure P1-11

Consider only the pollutant carbon monoxide and assume that the source of CO is from automobile exhaust and that, on the average, there are 400,000 cars operating in the basin at any one time. Each car gives off roughly 3000 standard cubic feet of exhaust each hour containing 2 mole percent carbon monoxide.

We shall perform an unsteady-state mole balance on CO as it is depleted from the basin area by a Santa Ana wind. Santa Ana winds are high velocity winds that originate in the Mojave Desert just to the northeast of Los Angeles. This clean desert air flows into the basin through a corridor assumed to be 20 miles wide and 2000 ft high (inversion height replacing the polluted air which flows out to sea or towards the south. The concentration of CO in the Santa Ana wind entering the basin is .08 ppm (2.04×10^{-10} lbmoles/ft^3).

1. How many pound moles of gas are in the system volume we have chosen for the L. A. basin if the temperature is 75°F and the pressure is 1 atm? (Values of the ideal gas constant may be found in Appendix B.)
2. What is the rate, $F_{CO,A}$, at which all of the autos emit carbon monoxide into the basin (lb mole CO/hr)?
3. What is the volumetric flow rate (cu ft/hr) of a 15 mph wind through the corridor which is 20 miles wide and 2000 ft high? (ANS. 1.67×10^{13} cu ft/hr)
4. At what rate, $F_{CO,s}$ does the Santa Ana wind bring carbon monoxide into the basin (lb mole/hr)?
5. Assuming that the volumetric flow rates entering and leaving the basin are identical, $v = v_0$, show that the unsteady mole balance on CO within the basin becomes

$$F_{CO,A} + F_{CO,s} - v_0 C_{CO} = V \frac{dC_{CO}}{dt} \quad \text{(P1-11-1)}$$

6. Verify that the solution to Equation (P1-11-1) is

$$t = \frac{V}{v_0} \ln\left[\frac{F_{CO,A} + F_{CO,s} - v_0 C_{CO,0}}{F_{CO,A} + F_{CO,s} - v_0 C_{CO}}\right] \quad \text{(P1-11-2)}$$

7. If the initial concentration of carbon monoxide in the basin before the Santa Ana wind starts to blow is 8 ppm (2.04×10^{-8} lbmole/ft^3), calculate the time required for the carbon monoxide to reach a level of 2. ppm.
8. Repeat Steps 2 through 7 for another pollutant, NO. The concentration of NO in the auto exhaust is 1500 ppm (3.84×10^{-6} lbmole/ft^3), and the initial NO concentration in the basin is 0.5 ppm. If there is no NO in the Santa Ana wind, calculate the time for the NO concentration to reach 0.1 ppm. What is the lowest concentration of NO that could be reached?

SUPPLEMENTARY READING

1. For further elaboration of the development of the general balance equation see Chapters 2 and 6 of

    HIMMELBLAU, D. M., *Basic Principles and Calculations in Chemical Engineering*, (2d ed.), Englewood Cliffs, N.J.: Prentice-Hall, Inc., 1967.

    and

    DIXON, D. C., *Chemical Engineering Science*, 25 (1970) 337.

2. An excellent description of the various types of commercial reactors used in industry is found in Chapter 11 of

    WALAS, S. M., *Reaction Kinetics for Chemical Engineers*. New York: McGraw-Hill Book Co., 1950.

    A somewhat different discussion of the usage, advantages, and limitations of various reactor types can be found in pp. 1–10 of

    DENBIGH, K. G. and J. C. R. TURNER, *Chemical Reactor Theory*, (2d ed.), London: Cambridge University Press, 1971.

3. A discussion of some of the most important industrial processes is presented by Stephenson. Here, not only is the type of reactor used in a particular process discussed, but also the type and arrangement of the associated equipment (e.g., separators, heat exchangers, etc.) is presented. A significant amount of kinetic data is also given.

    STEPHENSON, R. M., *Introduction to the Chemical Process Industries*. New York: Reinhold Publishing Corp., 1966.

    A similar book, which describes a larger number of processes but does not give as much kinetic data for each of the individual processes, is

    SHREVE, R. N., *Chemical Process Industries* (2d ed.), New York: McGraw-Hill Book Co., 1956.

2
Stoichiometric Relationships and Reactor Staging

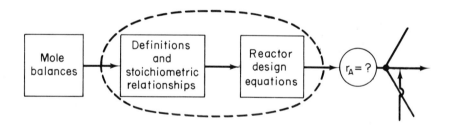

2.1–Stoichiometric Table

In Chapter 1 we stated that for a chemical species to be formed, some other species must lose their identity (i.e., disappear). This chapter will discuss how many molecules of one species will be formed during a chemical reaction when a given number of molecules of another species disappear. That is, our discussion will be concerned with the stoichiometric relationships between reacting molecules. These relationships will first be developed in a general form by considering the reaction

$$aA + bB \longrightarrow cC + dD \tag{2-1}$$

The capital letters represent chemical species and the lower case letters represent stoichiometric coefficients. One of the reacting species is usually chosen as the basis for the calculations and the other species are then related to this basis. Taking species A as our basis of calculation, we

usually find it more convenient to divide the reaction expression through by the stoichiometric coefficient of A, in order to arrange the reaction expression in the form

$$A + \frac{b}{a} B \longrightarrow \frac{c}{a} C + \frac{d}{a} D. \tag{2-2}$$

After writing the expression in this form we can easily see that for every gmole (6.02×10^{23} molecules) of A that is consumed, b/a moles of B are consumed. If 10 moles of B are consumed, how many moles of C will be produced?

(B1) _____

2.1A Batch System

For a batch system we define the conversion of A, X_A, as the number of moles of A that have reacted per mole of A initially in the reactor. In batch systems the conversion will be a function of time. If N_{A0} is the number of moles of A initially, then the total number of moles of A that have reacted after a time t is $N_{A0} X_A$. (For the sake of brevity, we shall eliminate the subscript A, and let $X \equiv X_A$. However, when considering simultaneous reactions and certain other reaction schemes, it is often more convenient to retain the subscript.)

Write the number of moles of A remaining in the reactor after a time t, N_A, in terms of N_{A0} and X.

(B2) _____

In order to determine the number of moles of each species remaining after $N_{A0} X$ moles of A have reacted, we form the Table 2-1. In this *stoichiometric* table the following information is presented: (1) the number of moles of each species initially present in the system, (2) the number of moles of each species that has reacted (i.e., the change in the number of moles of each species), and (3) the number of moles of each species remaining. As stated above, we shall take N_{A0} as the number of moles of A initially present in the reactor. Of these, $N_{A0} X_A$ moles of A were consumed in the system as a result of the chemical reaction, leaving $(N_{A0} - N_{A0} X)$ moles of A in the system. Therefore the number of moles of A remaining in the system is $N_A = N_{A0}(1 - X)$. The complete stoichiometric table is delineated in Table 2-1 for all species in the reaction,

(B1)

$$\left[\frac{(c \text{ moles C formed})/(a \text{ mole A reacted})}{(b \text{ moles B consumed})/(a \text{ mole A reacted})}\right] = \frac{c}{b} \frac{\text{moles C formed}}{\text{mole B consumed}}$$

$$[10 \text{ moles B consumed}]\left[\frac{c}{b} \frac{\text{moles C formed}}{\text{mole B consumed}}\right] = \left[10\frac{c}{b} \text{ moles C formed}\right]$$

(B2)

$$\begin{bmatrix}\text{Moles of A} \\ \text{in reactor} \\ \text{at time } t\end{bmatrix} = \begin{bmatrix}\text{Moles of A} \\ \text{initially in} \\ \text{reactor at} \\ t = 0\end{bmatrix} - \begin{bmatrix}\text{Moles of A that} \\ \text{have been con-} \\ \text{sumed by chemical} \\ \text{reaction}\end{bmatrix}$$

$$\begin{bmatrix}\text{Moles of A} \\ \text{consumed}\end{bmatrix} = \begin{bmatrix}\text{Moles of A} \\ \text{initially}\end{bmatrix}\begin{bmatrix}\text{Moles of A reacted} \\ \text{Mole of A initially}\end{bmatrix}$$

$$\text{Moles of A reacted} = N_{A0} \cdot X$$

The number of moles of A in the reactor after a conversion X has been achieved is

$$N_A = N_{A0} - N_{A0}X = N_{A0}(1 - X)$$

$$A + \frac{b}{a}B \rightarrow \frac{c}{a}C + \frac{d}{a}D. \quad (2\text{-}2)$$

TABLE 2-1

Species	Initially (moles)	Change (moles)	Remaining (moles)
A	N_{A0}	$-N_{A0}X$	$N_A = N_{A0} - N_{A0}X$
B	N_{B0}	$-\frac{b}{a}(N_{A0}X)$	$N_B = N_{B0} - \frac{b}{a}N_{A0}X$
C	N_{C0}	$+\frac{c}{a}(N_{A0}X)$	$N_C = N_{C0} + \frac{c}{a}N_{A0}X$
D	N_{D0}	$+\frac{d}{a}(N_{A0}X)$	$N_D = N_{D0} + \frac{d}{a}N_{A0}X$
Inerts, I	N_{I0}	—	$N_I = N_{I0}$
	Total $= N_{t0}$		Total $= N_{t0} + \left(\frac{d}{a} + \frac{c}{a} - \frac{b}{a} - 1\right)N_{A0}X$

The number of moles of A that have reacted is $N_{A0}X$. For every mole of A that reacts, (b/a) moles of B must react; therefore the total number of moles of B that have reacted is

$$\text{moles B reacted} = \frac{\text{moles B reacted}}{\text{moles A reacted}} \cdot \text{moles A reacted}$$

$$= \frac{b}{a}(N_{A0}X_A)$$

Since B is disappearing from the system, the sign of the "change" is negative. Therefore the number of moles of B remaining in the system, N_B, is given in the last column of Table 2-1 as

$$N_B = N_{B0} - \frac{b}{a}N_{A0}X$$

Discuss in detail (similar to the above discussion of species A and B) how each term or column in Table 2-1 is defined and formed for species C.

(B3)

We recall from Chapter 1 that the kinetic rate law (e.g., $-r_A = kC_A^2$) is a function solely of the intensive properties of the reacting materials (e.g., temperature, pressure, concentration, and catalysts, if any). The reaction rate, $-r_A$, usually depends on the concentration of the reacting species raised to some power. Consequently, to determine the reaction rate at any conversion X, we need to know the concentrations of the reacting species at that conversion.

The concentration of A is the number of moles of A per unit volume.

$$C_A = \frac{N_A}{V} \quad (2\text{-}3)$$

After writing similar equations for B, C, and D, we use the stoichiometric table to express the concentration of each compound in terms of the conversion X; i.e.,

$$C_A = \frac{N_A}{V} = \frac{N_{A0}(1-X)}{V}$$

$$C_B = \frac{N_B}{V} = \frac{N_{B0} - \frac{b}{a}N_{A0}X}{V} \quad (2\text{-}4)$$

(B3)

Column 1 gives the particular species, C.
Column 2 gives the number of moles of each species initially present. N_{C0} is the number of moles of C in the system at time $t = 0$.
Column 3 gives the change in the number of moles of C brought about by the reaction.

$$\text{moles of C formed} = \frac{\text{moles of C formed}}{\text{mole of A reacted}} \cdot (\text{moles of A reacted})$$

i.e.,

$$\text{change} = \frac{c}{a}(N_{A0}X)$$

Column 4 gives the number of moles of C in the system at any time t.

$$\begin{bmatrix} \text{Moles of C} \\ \text{in reactor} \\ \text{at time } t \end{bmatrix} = \begin{bmatrix} \text{Moles of C} \\ \text{initially} \\ \text{present} \end{bmatrix} + \begin{bmatrix} \text{Moles of C} \\ \text{generated by} \\ \text{chemical} \\ \text{reaction} \end{bmatrix}$$

$$N_C = N_{C0} + \frac{c}{a}(N_{A0}X)$$

$$C_C = \frac{N_C}{V} = \frac{N_{C0} + \frac{c}{a}N_{A0}X}{V}$$

$$C_D = \frac{N_D}{V} = \frac{N_{D0} + \frac{d}{a}N_{A0}X}{V}$$
(2-4)

What conditions and manipulations are necessary in order to write the concentration of B in the form

$$C_B = C_{A0}\left(\theta_B - \frac{b}{a}X\right)$$
(2-5)

Also, to what is θ_B equal?

(B4)

2.1B Flow Systems

For a flow system, the conversion of species A is defined as the moles of A reacted per mole of A fed to the system. Normally, the conversion increases with the time the reactants spend in the reactor. For continuous-flow systems, this time usually increases with increasing reactor volume; consequently, the conversion X is a function of the reactor volume V. If F_{A0} is the molar flow rate of A fed to a system operated at steady state, the molar rate at which A is reacting within the entire system will be $F_{A0}X$. The molar feed rate to the system minus the rate of reaction of A within the system will be equal to the molar flow rate of A leaving the system F_A. Write the preceding sentence in the form of a mathematical statement.

(B5)

What modifications of Table 2-1 would be necessary in order to write a stoichiometric table for a flow system? Taking A as your basis, try to set up a stoichiometric table for a flow system for the reaction $aA + bB \rightarrow cC + dD$ without referring to Table 2-1 for a batch system.

(B4)

From Equation (2-4)

$$C_B = \frac{N_{B0}}{V} - \frac{\frac{b}{a} N_{A0} X}{V}$$

For constant volume, $V = V_0$, $C_{B0} = N_{B0}/V_0$,

$$C_B = C_{B0} - \frac{b}{a} C_{A0} X$$

$$= C_{A0} \left[\frac{C_{B0}}{C_{A0}} - \frac{b}{a} X \right]$$

θ_B is defined by

$$\theta_B = \frac{N_{B0}}{N_{A0}} \qquad \text{for batch systems}$$

$$\theta_B = \frac{F_{B0}}{F_{A0}} \qquad \text{for flow systems}$$

Let $\theta_B = C_{B0}/C_{A0}$:

$$C_B = C_{A0} \left(\theta_B - \frac{b}{a} X \right)$$

Therefore the concentration of B can be given by Equation (2-5) *only* for the conditions of constant volume.

(B5)

$$[F_{A0}] \cdot [X] = \frac{\text{moles of A fed}}{\text{time}} \cdot \frac{\text{moles of A reacted}}{\text{mole of A fed}}$$

$$F_{A0} \cdot X = \frac{\text{moles of A reacted}}{\text{time}}$$

$$\begin{bmatrix} \text{Molar flow rate} \\ \text{at which A is} \\ \text{fed to the system} \end{bmatrix} - \begin{bmatrix} \text{Molar rate at} \\ \text{which A is} \\ \text{consumed} \\ \text{within the} \\ \text{system} \end{bmatrix} = \begin{bmatrix} \text{Molar flow rate} \\ \text{at which A leaves} \\ \text{the system} \end{bmatrix}$$

$$F_{A0} \quad - \quad F_{A0} X \quad = \quad F_A$$

Rearranging

$$F_A = F_{A0}(1 - X)$$

(B6)

Species				

For a flow system, the concentration C_A at a given point is defined in terms of F_A and the volumetric flow rate v at that point. The units of v are usually given in terms of liters per second or cubic feet per minute:

$$C_A = \frac{F_A}{v} = \frac{\text{mole/time}}{\text{liters/time}} = \frac{\text{moles}}{\text{liter}}$$

Write the concentrations of A, B, C, and D in terms of the entering molar flow rates (F_{A0}, F_{B0}, F_{C0}, F_{D0}); the conversion of A, X, and the volumetric flow rate, v; for the general reaction given by Equation (2-1).

(B7)

For a flow system, under what conditions does C_B, given by Equation (B7-2), reduce to Equation (2-5) (i.e., $C_B = C_{A0}[\theta_B - (b/a)X]$)?

(B8)

We also note that for almost all liquid phase reactions the fluid density is almost constant; therefore the assumptions of constant fluid volume V (batch) and constant volumetric flow rate v (flow) for liquid phase reactions are most often useful approximations (i.e., $V = V_0$ and $v = v_0$).

The ratio of stoichiometric coefficients can also be used to obtain the relative rates of reactions of the various species involved in the reaction. For the reaction

$$A + \frac{b}{a}B \longrightarrow \frac{c}{a}C + \frac{d}{a}D$$

we see that for every mole of A that is consumed, c/a moles of C appear:

$$\text{Rate of formation of C} = \frac{c}{a}(\text{Rate of disappearance of A})$$

Write this statement in terms of r_C, r_A, and the stoichiometric coefficients. Also write an equation relating the rates of formation of C and D.

(B6) For continuous-flow systems in which the same single chemical reaction is taking place, we can simply replace N_{j0} by F_{j0} and N_j by F_j. Taking A as our basis, divide Equation (2-1) through by the stoichiometric coefficient of A as in Equation (2-2) to obtain

$$A + \frac{b}{a}B \longrightarrow \frac{c}{a}C + \frac{d}{a}D \tag{2-2}$$

TABLE 2-1(a)

Species	Feed rate to reactor (moles/time)	Change within reactor (moles/time)	Effluent rate from reactor (moles/time)
A	F_{A0}	$-F_{A0}X$	$F_A = F_{A0}(1-X)$
B	F_{B0}	$-\frac{b}{a}F_{A0}X$	$F_B = F_{B0} - \frac{b}{a}F_{A0}X$
C	F_{C0}	$\frac{c}{a}F_{A0}X$	$F_C = F_{C0} + \frac{c}{a}F_{A0}X$
D	F_{D0}	$\frac{d}{a}F_{A0}X$	$F_D = F_{D0} + \frac{d}{a}F_{A0}X$
I	F_{I0}	—	F_{I0}
	F_{t0}		$F_t = F_{t0} + \left(\frac{d}{a} + \frac{c}{a} - \frac{b}{a} - 1\right)F_{A0}X$

(B7)
$$aA + bB \longrightarrow cC + dD \tag{2-1}$$

Taking A as our basis,

$$A + \frac{b}{a}B \longrightarrow \frac{c}{a}C + \frac{d}{a}D \tag{2-2}$$

$$C_A = \frac{F_A}{v} = \frac{F_{A0}}{v}(1-X) \quad \text{(B7-1)} \qquad C_B = \frac{F_B}{v} = \frac{F_{B0} - \frac{b}{a}F_{A0}X}{v} \quad \text{(B7-2)}$$

$$C_C = \frac{F_C}{v} = \frac{F_{C0} + \frac{c}{a}F_{A0}X}{v} \quad \text{(B7-3)} \qquad C_D = \frac{F_D}{v} = \frac{F_{D0} + \frac{d}{a}F_{A0}X}{v} \quad \text{(B7-4)}$$

The flow rates F_A, F_B, F_C, and F_D are found in Table 2-1(a), Frame (B6).

(B8)
$$C_B = \frac{F_{B0} - \frac{b}{a}F_{A0}X}{v} = \frac{F_{A0}\left(\frac{F_{B0}}{F_{A0}} - \frac{b}{a}X\right)}{v}$$

For constant volumetric flow rate $v = v_0$, $C_{A0} = F_{A0}/v_0$

$$C_B = C_{A0}\left(\theta_B - \frac{b}{a}X\right)$$

$$\theta_B = \frac{F_{B0}}{F_{A0}} = \frac{C_{B0}v_0}{C_{A0}v_0} = \frac{C_{B0}}{C_{A0}} = \frac{y_{B0}}{y_{A0}}$$

the same as in Frame (B4) (where y_{B0} and y_{A0} are the initial mole fractions of B and A respectively).

(B9)

A similar relationship can be written between A and B:

$$r_B = \frac{b}{a} r_A \tag{2-6}$$

and between C and D:

$$r_C = \frac{c}{d} r_D \tag{2-7}$$

2.2–Design Equations (Mole Balances)

2.2A Batch

The mole balance on species A for a batch system in which there are no spatial variations in temperature or concentration results in the following equation. [Review Frame (A12).]

$$\frac{dN_A}{dt} = r_A V \tag{2-8}$$

This equation holds, whether or not the reactor volume is constant. In the general reaction

$$A + \frac{b}{a} B \longrightarrow \frac{c}{a} C + \frac{d}{a} D \tag{2-2}$$

reactant A is disappearing; therefore we multiply both sides of Equation (2-8) by -1 to obtain the mole balance for the batch reactor in the form

$$-\frac{dN_A}{dt} = (-r_A)V \tag{2-9}$$

The rate of disappearance of A, $-r_A$, in this reaction might be given by a rate law similar to Equation (1-2), such as $-r_A = k C_A C_B$.

For batch reactors we are interested in determining how long we should leave the reactants in the reactor in order to achieve a certain conversion X.

After referring back to Table 2-1, rewrite the differential on the left-hand side of Equation (2-8) in terms of the conversion and the number of moles of A initially present in the system.

(B10)

The differential forms of the design equations are usually used to interpret reaction rate data. For a *batch* reactor, the design equation in differential form is

$$\boxed{N_{A0} \frac{dX}{dt} = -r_A V} \tag{B10-4}$$

(B9)

Make sure you did not leave out the minus sign in writing the rate of disappearance.

$$r_C = \frac{c}{a}(-r_A) = -\frac{c}{a}r_A$$

The relationship between the rate of formation of C and D is

$$\left.\begin{array}{l} r_C = -\dfrac{c}{a}r_A \\ r_D = -\dfrac{d}{a}r_A \end{array}\right\} \quad r_C = \dfrac{c}{d}r_D$$

Actually this manipulation is not at all necessary, since one can write the relationship directly from the stoichiometry of the reaction.

$$aA + bB \longrightarrow cC + dD$$

for which

$$\boxed{\dfrac{-r_A}{a} = \dfrac{-r_B}{b} = \dfrac{r_C}{c} = \dfrac{r_D}{d}}$$

Then

$$r_C = \frac{c}{d}r_D$$

$$r_D = -\frac{d}{a}r_A$$

⋮

(B10)

$$N_A = N_{A0} - N_{A0}X \tag{B10-1}$$

Differentiating with respect to time and remembering that N_{A0} is the number of moles initially present and is therefore a constant with respect to (w.r.t.) time,

$$\frac{dN_A}{dt} = 0 - N_{A0}\frac{dX}{dt} \tag{B10-2}$$

Combining with Equation (2-8),

$$-N_{A0}\frac{dX}{dt} = r_A V \tag{B10-3}$$

$$N_{A0}\frac{dX}{dt} = -r_A V \tag{B10-4}$$

In the special case of a constant-volume batch reactor, Equation (2-9) can be arranged in the form

$$-\frac{1}{V}\frac{dN_A}{dt} = -\frac{d\frac{N_A}{V}}{dt} = \frac{-dC_A}{dt} = -r_A \qquad (2\text{-}10)$$

For reaction systems in which the volume varies during the reaction, it will be shown that we can express V and $-r_A$ as functions of X alone, for either adiabatic or isothermal reactors. Consequently, we can separate the variables of the differential equation in Equation (B10-4):

$$dt = \frac{dX}{-r_A V} N_{A0}$$

Integrating this equation with the limits $t = 0$, $X = 0$, we obtain an equation giving the time t necessary to achieve a specified conversion X.

$$t = N_{A0} \int_0^X \frac{dX}{(-r_A)V} \qquad (2\text{-}11)$$

Equations (B10-4) and (2-11) are differential and integral forms, respectively, of the design equations for a batch reactor. The differential form is most often used in the interpretation of laboratory rate data.

2.2B CSTR or Backmix Reactor

The equation resulting from a mole (mass) balance on species A for the reaction

$$A + \frac{b}{a}B \longrightarrow \frac{c}{a}C + \frac{d}{a}D$$

Equation (2-2), occurring in a CSTR was previously given in Chapter 1 by Equation (A15-4):

$$F_{A0} - F_A = -r_A V \qquad (2\text{-}12)$$

Rewrite Equation (2-12) utilizing the stoichiometric table [Table 2-1(a)] to substitute for the exiting molar flow rate, F_A, of A in terms of the conversion X and the entering molar flow rate F_{A0}, of A.

(B11)

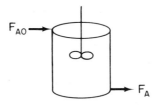

Figure 2–1

(B11)

Equation (2-12):
$$F_{A0} - F_A = -r_A V \qquad (2\text{-}12)$$

But by Equation (B5) we have
$$F_A = F_{A0} - F_{A0}X \qquad (B5)$$

Combining Equations (2-12) and (B5),
$$F_{A0}X = -r_A V \qquad (B11)$$

We can rearrange Equation (B11) to determine the backmix reactor volume necessary to achieve a specified conversion X.

$$V = \frac{F_{A0}X}{-r_A} \qquad (2\text{-}13)$$

2.2C Tubular Flow Reactor

After multiplying both sides of the tubular reactor design Equation (1-5) by -1, we express the mole balance equation for species A in the reaction given by Equation (2-2)

$$-\frac{dF_A}{dV} = -r_A \qquad (2\text{-}14)$$

For a flow system, F_A has previously been given in terms of the entering molar flow rate F_{A0} and the conversion X.

$$F_A = F_{A0} - F_{A0}X \qquad (2\text{-}15)$$

Combining Equations (2-14) and (2-15) we obtain the differential form of the design equation of a plug flow reactor.

$$F_{A0}\frac{dX}{dV} = -r_A \qquad (2\text{-}16)$$

We now separate the variables and integrate with the limits $V = 0$ when $X = 0$ to obtain the plug flow reactor volume necessary to achieve a specified conversion X:

$$V = F_{A0}\int_0^X \frac{dX}{-r_A} \qquad (2\text{-}17)$$

In order to carry out the integrations in the batch and plug flow reactor design Equations (2-11) and (2-17), as well as the evaluation of the CSTR design equation, we need to know how the reaction rate $-r_A$ varies with the concentration (hence conversion) of the reacting species. This relationship between reaction rate and concentration will be developed in Chapter 3. However, before proceeding to Chapter 3, let us work through the problems in Section 2.3 to ensure that the concepts presented in the preceding sections are clear.

2.3–PROBLEMS

2.3A–Stoichiometric Table

Soap consists of the sodium and potassium salts of various fatty acids such as oleic, stearic, palmitic, lauric, and myristic. The saponification reaction for the formation of soap from aqueous caustic soda and glyceryl stearate is

$$3\text{NaOH(aq.)} + (C_{17}H_{35}COO)_3C_3H_5 \longrightarrow 3C_{17}H_{35}COONa + C_3H_5(OH)_3$$

Letting X represent the conversion of sodium hydroxide (the moles of sodium hydroxide reacted per mole sodium hydroxide initially present), set up a stoichiometric table expressing

the concentration of each species in terms of its initial concentration and the conversion X. (*Hint:* Remember the relationship between V and V_0 and v and v_0 for a liquid phase reaction.) Work this out for a batch reactor.

(B12)

Species			

If the initial mixture consists solely of sodium hydroxide at a concentration of 10 gmoles per liter and of glyceryl stearate at a concentration of 2 gmoles per liter, what is the concentration of glycerine when the conversion of sodium hydroxide is (1) 20% and (2) 90%? Be careful with part 2.

(B13)

2.3B Applications of the Design Equations

The rate of disappearance of A, $-r_A$, is a function of concentration of the various species, each of which can be expressed as a function of the conversion X for a single reaction; consequently $-r_A$ can be expressed as a function of X. Make a sketch showing how $(1/-r_A)$ (y axis) might vary with X (x axis). Explain your reasoning for sketching the curve the way you did. (*Hint:* A first-order rate equation may sometimes take the form $-r_A = kC_{A0}(1 - X)$ when C_A is expressed in terms of X.)

(B14)

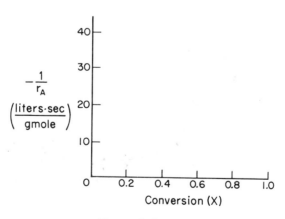

Figure 2–2

Equation (2-13) gives the volume of a CSTR as a function of F_{A0}, X, and $-r_A$; i.e.,

$$V = \frac{F_{A0} X}{-r_A}$$

51

(B12)

Since we are taking sodium hydroxide as our basis, we divide through by the stoichiometric coefficient of sodium hydroxide to put the reaction expression in the form

$$NaOH + \tfrac{1}{3}(C_{17}H_{35}COO)_3 C_3H_5 \longrightarrow C_{17}H_{35}COONa + \tfrac{1}{3}C_3H_5(OH)_3$$

$$A \quad + \tfrac{1}{3}B \quad\quad\quad\quad \longrightarrow C \quad\quad + \tfrac{1}{3}D$$

Species	Initially	Change	Remaining	Concentration
NaOH	N_{A0}	$-N_{A0}X$	$N_{A0}(1-X)$	$C_{A0}(1-X)$
$(C_{17}H_{35}COO)_3C_3H_5$	N_{B0}	$-\tfrac{1}{3}N_{A0}X$	$N_{A0}\left(\theta_B - \tfrac{X}{3}\right)$	$C_{A0}\left(\theta_B - \tfrac{X}{3}\right)$
$C_{17}H_{35}COONa$	N_{C0}	$N_{A0}X$	$N_{A0}(\theta_C + X)$	$C_{A0}(\theta_C + X)$
$C_3H_5(OH)_3$	N_{D0}	$\tfrac{1}{3}N_{A0}X$	$N_{A0}\left(\theta_D + \tfrac{X}{3}\right)$	$C_{A0}\left(\theta_D + \tfrac{X}{3}\right)$
Water (inert solvent)	N_{I0}	—	—	C_{I0}
Total	N_{t0}		N_{t0}	

Since this is a liquid phase reaction, the density ρ is considered to be constant; therefore $V = V_0$.

$$C_A = \frac{N_A}{V} = \frac{N_A}{V_0} = \frac{N_{A0}(1-X)}{V_0} = C_{A0}(1-X)$$

$$\theta_B = \frac{C_{B0}}{C_{A0}}, \quad \theta_C = \frac{C_{C0}}{C_{A0}}, \quad \theta_D = \frac{C_{D0}}{C_{A0}}$$

(B13)

Only the reactants NaOH and $(C_{17}H_{35}COO)_3C_3H_5$ are initially present; therefore $\theta_C = \theta_D = 0$.

1. For 20% conversion:

$$C_D = C_{A0}\frac{X}{3} = (10)\left(\frac{.2}{3}\right) = .66 \text{ gmole}/l$$

$$C_B = C_{A0}\left(\theta_B - \frac{X}{3}\right) = 10\left(\frac{2}{10} - \frac{.2}{3}\right) = 10(.133) = 1.33 \text{ gmole}/l$$

2. For 90% conversion:

$$C_D = C_{A0}\left(\frac{X}{3}\right) = 10\left(\frac{.9}{3}\right) = 3 \text{ gmoles}/l$$

Let us find C_B:

$$C_B = 10\left(\frac{2}{10} - \frac{.9}{3}\right) = 10(.2 - .3) = -1 \text{ gmole}/l$$

Negative concentration—impossible!
90% conversion of NaOH is not possible, since glyceryl stearate is the limiting reactant.

(B14)

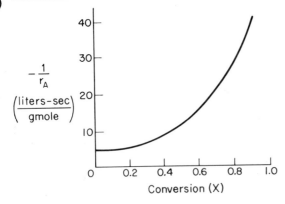

For most reactions, the reaction rate is directly proportional to some power of the concentrations of the reactants. The reaction rate is greatest when negligible conversion has been achieved ($X \simeq 0$) and the concentration of the reactants is greatest. (Hence $1/-r_A$ will be small.) When the conversion is large, the reactant concentration will be small, as will be the reaction rate. (Consequently $(1/-r_A)$ is large.) A sketch of a typical reciprocal rate conversion curve is shown in Figure 2-3.

Figure 2–3

Shade the area on Figure 2-2 which, when multiplied by F_{A0}, would give the volume of a CSTR necessary to achieve 80% conversion (i.e., $X = .8$).

(B15) _____

_____

The design equation for a plug flow reactor is

$$V = F_{A0} \int_0^X \frac{dX}{-r_A} \qquad (2\text{-}17)$$

Assuming that you know F_{A0} and you know $-r_A$ as a function of X, how would you plot $-r_A$ as a function of X to determine, by graphical integration, the tubular reactor volume necessary to achieve a specified conversion of A?

(B16) _____

_____

Shade the area under the curve which is equal to the integral

$$I = \int_0^{.8} \frac{dX}{-r_A}$$

Also state how you would determine the tubular reactor volume for a given F_{A0}.

(B17) _____

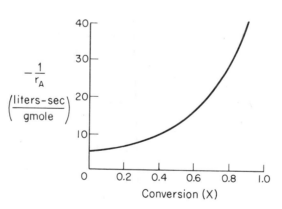

Figure 2–5

_____

Shown in Frame (B18), Figure 2-7, is a plot of the reciprocal reaction rate as a function of conversion. For a given flow rate, F_{A0}, which reactor would require the smaller volume to achieve a conversion of 60%; a CSTR or a tubular reactor? Explain.

(B15)

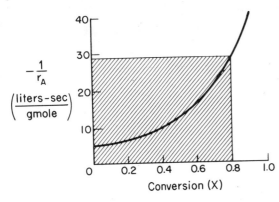

Figure 2-4

Algebraically,

$$V = F_{A0}\left(\frac{1}{-r_A}\right)X, \quad \frac{V}{F_{A0}} = \left(\frac{1}{-r_A}\right)(.8)$$

In a CSTR the effluent stream is identical to the conversion within the reactor since perfect mixing is assumed. Therefore we need to find the value of $-r_A$ (or reciprocal thereof) when $X = .8$. From Figure 2-4 we see that when $X = .8$, then $1/-r_A = 27.5$

$$\frac{V}{F_{A0}} = (27.5)(.8) = 22 \text{ liters-sec/gmole}$$

In Figure 2-4 this value of V/F_{A0} is equal to the area of a rectangle with a height $1/-r_A = 27.5$ and a base $X = .8$. This rectangle is shaded in Figure 2-4.

If the entering molar flow rate, F_{A0}, were 2 gmoles/sec, the CSTR reactor volume necessary to achieve 80% conversion would be

$$V = 2\frac{\text{gmoles}}{\text{sec}} \times 22\frac{l \cdot \text{sec}}{\text{gmoles}}$$

$$V = 44 \, l$$

(B16) The integral in Equation (2-17) can be evaluated from the area under the curve of a plot of $1/-r_A$ vs. X. The product of this area and F_{A0} will give the tubular reactor volume necessary to achieve the specified conversion of A, X.

(B17)

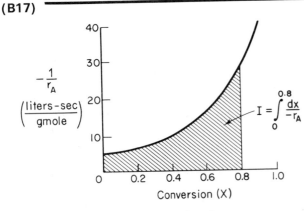

Figure 2-6

The shaded area is roughly equal to 10 l-sec per gmole. The tubular reactor volume can be determined by multiplying this area (in l-sec per gmole) by F_{A0} (gmole per sec). Short cuts for evaluating the area under the curve in graphical integration are given in Appendix A-1.

(B18)

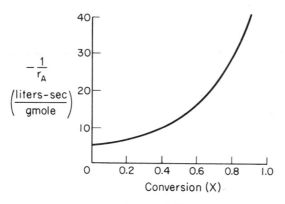

Figure 2-7

2.3C Reactor Staging

The gas phase reaction $A \rightleftharpoons 3B$ is carried out isothermally. Table 2-2 gives the conversion as a function of chemical reaction rate at a temperature of 300°F and a pressure of 10 atm. The initial charge was an equal molar mixture of A and inerts.

TABLE 2-2

X	$-r_A$ (gmoles/l/sec)
0	.0053
.1	.0052
.2	.0050
.3	.0045
.4	.0040
.5	.0033
.6	.0025
.7	.0018
.8	.00125
.85	.00100

How would you plot these data so that they could most conveniently be used to determine the reactor volume necessary to achieve a specified conversion for either a CSTR or a plug flow reactor? Also, what additional information, if any, is necessary to determine the time required to achieve 70% conversion in a batch reactor?

(B19)

The volumetric feed rate is 2070 l/hr. The gas mixture, which consists of 50% A and 50% inerts at 10 atm, enters the reactor at 300°F. What is the concentration, C_{A0}, of A initially? The entering molar flow rate of A, F_{A0}? The ideal gas constant $= R = .082$ l-atm/gmole-°K.

(B18)

For the CSTR:

$$\frac{V}{F_{A0}} = \frac{1}{-r_A} \cdot X = (16)\cdot(.6) = 9.6 \frac{l\text{-sec}}{gmole}$$

This is also the area of the rectangle with vertices $(X, 1/-r_A)$ of $(0, 0)$, $(0, 16)$, $(.6, 16)$, and $(.6, 0)$.

In the plug flow (tubular reactor)

$$\frac{V}{F_{A0}} = \int_0^{.6} \frac{dX}{-r_A} = \text{area under the curve between } X = 0 \text{ and } X = .6.$$

$$\simeq 5.1 \frac{l\text{-sec}}{gmole}$$

For the same flow rate F_{A0}, the plug flow reactor requires a smaller volume than the CSTR to achieve a conversion of 60%.

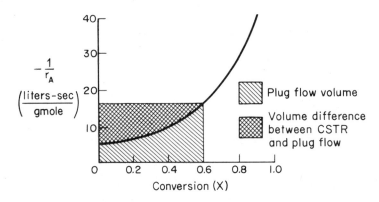

Figure 2–8

(B19)

Plot $1/-r_A$ vs. X. First we need to calculate $1/-r_A$ from Table 2-2:

$X =$	0	.1	.2	.3	.4	.5	.6	.7	.8	.85
$-r_A =$	.0053	.0052	.0050	.0045	.0040	.0033	.0025	.0018	.00125	.001
$\dfrac{1}{-r_A} =$	189	192	200	222	250	303	400	556	800	1000

For a batch reactor, the time necessary to achieve a specified conversion (e.g., 70%) is

$$t = N_{A0} \int_0^X \frac{dX}{(-r_A V)}$$

For a variable volume we cannot evaluate the integral unless we know V as a function of X. However, as we shall see later, this relationship can be obtained from knowledge of the initial concentration and the reaction stoichiometry.

(B20)

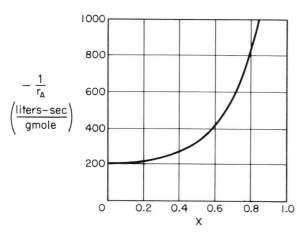

Figure 2-9

The reaction rates in Table 2-2 have been converted to reciprocal rates, $1/-r_A$, and plotted as a function of conversion in Figure 2-9. *This figure may be used in solving Frames (B21)–(B27).*

What is the reactor volume necessary to achieve an 80% conversion for (1) a plug flow reactor and (2) a CSTR?

(B21)

Consider two CSTRs in series shown in Figure 2-11.

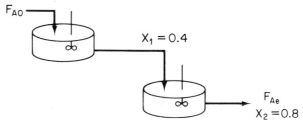

Figure 2–11

If a 40% conversion is achieved in the first reactor, what total reactor volume is necessary to accomplish an 80% overall conversion of species A entering reactor 1? ($F_{Ae} = .2\, F_{A0}$.) F_{Ae} is the molar flow rate of A exiting from the last reactor of the scheme.

(B20)

P_{A0} = Initial partial pressure
Π_0 = Initial total pressure
y_{A0} = Initial mole fraction of A

Ideal Gas Law

$$C_{A0} = \frac{P_{A0}}{RT_0} = \frac{y_{A0}\Pi_0}{RT_0}, \qquad (B20\text{-}1)$$

$T_0 = 300°\text{F} = 423°\text{K}$

$C_{A0} = \dfrac{(.5)(10 \text{ atm})}{(.082 \text{ } l\text{-atm/gmole/°K})(423°\text{K})}$

$C_{A0} = .1445 \text{ gmole}/l$

$F_{A0} = C_{A0}v_0 = .1445 \text{ gmole}/l \cdot 2070 \text{ } l/\text{h} = 299 \text{ gmoles/h} = .083 \text{ gmoles/sec}$

(B21)

For the CSTR (when $X = .8$, then $(1/-r_A) = 800 \, l\cdot\text{sec/gmole}$)

$$V = F_{A0}\frac{1}{-r_A}X = 299 \frac{\text{gmoles}}{\text{h}} \cdot 800 \frac{l\cdot\text{sec}}{\text{gmole}} \cdot .8 \cdot \frac{1 \text{ h}}{3600 \text{ sec}}$$

$V = 53$ liters

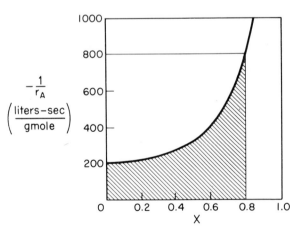

Figure 2–10

For the plug flow reactor

$$V = F_{A0}\int_0^{.8} \frac{dX}{-r_A} = F_{A0}I$$

The shaded area in Figure 2-10 is equal to the integral I above.

$I = 265 \; l\text{-sec/gmole}$

$V = 299 \text{ gmoles/h} \cdot 265 \; l\text{-sec/gmole} \cdot 1 \text{ h}/3600 \text{ sec}$

$V = 22$ liters

(B22)

For an overall conversion of 80%, how many cubic feet of reactor volume are saved by having two CSTRs in series with an intermediate conversion of 40% rather than one CSTR in which 80% conversion is achieved? (1 cu ft = 28.32 l.)

(B23)

(B22)

$$V_1 = F_{A0} \cdot \frac{1}{-r_{A1}} \cdot X$$

$$V_2 = \frac{F_{A2} - F_{A1}}{-r_{A2}} = \frac{F_{A0}(1-X_2) - F_{A0}(1-X_1)}{-r_{A2}} = \frac{F_{A0}(X_2 - X_1)}{-r_{A2}}$$

$F_{A0} = 299 \text{ gmoles/hr} = .083 \text{ gmole/sec}$

Reactor 1

$$V_1 = .083\left(\frac{1}{-r_{A1}}\right)(.4) = (.083)(250)(.4) = 8.30 \; l$$

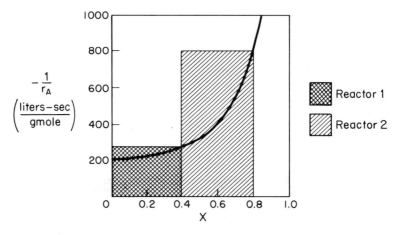

Figure 2-12

Reactor 2

$$V_2 = (.083)\left(\frac{1}{-r_{A2}}\right)(.8 - .4) = .083(800)(.4) = 26.6 \; l$$

where the rate $-r_{A1}$ is evaluated at a conversion of .4 and rate $-r_{A2}$ is evaluated at a conversion of .8.
Total volume:

$$V = V_1 + V_2 = 34.9 \text{ liters}$$

(B23)

From Frame (B21), we found that a CSTR volume of 53 liters was necessary to achieve 80% conversion when one reactor was used. However, by using two CSTRs in series, the total reactor volume necessary to achieve this conversion drops to 34.9 liters.

The saving in reactor volume is

$$V = (53 - 34.9) \; l \times \frac{1 \text{ ft}^3}{28.32 \; l} = .64 \text{ ft}^3$$

Consider two tubular reactors in series.

How many cubic feet of reactor volume would be saved by the arrangement in Figure 2-13 showing two tubular reactors, over a system in which there is only one tubular reactor?

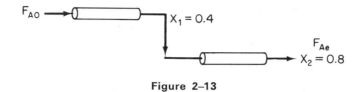

Figure 2-13

(B24)

For the reaction data given in Table 2–2, consider the series arrangements of a CSTR and a tubular flow reactor shown in Figure 2-15. If the intermediate conversion X_1 is .7, which reactor scheme (A or B) should be used to obtain the smaller total reactor volume? Support your decision with calculations.

SCHEME A

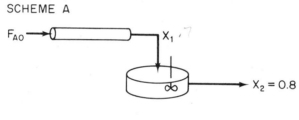

SCHEME B

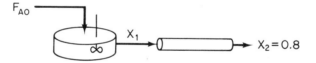

Figure 2-15

(B25)

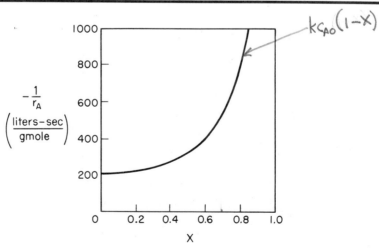

Figure 2-16

(B24) We can see from Figure 2-14 that no savings in reactor volume results from arranging the plug flow reactors in series. We also note

$$\int_0^{.8} \frac{dX}{-r_A} \equiv \int_0^{.4} \frac{dX}{-r_A} + \int_{.4}^{.8} \frac{dX}{-r_A}$$

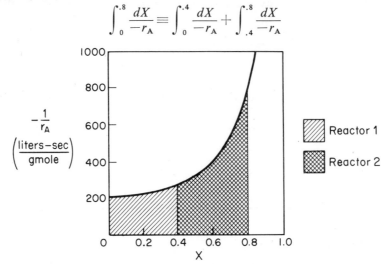

Figure 2-14

(B25) *Scheme A:* Plug flow:

$$V_1 = F_{A0} \int_0^{.7} \frac{dX}{-r_A} = .083 \frac{\text{gmole}}{\text{sec}} \cdot 200 \frac{l\text{-sec}}{\text{gmole}} = 16.6 \, l$$

CSTR:

$$V_2 = F_{A0} \frac{(X_2 - X_1)}{-r_{A2}} = .083(.8 - .7)(800) = 6.65 \, l$$

$$V_{\text{total}} = V_1 + V_2 = 23.25 \text{ liters}$$

Scheme B: CSTR:

$$V_1 = F_{A0} \frac{X_1}{-r_{A1}} = .083(.7)(556) = 32.4 \, l$$

Plug flow:

$$V_2 = F_{A0} \int_{.7}^{.8} \frac{dX}{-r_A} = .083(68) = 5.65 \, l$$

$$V_{\text{total}} = V_1 + V_2 = 38.05 \text{ liters}$$

System A will give the smallest total reactor volume for an intermediate conversion of 70%.

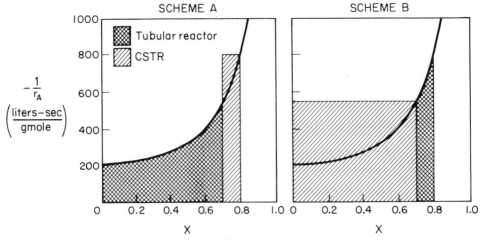

Figure 2-17

Can you extrapolate these results to make the statement that "if you have a CSTR and a tubular reactor in series, you should always place the plug flow reactor first to achieve the smallest total reactor volume for a given overall conversion"?

(B26)

If the intermediate conversion X_1 is .5, which system (A or B) will give the smaller total reactor volume?

(B27)

In the previous examples we observed that if we know the molar flow rate into the reactor, F_{A0}, and we know $-r_A$ as a function of X, we could calculate the reactor volume necessary to achieve a specified conversion. However, $-r_A$ is not a unique function of conversion. It also depends on the initial concentration of the reactants as well as the temperature. Consequently the experimental data obtained in the laboratory and presented in Table 2-2 in terms of reaction rate $-r_A$ and conversion X are useful only in the design of full-scale reactors that are to be operated at the same conditions (temperature, initial reactant concentrations) as the laboratory experiments from which the data were obtained.

For reactions in which the rate depends only on the concentration of one species, i.e., $-r_A = \text{fn}(C_A)$, it is usually convenient to report $-r_A$ as a function of concentration rather than conversion. Rewrite the design equation for a plug flow reactor [Equation (2-17)] in terms of the concentration of C_A rather than in terms of conversion, for the special case when $v = v_0$.

(B26) No; it depends upon the intermediate conversion.

(B27) *Scheme A*: Plug flow:

$$V_1 = F_{A0} \int_0^{.5} \frac{dX}{-r_A} = .083(107.5) = 8.9\, l$$

CSTR:

$$V_2 = F_{A0} \frac{(X_2 - X_1)}{-r_{A2}} = .083(.8 - .5)(800) = 19.9\, l$$

$$V_{total} = V_1 + V_2 = 28.8\, l$$

Scheme B:

$$V_1 = F_{A0} \frac{X_1}{-r_{A1}} = .083(.5)(303) = 12.80\, l$$

$$V_2 = F_{A0} \int_{.5}^{.8} \frac{dX}{-r_A} = .083(153) = 12.7\, l$$

$$V_{total} = 25.5 \text{ liters}$$

Scheme B will give the smallest total reactor volume for an intermediate conversion of 50%.

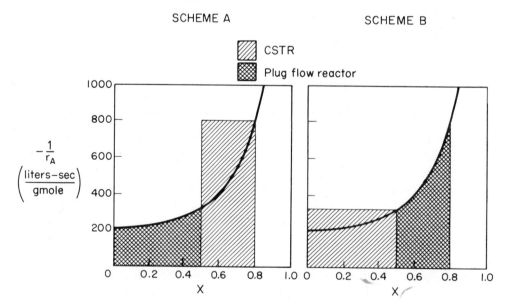

Figure 2–18

(B28)

If we divide both sides of Equation (2-17) by the entering volumetric flow rate v_0 and define the *space time* τ by the ratio

$$\tau = \frac{V}{v_0}, \qquad (2\text{-}18)$$

the design equation can be written in the form

$$\tau = C_{A0} \int_0^X \frac{dX}{-r_A} \qquad (2\text{-}19)$$

Either by dividing Equation (B28-6) by v_0 or by taking C_{A0} into the integral of Equation (2-19), the design equation for *no volume change* on reaction takes the form

$$\tau = \int_{C_A}^{C_{A0}} \frac{dC_A}{-r_A} \qquad (2\text{-}20)$$

The *space time* τ is the time needed to feed into the reactor an amount of material equivalent to that contained in one reactor volume at the entrance conditions. The space velocity SV is the reciprocal of the space time τ.

$$SV = \frac{1}{\tau} = \frac{v_0}{V} \qquad (2\text{-}21)$$

The two space velocities commonly referred to in industry are the liquid-hourly space velocity, *LHSV*, and the gas hourly space velocity, *GHSV*. The GHSV is normally reported at STP.

Equation (2-20) is a form of the design equation for constant volumetric flow rate v that may prove more useful in determining the space time or reactor volume for reaction rates that depend only on the concentration of one species. Consider such a system and sketch the reciprocal rate as a function of the concentration of A, and explain how you would determine the space time needed to reduce the concentration of A from C_{A0} to C_{A1}. Include in your explanation the shading of the area under the curve that is appropriate when graphical techniques are used in evaluating the design equation.

(B28)

$$V = F_{A0} \int_0^X \frac{dX}{-r_A} \qquad \text{(B28-1)}$$

$$F_{A0} = v_0 C_{A0} \qquad \text{(B28-2)}$$

$$X = \frac{F_{A0} - F_A}{F_{A0}} \qquad \text{(B28-3)}$$

With $v = v_0$

$$X = \frac{C_{A0} - C_A}{C_{A0}} \qquad \text{(B28-4)}$$

$$dX = \frac{-dC_A}{C_{A0}} \qquad \text{(B28-5)}$$

when $X = 0$, $\quad C_A = C_{A0}$
when $X = X$, $\quad C_A = C_A$

$$V = v_0 \int_{C_A}^{C_{A0}} \frac{dC_A}{-r_A} \qquad \text{(B28-6)}$$

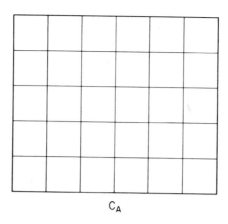

Figure 2-19

In summary of these last examples, we have seen that, in the design of reactors that are to be operated at conditions (e.g., temperature, initial concentration) identical to those at which the reaction rate data were obtained, detailed knowledge of the kinetic rate law $-r_A$ is not always necessary. In some instances it may be possible to scale-up a laboratory-bench or pilot-plant reaction system solely from knowledge of $-r_A$ as a function of X or C_A. Unfortunately for most reactor systems, a scale-up process cannot be achieved simply from a knowledge of $-r_A$ as a function of C_A. In Chapter 3 we shall present elementary forms of the kinetic rate equation from which the design equations can be evaluated, either by graphical integration or with the aid of a *table of integrals*. A short table of integrals that will be useful in the evaluation of the reactor design equations can be found in Appendix A-2.

SUMMARY

1. The stoichiometric table for the reaction

$$A + \frac{b}{a} B \longrightarrow \frac{c}{a} C + \frac{d}{a} D \tag{S2-1}$$

being carried out in a flow system is:

Species	Entering	Change	Leaving
A	F_{A0}	$-F_{A0}X$	$F_{A0}(1-X)$
B	F_{B0}	$-\frac{b}{a}F_{A0}X$	$F_{A0}\left(\theta_B - \frac{b}{a}X\right)$
C	F_{C0}	$\frac{c}{a}F_{A0}X$	$F_{A0}\left(\theta_C + \frac{c}{a}X\right)$
D	F_{D0}	$\frac{d}{a}F_{A0}X$	$F_{A0}\left(\theta_D + \frac{d}{a}X\right)$
I	F_{I0}	—	F_{I0}

where

$$\theta_i = \frac{F_{i0}}{F_{A0}} \quad \left(\text{For batch systems, } \theta_i = \frac{N_{i0}}{N_{A0}}\right)$$

67

(B29)

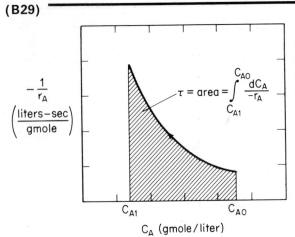

Figure 2-20

Figure 2-20 shows a typical curve of the reciprocal reaction rate as a function of concentration for an isothermal reaction carried out at constant volume. For reaction orders greater than zero, the rate decreases as the concentration decreases.

2. Relative rates of reaction for the reaction given by Equation (S2-1) are

$$\frac{-r_A}{a} = \frac{-r_B}{b} = \frac{r_C}{c} = \frac{r_D}{d} \tag{S2-2}$$

3. The differential and integral design equations for a batch system are, respectively,

$$N_{A0}\frac{dX}{dt} = -r_A V \tag{S2-3}$$

$$t = N_{A0}\int_0^X \frac{dX}{(-r_A)V} \tag{S2-4}$$

For constant volume batch reactors, i.e., $V = V_0$

$$C_{A0}\frac{dX}{dt} = -\frac{dC_A}{dt} = -r_A \tag{S2-5}$$

$$t = C_{A0}\int_0^X \frac{dX}{-r_A} \tag{S2-6}$$

4. The design equation for a CSTR can be expressed in the following forms:

$$V = \frac{F_{A0}X}{-r_A} \tag{S2-7}$$

$$V = \frac{v_0(C_{A0} - C_A)}{-r_A} \quad \text{(only for } v = v_0\text{)} \tag{S2-8}$$

$$\tau = \frac{C_{A0} - C_A}{-r_A} = \frac{C_{A0}X}{-r_A} \tag{S2-9}$$

where
τ = space time = V/v_0
SV = space velocity = $v_0/V = 1/\tau$
LHSV = Liquid Hourly Space Velocity
GHSV = Gas Hourly Space Velocity

5. The differential and integral forms of the design equation for a plug flow reactor are, respectively,

$$\frac{dX}{d\tau} = v_0\frac{dX}{dV} = \frac{-r_A}{C_{A0}} \tag{S2-10}$$

$$-\frac{dC_A}{d\tau} = -r_A \tag{S2-11}$$

$$V = F_{A0}\int_0^X \frac{dX}{-r_A} \tag{S2-12}$$

$$\tau = C_{A0}\int_0^X \frac{dX}{-r_A} \tag{S2-13}$$

For *no* volume change from reaction, Equation (S2-13) can be put in the form

$$\tau = \int_{C_{A1}}^{C_{A0}} \frac{dC_A}{-r_A} \tag{S2-14}$$

QUESTIONS AND PROBLEMS

P2-1. Fill in the intermediate steps, going from Equation (S2-7) to Equation (S2-9).

P2-2. Without referring to the text, write the differential and integral design equations for a batch reactor, a plug flow reactor, and a CSTR. When would you expect to use the differential form of the mole balance (design) equation?

P2-3. The space time necessary to achieve 80% conversion in a CSTR is 5 hours. Determine (if possible) the reactor volume required to process 2 cu ft/per min. What is the space velocity for this system?

P2-4. Phthalic anhydride can be produced by the partial oxidation of naphthalene in either a fixed or fluidized catalytic bed.

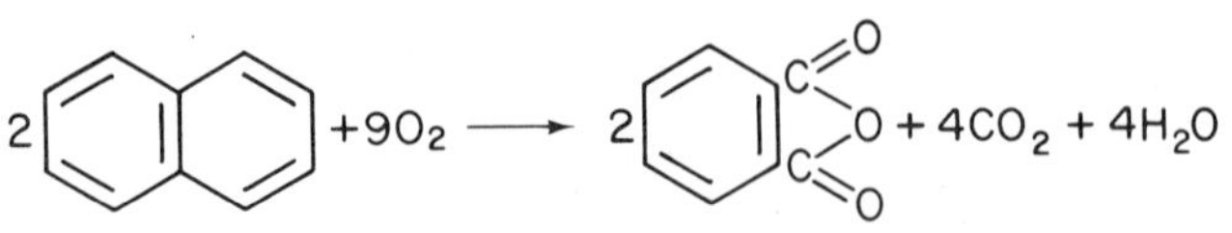

$$2 \text{(naphthalene)} + 9O_2 \longrightarrow 2 \text{(phthalic anhydride)} + 4CO_2 + 4H_2O$$

Figure P2-4

Set up a stoichiometric table for this reaction for an initial mixture of 15% naphthalene and 85% air (mole %), and use this table to develop the relations listed below.

    a. For a constant-volume isothermal batch reaction, determine each of the following as a function of the conversion of naphthalene, X_N.
        (1) The partial pressure of O_2, P_{O_2}.
        (2) The concentration of O_2, C_{O_2}.
        (3) The total pressure Π.
    b. For a constant-pressure isothermal batch reactor, determine each of the following as a function of the conversion of naphthalene, X_N.
        (1) The partial pressure of O_2, P_{O_2}.
        (2) The concentration of O_2, C_{O_2}.
        (3) The total volume V.
    c. Would your results in A and B be changed if the water was condensed out of the system, and if so, how?

[For explosive limits of this reaction, see *Chem. Engr. Prog.* 66 (1970), 49.]

Sample Answers—Part a	Sample Answers—Part b
$C_{O_2} = C_{N,0}(1.19 - 4.5X_N)$	$P_{O_2} = y_{O_2} \cdot \Pi_0 = \left\{\dfrac{1.185 - 4.5X_N}{6.66 - .5X_N}\right\} \Pi_0$
Π = total pressure	$P_{O_2} = P_{N,0} \left\{\dfrac{1.19 - 4.5X_N}{1 - .075X_N}\right\}$
y = mole fraction	$P_{N,0} = \Pi_0/6.66$

P2-5. Show that for CSTRs in series, the volume of the nth reactor is

$$V_n = F_{A0} \frac{(X_n - X_{n-1})}{-r_{An}}$$

where X_{n-1} is the conversion exiting from the $n-1$ reactor and X_n is the conversion exiting

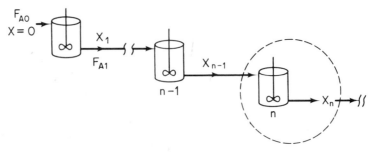

Figure P2-5

from reactor n. F_{A0} is the molar flow rate of A entering the first reactor in the series. (See Figure P2-5.) What type of reactor does this system approach in the limit when the volume of each reactor becomes small, $V_i \to 0$, and the number of CSTRs becomes large, $n \to \infty$?

P2-6. For the same total initial feed rate F_{A0}, which will give the smallest total reactor volume; two equal sized plug flow reactors connected in series or connected in parallel? For the parallel system, consider the molar flow rates to each reactor the same.

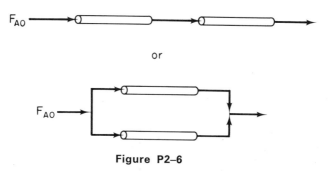

Figure P2-6

P2-7. Pure A is fed at a volumetric flow rate of 1000 cu ft/hr and at a concentration of .005 lb-moles/cu ft to an existing CSTR, which is connected in series to an existing tubular reactor. If the volume of the CSTR is 1200 cu ft and the tubular reactor volume is 600 cu ft, what are the intermediate and final conversions that can be achieved with the existing system? The reciprocal rate is plotted in Figure P2-7 as a function of conversion for the conditions at which the reaction is to be carried out.

$$A \longrightarrow \text{Products}$$

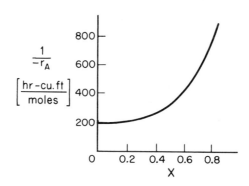

Figure P2-7

P2-8. Cumene decomposes over a solid catalyst to form benzene and propylene:

$$C_6H_5CH(CH_3)_2 \longrightarrow C_6H_6 + C_3H_6$$

For an equal molar feed of cumene and nitrogen, set up a stoichiometric table and then express the concentration of each species in terms of conversion and the initial concentration of cumene when the reaction is carried out:
(a) under constant pressure and
(b) in a constant volume reactor.

P2-9. Figure P2-9a shows $C_{A0}/-r_A$ vs. X_A for a nonisothermal, nonelementary, multiple reaction reactor.
   (a) Consider the two systems shown in Figure P2-9b in which a CSTR and a plug flow reactor are connected in series. The intermediate conversion is .3 and the final conversion is .7. How should the reactors be arranged to obtain the minimum total reactor volume? Explain.

(b) If the volumetric flow rate is 50 *l*/min, what is the minimum total reactor volume?

(c) Is there a better means of achieving 70% conversion other than either of the systems proposed above?

(Ans.: 525 *l*)

(d) At what conversion(s) would the required reactor volume be identical for either a CSTR or a tubular plug-flow reacter (PFR)?

Ans. ($X = .45$, and $X = ?$)

(e) Using the information in Figure P2-9a along with the CSTR design equation, make a plot of τ vs X. If the reactor volume is 700 liters and the volumetric flow rate 50 liters/min what are the possible outlet conversions (i.e. multiple steady states) for this reactor?

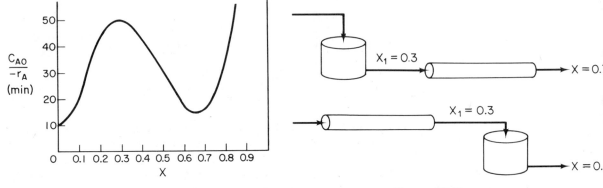

Figure P2-9a

Figure P2-9b

SUPPLEMENTARY READING

1. Further discussion of stoichiometry may be found in Chapter 2 of

    HIMMELBLAU, D. M., *Basic Principles and Calculations in Chemical Engineering*, 2d ed., Englewood Cliffs, N.J.: Prentice-Hall, 1967.

2. Further discussion of the proper staging of reactors in series for various rate laws, in which a plot of $-1/r_A$ vs. X is given, is presented in Chapter 6 (especially pp. 139-156) of

    LEVENSPIEL, O., *Chemical Reaction Engineering*, 2d ed., New York: John Wiley and Sons, 1972.

3
Elementary Forms of The Rate Law

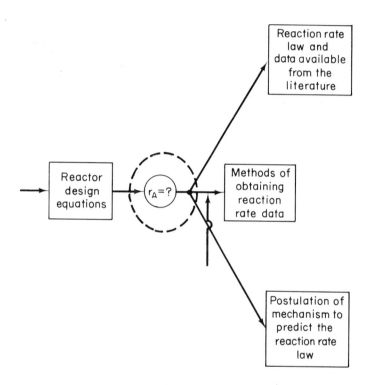

We have shown that in order to evaluate the time necessary to achieve a given conversion X in a batch system, or to evaluate the reactor volume necessary to achieve a given conversion X in a flow system, we need to know the reaction rate as a function of the concentrations of the reacting species. In the following chemical reactions we shall take as our basis of calculation species A, which is one of the reactants that is disappearing as a result of a chemical reaction. The rate of disappearance of A, $-r_A$, can be written as the product of the *specific reaction rate* k and a function of the concentration of the various species involved in the reaction, i.e.,

$$-r_A = k \, \text{fn}(C_A, C_B, \cdots) \tag{3-1}$$

The algebraic equation that relates $-r_A$ to the species concentrations [e.g., Equation (3-1)] is called the kinetic expression or *rate law*.

3.1–Basic Definitions

The specific reaction rate is usually a function only of the temperature of the reacting mixture. In liquid systems k can be a function of other parameters such as ionic strength, solvent, or total pressure. Recent experiments on liquid phase reaction rates carried out under very high pressures showed that k increased significantly with increasing pressure.* However, for purposes of the material presented here, and as is the case in most laboratory and industrial reactions, we assume that k depends only on temperature. The temperature dependence of the specific reaction rate can usually be correlated by the Arrhenius equation

$$k(T) = Ae^{-E/RT}, \qquad (3\text{-}2)$$

where

A = frequency factor
E = activation energy
R = ideal gas constant
T = absolute temperature

although other similar expressions exist. One such expression is the temperature dependence derived from transition-state theory, which takes a form similar to Equation (3-2):

$$k(T) = A'T^n e^{-E'/RT} \qquad (3\text{-}3)$$

where $0 \leq n \leq 1$.

In gas phase reactions the molecules must collide with a certain minimum energy in order to react. The exponential factor $e^{-E/RT}$ gives the fraction of the collisions between molecules that have sufficient energy to result in reaction. The activation energy E is the energy that the sum of the energies of the colliding molecules must exceed for the reaction to occur.[†]

If Equations (3-2) and (3-3) are used to describe the temperature dependence for the same reaction, it will be found that the activation energies E and E' differ. The verification of this point is left as an exercise in Problem P3-3.

The functional dependence of the reaction rate $-r_A$ on the concentration of the reacting species, $\text{fn}(C_j)$, is determined by *experimental observation*. One of the most common expressions of this function, especially in the absence of the kinetic mechanism, is the product of concentrations of the individual reacting species, each of which is raised to a power.

$$-r_A = kC_A^\alpha C_B^\beta \qquad (3\text{-}4)$$

The *reaction order* is the power to which the concentration is raised in the kinetic rate law. In Equation (3-4) the reaction is α-order with respect to A, and β-order with respect to B. The overall order of the reaction, n, is

$$n = \alpha + \beta$$

The activation energy, frequency factor, and reaction orders for a large number of gas and liquid phase reactions can be found in the National Bureau of Standards circular and supplements.[‡]

Methane is the major constituent (approximately 95%) of *natural gas*. When methane is

---

*R. A. Grieger, and C. A. Eckert, *A.I.Ch.E. J.* 16, (1970), 766.

[†]For further discussion of the collision and transition state theories, see Chapter 3 of *Chemical Kinetics of Gas Reactions* by V. N. Kondrat'ev, London: Pergamon Press, 1964, or Chapters 13 and 14 of *Physical Chemistry* by D. F. Eggers et al., New York: John Wiley and Sons, Inc., 1964.

[‡]*Tables of Chemical Kinetics Homogeneous Reactions*, National Bureau of Standards, Circular 510 (September 28, 1951); Supplement 1 (November 14, 1956); Supplement 2 (August 5, 1960); Supplement 3 (September 15, 1961) (Washington, D.C.).

used as a fuel, the primary reaction that occurs is

$$CH_4 + 2O_2 \longrightarrow CO_2 + 2H_2O + 213 \frac{kcal}{gmole\ methane}$$

What is the reaction order (1) with respect to methane, (2) with respect to oxygen and (3) overall?

(C1)

The reaction of methane with oxygen is *exothermic*. That is, heat will be given off by the reacting mixture when the system is maintained in an isothermal condition. The heat of reaction, ΔH_R, is negative for an exothermic reaction. The heat of reaction at 25°C for the combustion of methane is

$$\Delta H_R(298°K) = -213 \frac{kcal}{gmole\ CH_4}$$

In an *endothermic reaction*, heat is absorbed from the surroundings by the reacting mixture when the system is maintained in an isothermal condition. In this instance, the heat of reaction is positive.

Although we are not able to determine the reaction order for the methane combustion reaction from the given information, there are a number of reactions for which the reaction order can be deduced from the stoichiometric reaction expression. An *elementary reaction* is one in which the reaction order of each species turns out to be identical to the stoichiometric coefficient of that species. In addition, elementary reaction must be either first or second order for any particular species. If the reaction in Equation (2-1), $(aA + bB \rightarrow cC + dD)$, is elementary as written, what is the order with respect to A and to B? What is the overall order of reaction? Choose the correct answer from those given in Frame (C2).

(C2)
1. First order with respect to (w.r.t.) A and first order w.r.t. B. Overall, second order.
2. ath order w.r.t. A, bth order w.r.t. B. Overall order, $n = a + b$.
3. Cannot determine the reaction order from the information given.

Before proceeding further we shall define a few more terms pertaining to reacting mixtures.

A *homogeneous reaction* is one that takes place only in one phase.

A *heterogeneous reaction* takes place at the interface between two or more phases.

An *irreversible reaction* is one that proceeds only in one direction and continues in that direction until the reactants are exhausted. It behaves as if no equilibrium condition exists. Strictly speaking, no chemical reaction is completely irreversible. However, in a number of chemical reactions the equilibrium condition lies so far to the right that they are treated as irreversible reactions.

The reaction between methyl bromide and sodium hydroxide is classified as a nucleophilic aliphatic substitution*:

$$NaOH + CH_3Br \longrightarrow CH_3OH + NaBr$$

This elementary reaction is carried out in aqueous ethanol. Therefore like almost all liquid

*R. T. Morrison and R. N. Boyd, *Organic Chemistry*, 2nd ed. (Boston: Allyn and Bacon, 1966), p. 468.

(C1)

The reaction order is determined from experimental observation. There is no way to determine a priori the reaction order from the stoichiometric relationships.

(C2)

Since the reaction is an elementary reaction as written, the stoichiometric coefficients are identical with the reaction order.

$$\alpha = a$$
$$\beta = b$$
$$n = a + b = \text{overall order}$$
$$-r_A = k C_A^a C_B^b$$

It is ath order w.r.t. A and bth order w.r.t. B.

phase reactions, the density remains relatively constant throughout the reaction. Consequently, for liquid phase reactions, the volume V for a batch reaction and the volumetric flow rate v for a continuous flow system will be constant during the course of the chemical reaction. Write the rate of disappearance of methyl bromide, $-r_{MB}$, in terms of the appropriate concentrations and the specific reaction rate k. Also state the overall reaction order as well as the order w.r.t. each of the reactants.

(C3) ───

───

Another nucleophilic aliphatic substitution is the reaction of sodium hydroxide and tert-butyl bromide (TBB).

$$NaOH + CH_3-\underset{\underset{Br}{|}}{\overset{\overset{CH_3}{|}}{C}}-CH_3 \longrightarrow CH_3-\underset{\underset{OH}{|}}{\overset{\overset{CH_3}{|}}{C}}-CH_3 + NaBr$$

State the order of reaction for each reactant and write the rate of formation of tert-butyl alcohol, r_{TBA}, in terms of the conversion X of TBB the specific reaction rate k, and the initial concentrations of NaOH and TBB, (C_{NaOH_0}, C_{TBB_0}).

(C4) ───

───

We shall now describe the reaction in Frame (C3) using the following statement:

The reaction of methyl bromide with sodium hydroxide is (1) homogeneous, liquid phase; (2) irreversible; (3) first order w.r.t. methyl bromide; (4) first order w.r.t. sodium hydroxide; (5) overall second order; and (6) an elementary reaction.

Write a similar description for the reaction of tert-butyl bromide with sodium hydroxide.

(C5) ───

───

The irreversible *gas phase* reaction shown in Equation (3-5) is first order in A and second order in B:

$$2A + B \longrightarrow 2C \tag{3-5}$$

Which of the kinetic rate laws shown in Frame (C6) describes the reaction given by Equation (3-5) and the accompanying description?

(C3)

$$1\text{NaOH} + 1\text{CH}_3\text{Br} \longrightarrow 1\text{CH}_3\text{OH} + 1\text{NaBr}$$

$\alpha = 1,$ first order w.r.t. NaOH.

$\beta = 1,$ first order w.r.t. methyl bromide (MB).

$$-r_{\text{MB}} = kC_{\text{NaOH}}C_{\text{CH}_3\text{Br}}$$

Overall, this reaction is second order.

(C4)

Since it was not stated that the reaction is elementary and there is no experimental rate data presented for this reaction, we cannot determine the reaction order. From p. 475 of *Organic Chemistry* by Morrison and Boyd (See footnote on p. 77) we find that the reaction order is first order w.r.t. tert-butyl bromide and zero order with respect to sodium hydroxide.

$$-r_{\text{TBB}} = r_{\text{TBA}} = kC_{\text{TBB}}$$

TBA = Tert-butyl alcohol

TBB = Tert-butyl bromide $\equiv$ A (Basis of Calculation)

Since this is a liquid phase reaction, we can assume that there is *no volume change* during the course of the reaction:

$$V = V_0, \qquad C_A = \frac{N_{A0}(1-X)}{V_0} = C_{A0}(1-X)$$

$$r_{\text{TBA}} = kC_{A0}(1-X) = -r_{\text{TBB}}$$

Species	Initial	Change	Remaining	Concentration
$(CH_3)_3CBr$	N_{A0}	$-XN_{A0}$	$N_{A0}(1-X)$	$C_{A0}(1-X)$
NaOH	$N_{\text{NaOH},0}$	$-XN_{A0}$	$N_{A0}(\theta_{\text{NaOH}} - X)$	$C_{A0}(\theta_{\text{NaOH}} - X)$
$(CH_3)_3COH$	$N_{\text{TBA},0}$	$+XN_{A0}$	$N_{A0}(\theta_{\text{TBA}} + X)$	$C_{A0}(\theta_{\text{TBA}} + X)$
NaBr	$N_{\text{NaBr},0}$	$+XN_{A0}$	$N_{A0}(\theta_{\text{NaBr}} + X)$	$C_{A0}(\theta_{\text{NaOH}} + X)$

(C5)

(1) Homogeneous, liquid phase; (2) irreversible; (3) first order w.r.t. tert-butyl bromide; (4) zero order w.r.t. sodium hydroxide; (5) overall first order; (6) nonelementary reaction.

(C6)

1. $-r_A = kC_A^2 C_B$.
2. $-r_A = kC_A C_B$.
3. $-r_A = kC_A C_B^2$.
4. $-r_A = kC_A C_B^{1/2}$.

Taking A as your basis, set up a stoichiometric table for the reaction given by Equation (3-5). Then write $-r_A$ in terms of k, X, V, N_{A0}, and N_{B0}. Assume that there are no inerts or product initially present in the system.

(C7)

3.2–Volume Change with Reaction

In our previous discussions we primarily considered systems in which either the volume is constant in a batch system or the volumetric flow rate is constant in a flow system. We shall develop relationships that will enable us to easily design systems in which V or v vary during the reaction. As you will recall, we arranged the general reaction given by Equation (2-1) in the form

$$A + \frac{b}{a}B \longrightarrow \frac{c}{a}C + \frac{d}{a}D$$

The total number of moles is given in Table 2-1 as

$$N_t = N_{t0} + \left(\frac{c}{a} + \frac{d}{a} - \frac{b}{a} - 1\right) N_{A0} X \tag{3-6}$$

Let δ be the increase in the total number of moles of the gas phase per mole of A reacted. Now write N_t in terms of δ, X, and any other appropriate terms.

(C8)

The equation of state that we shall use is

$$\Pi V = Z N_t R T \tag{3-7}$$

where V and N_t are defined as before and

$T =$ temperature
$\Pi =$ total pressure
$Z =$ compressibility factor
$R =$ ideal gas constant

This equation is valid at any point in the system at any time t. At time $t = 0$, i.e., the time the reaction is initiated, Equation (3-7) becomes

$$\Pi_0 V_0 = Z_0 N_{t0} R T_0 \tag{3-8}$$

(C6)

$$-r_A = kC_A^\alpha C_B^\beta$$

Since the reaction is first order with respect to A, ($\alpha = 1$) and second order with respect to B, ($\beta = 2$), the rate equation is

$$-r_A = kC_A C_B^2 \qquad \text{(Ans.: 3)}$$

(C7)

Once having determined the reaction order, we now take A as our basis and divide by the stoichiometric coefficient of A and set up our stoichiometric table.

$$A + \tfrac{1}{2}B \longrightarrow C$$

Species	Initially	Change	Remaining
A	N_{A0}	$-N_{A0}X$	$N_A = N_{A0}(1 - X)$
B	N_{B0}	$-\tfrac{1}{2}N_{A0}X$	$N_B = N_{A0}\left(\dfrac{N_{B0}}{N_{A0}} - .5X\right)$
			$= N_{A0}(\theta_B - .5X)$
C	0	$N_{A0}X$	$N_C = XN_{A0}$
Inerts	0	0	0
Total	N_{t0}		$N_t = N_{t0} - .5N_{A0}X$

$$C_A = \frac{N_{A0}(1-X)}{V}, \quad C_B = \frac{N_{A0}}{V}(\theta_B - .5X), \quad \text{where } \theta_B = \frac{N_{B0}}{N_{A0}}$$

$$-r_A = kC_A C_B^2 = \frac{kN_{A0}^3}{V^3}(1-X)(\theta_B - .5X)^2 \qquad \text{(Ans.)}$$

(C8)

Since we are taking A as our basis, we must rewrite the generalized reaction

$$aA + bB \longrightarrow cC + dD \qquad (2\text{-}1)$$

in the form

$$A + \frac{b}{a}B \longrightarrow \frac{c}{a}C + \frac{d}{a}D \qquad (2\text{-}2)$$

in order to more readily form our stoichiometric relationships. The sum of the stoichiometric coefficients of the products minus the sum of the stoichiometric coefficients of the reactants will give the net increase in the total number of moles *per mole of A reacted*, which is represented by the symbol, δ:

$$\delta = \left(\frac{c}{a} + \frac{d}{a}\right) - \left(\frac{b}{a} + 1\right) = \frac{c}{a} + \frac{d}{a} - \frac{b}{a} - 1 \qquad (C8\text{-}1)$$

The number of moles of A reacted is $N_{A0}X$; therefore the increase in the total number of moles is $N_{A0}X\delta$. The total number of moles is

$$N_t = N_{t0} + N_{A0}X\delta \qquad (C8\text{-}2)$$

In using Equation (C8-1) to calculate δ, *we are assuming that each of the species involved in the reaction, (i.e., A, B, C, and D) is in the gaseous state.*

Dividing Equation (3-7) by Equation (3-8) and rearranging, we obtain

$$V = V_0 \left(\frac{\Pi_0}{\Pi}\right)\left(\frac{T}{T_0}\right)\left(\frac{Z}{Z_0}\right)\left(\frac{N_t}{N_{t0}}\right) \tag{3-9}$$

We now want to express the volume V as a function of the conversion X. Recalling Equation (C8-2),

$$N_t = N_{t0} + \delta N_{A0} X \tag{3-10}$$

we divide through by N_{t0}:

$$\frac{N_t}{N_{t0}} = 1 + \delta y_{A0} X \tag{3-11}$$

What is the definition of the symbol y_{A0}?

(C9)

1. It is the fraction of the total moles reacted.
2. It is the initial mole fraction of species A.
3. It is the fractional change in volume per mole of A reacted.

Equation (3-11) is further simplified by letting

$$\boxed{\epsilon = y_{A0}\delta} \tag{3-12}$$

If all the species in the generalized reaction are in the gas phase, then

$$\delta = \frac{d}{a} + \frac{c}{a} - \frac{b}{a} - 1 \tag{C8-1}$$

Equation (3-9) now becomes

$$V = V_0 \left(\frac{\Pi_0}{\Pi}\right)\left(\frac{T}{T_0}\right)\left(\frac{Z}{Z_0}\right)(1 + \epsilon X) \tag{3-13}$$

In most gas phase systems that we shall be studying, the temperatures and pressures are such that the compressibility factor will not change significantly during the course of the reaction; hence $Z_0 = Z$. In addition, f will denote the ratio of the temperatures T/T_0. Using these two statements, rearrange Equation (3-13) to give the pressure Π as a function of the variables discussed above.

(C10)

For a batch system the volume of gas at any time t is

$$\boxed{V = V_0 \left(\frac{\Pi_0}{\Pi}\right) f(1 + \epsilon X)} \tag{3-14}$$

A similar expression for the gas volumetric flow rate in a flow system is

$$\boxed{v = v_0 \left(\frac{\Pi_0}{\Pi}\right) f(1 + \epsilon X)} \tag{3-15}$$

(C9)
$$N_t = N_{t0} + \delta N_{A0} X$$

Divide by N_{t0}:
$$\frac{N_t}{N_{t0}} = 1 + \frac{N_{A0}}{N_{t0}} \delta X$$

$$\frac{N_t}{N_{t0}} = 1 + \delta y_{A0} X, \qquad y_{A0} = \frac{N_{A0}}{N_{t0}}, \qquad \epsilon = y_{A0}\delta$$

y_{A0} is the initial mole fraction of A.

(C10)
$$V = V_0 \frac{\Pi_0 T Z}{\Pi T_0 Z_0}(1 + \epsilon X)$$

By rearranging and letting $f = T/T_0$ and $Z = Z_0$,
$$\Pi = \Pi_0 \left(\frac{V_0}{V}\right) f(1 + \epsilon X)$$

For a batch system, the subscript zero in Equation (3-14) refers to the initial conditions. To what conditions does the subscript zero refer in Equation (3-15)?

(C11)

As previously noted, f is defined by

$$f = \frac{T}{T_0} \qquad (3\text{-}16)$$

3.3–Expressing $-r_A$ as a Function of X

3.3A Irreversible Reactions

The gas phase reaction discussed in Equation (3-5),

$$2A + B \longrightarrow 2C$$

is carried out isothermally at constant pressure in a batch system. Write the rate law, $-r_A$, in terms of k, C_{A0}, θ_B, ϵ, and X. The reaction is first order in A and second order in B. (*Hint:* use Equation (3-14) in conjunction with the definition $C_A = N_A/V$).

(C12)

Calculate ϵ when reactants A and B are fed to the reactor in a stoichiometric ratio in a mixture containing no inerts. The correct answer is one of the choices in Frame (C13).

(C13)

1. $\epsilon = -4/3$
2. $\epsilon = -1/4$
3. $\epsilon = +1/3$
4. $\epsilon = -1/3$
5. $\epsilon = +4/3$
6. $\epsilon = -1$

Calculate ϵ when the initial mixture consists of 25% A, 25% B, and 50% inerts.

(C14)

For the initial composition of 25% A, 25% B, and 50% inerts, a value of k of 10^5 (gmoles/l)$^{-2}$/sec and an initial concentration of A of 10^{-3} gmole/l, evaluate all possible symbols (ϵ, k, θ_B) in Equation (C12-4) in order to express $-r_A$ as a function of X alone. The reaction is carried out isothermally under constant pressure.

(C11)

The subscript zero in Equation (3-15) refers to the entrance condition of the reactor.

(C12)

$$-r_A = kC_A C_B^2$$

$$-r_A = k\left(\frac{N_A}{V}\right)\left(\frac{N_B}{V}\right)^2 \tag{C12-1}$$

The reaction is carried out isothermally ($f = 1$) under constant pressure (variable volume); therefore
$V = V_0(1 + \epsilon X)$

$$= k\left(\frac{N_{A0}(1 - X)}{V_0(1 + \epsilon X)}\right)\left(\frac{N_{A0}(\theta_B - .5X)}{V_0(1 + \epsilon X)}\right)^2 \tag{C12-2}$$

$$= kC_{A0}^3 \left(\frac{1 - X}{1 + \epsilon X}\right)\left(\frac{\theta_B - .5X}{1 + \epsilon X}\right)^2 \tag{C12-3}$$

$$-r_A = \frac{kC_{A0}^3}{(1 + \epsilon X)^3}(1 - X)(\theta_B - .5X)^2 \tag{C12-4}$$

(C13)

For stoichiometric feed 2A to 1B, $y_{A0} = \frac{2}{3}$. Now, $\epsilon = y_{A0}\delta =$ the net change in the total number of moles per mole of A reacted. By transposing A and $\frac{1}{2}$B to the right hand side of the arrow in the reaction

$$A + \tfrac{1}{2}B \longrightarrow C$$

the reaction equation can be written in terms of the sum of the stoichiometric coefficient products minus the sum of the stoichiometric coefficient reactants, i.e.,

$$C - \tfrac{1}{2}B - A$$
$$\delta = 1 - \tfrac{1}{2} - 1 = -\tfrac{1}{2}$$
$$y_{A0} = \tfrac{2}{3}$$
$$\epsilon = y_{A0}\delta = \tfrac{2}{3}(-\tfrac{1}{2}) = -\tfrac{1}{3} = -.33$$

(C14)

$$\delta = -\tfrac{1}{2}$$
$$\epsilon = (.25)(-\tfrac{1}{2}) = -\tfrac{1}{8} = -.125$$

(C15)

Equation (C15-2) gives the rate of disappearance of A, $-r_A$, as a function of X alone. This relationship can now be substituted directly into any one of the design equations.

$$\text{Batch:} \qquad t = N_{A0} \int_0^X \frac{dX}{-r_A V} \qquad (2\text{-}11)$$

$$\text{Backmix (CSTR):} \qquad V = \frac{F_{A0} X}{-r_A} \qquad (2\text{-}13)$$

$$\text{Tubular (plug flow):} \qquad V = F_{A0} \int_0^X \frac{dX}{-r_A} \qquad (2\text{-}17)$$

Since we now can express $-r_A$ as a function of X, all that is needed to evaluate the "batch or tubular design equations" is an integral table or some other means such as graphical or numerical integration.

For a 90% conversion of the reaction we have been discussing, i.e.,

$$2A + B \longrightarrow 2C$$

along with the conditions given in Frame (C15), substitute numerical values for all symbols wherever possible in a batch reactor design equation. Then explain how you would determine the time necessary to achieve a 90% conversion.

(C16)

For the same conditions as those of Frame (C15), plus an entering volumetric flow rate of 6 l/sec, evaluate the backmix reactor (CSTR) volume that is necessary to achieve the conversion specified above. (*Hint:* first calculate F_{A0}).

(C15)

$$\theta_B = \frac{C_{B0}}{C_{A0}} = \frac{.25}{.25} = 1$$

$$\epsilon = -.125$$

$$C_{A0} = 10^{-3} \text{ gmole}/l$$

Since the reaction is carried out isothermally, the specific reaction rate is constant and equal to

$$k = 10^5 \text{ (gmoles}/l)^{-2}/\text{sec}$$

$$-r_A = \frac{10^5 \left(\frac{\text{gmoles}}{l}\right)^{-2}}{\sec} \frac{\left[10^{-3} \left(\frac{\text{gmole}}{l}\right)\right]^3 (1-X)(1-.5X)^2}{(1-.125X)^3} \tag{C15-1}$$

$$-r_A = \frac{10^{-4}(1-X)(1-.5X)^2}{(1-.125X)^3} \frac{\text{gmoles}}{l\text{-sec}} \tag{C15-2}$$

(C16)

$$t = N_{A0} \int_0^X \frac{dX}{-r_A V} = C_{A0} \int_0^X \frac{dX}{-r_A(1+\epsilon X)}$$

From the conditions of Frame (C14):

$$\epsilon = -.125, \quad C_{A0} = 10^{-3} \text{ gmoles}/l$$

$$t = C_{A0} \int_0^{.9} \frac{dX}{-r_A(1-.125X)}$$

$$t = C_{A0} \int_0^{.9} \frac{dX}{\frac{10^{-4}(1-X)(1-.5X)^2}{(1-.125X)^3}(1-.125X)}$$

$$t = 10 \int_0^{.9} \frac{(1-.125X)^2 \, dX}{(1-X)(1-.5X)^2} \text{ seconds.}$$

This integral is graphically evaluated in Figure 3–1 from which the time to achieve 90% conversion is found to be 50 seconds.

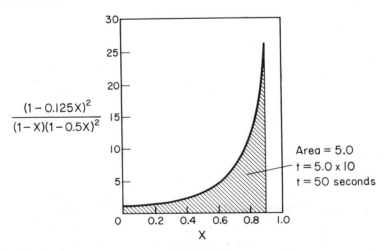

Figure 3–1

(C17)

For a tubular reactor with the same volumetric flow rate as the CSTR, evaluate all possible symbols and outline how you would evaluate the tubular reactor design integral.

(C18)

In the study of reaction orders and kinetic mechanisms reference is sometimes made to the molecularity of a reaction. The molecularity is the number of atoms, ions, or molecules involved (colliding) in the rate-determining step of the reaction. The terms *unimolecular*, *bimolecular*, and *termolecular* refer to reactions involving, respectively, one, two, or three atoms (or molecules) interacting or colliding in any one reaction step.

The most common example of a unimolecular reaction is radioactive decay, such as the spontaneous emission of an alpha particle from uranium 238 to give thorium:

$$_{92}U^{238} \longrightarrow {}_{90}Th^{234} + {}_{2}He^{4}$$

It had been previously thought that the reaction between hydrogen and iodine was bimolecular:

$$H_2 + I_2 \longrightarrow 2HI$$

However, in recent years research on this reaction has revealed that even though the reaction is first order in H_2, first order in I_2, and overall second order, i.e.,

$$r_{HI} = k C_{H_2} C_{I_2} \tag{3-17}$$

the reaction mechanism is not bimolecular.* Currently it is believed that the reaction is termolecular and proceeds via the sequence,

$$I_2 \underset{k_2}{\overset{k_1}{\rightleftharpoons}} 2I \tag{3-18}$$

$$2I + H_2 \xrightarrow{k_3} 2HI \tag{3-19}$$

By using a technique that involves the pseudo steady-state hypothesis discussed in Chapter 7, it can be shown that the rate law in Equation (3-17) is consistent with the reaction steps given by Equations (3-18) and (3-19).

*J. Chem. Phys. 30 (1959): 1292; J. Chem. Phys. 46 (1967): 73; Chemistry 40, no. 4 (1967): 29.

(C17)

$$F_{A0} = C_{A0}v_0 = 10^{-3} \cdot 6 = 6 \times 10^{-3} \text{ gmoles A/sec}$$

$$F_{A0}X = 6 \times 10^{-3} \times .9 = 5.4 \times 10^{-3} \text{ gmoles A/sec}$$

$$V = \frac{F_{A0}X}{-r_A}$$

$$-r_A = \frac{10^{-4}(1-X)(1-.5X)^2}{(1-.125X)^3} \text{ gmoles A}/l/\text{sec}$$

When $X = .9$,

$$-r_A = \frac{10^{-4}(.1)(.55)^2}{(.887)^3} = 4.33 \times 10^{-6} \text{ gmoles A}/l/\text{sec}$$

$$V = \frac{5.4 \times 10^{-3}}{4.33 \times 10^{-6}} = 1250 \text{ liters}$$

(C18)

$$V = F_{A0} \int_0^{.9} \frac{dX}{-r_A}$$

$$= 6.0 \times 10^{-3} \int_0^{.9} \frac{dX}{10^{-4} \frac{(1-X)(1-.5X)^2}{(1-.125X)^3}} \quad \text{(C18-1)}$$

$$= 60 \int_0^{.9} \frac{(1-.125X)^3 \, dX}{(1-X)(1-.5X)^2} \quad \text{(C18-2)}$$

One could find the integral in a table of integrals, evaluate it numerically, or evaluate it graphically. To evaluate Equation (C18-2) graphically, plot as shown in Figure 3-2. From this plot we can see that the tubular reactor volume necessary to achieve 90% conversion is 294 liters.

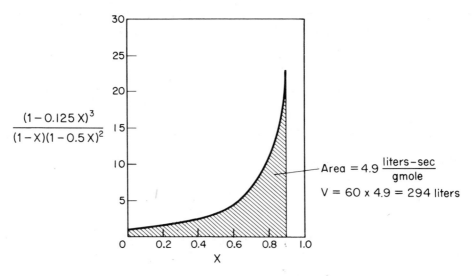

Figure 3–2

3.3B Reversible Reactions

Another example of a second-order, gas-phase reaction is the combination of two benzene molecules to form one molecule of hydrogen and one of diphenyl. This reaction is elementary and reversible:

$$2C_6H_6 \underset{k_2}{\overset{k_1}{\rightleftharpoons}} C_{12}H_{10} + H_2 \tag{3-20}$$

The forward and reverse *specific reaction rates*, k_1 and k_2 respectively, will be defined with respect to benzene.

Benzene (B) is being deleted by the forward reaction

$$2C_6H_6 \xrightarrow{k_1} C_{12}H_{10} + H_2$$

in which the rate of disappearance of benzene is

$$-r_{B,\text{forward}} = k_1 C_B^2$$

If we multiply both sides of this equation by -1, we obtain the expression for the rate of formation of benzene for the forward reaction.

$$r_{B,\text{forward}} = -k_1 C_B^2 \tag{3-21}$$

For the reverse reaction between diphenyl (D) and hydrogen (H)

$$C_{12}H_{10} + H_2 \xrightarrow{k_2} 2C_6H_6$$

The rate of formation of benzene is given by Equation (3-22):

$$r_{B,\text{reverse}} = k_2 C_D C_H \tag{3-22}$$

The net rate of formation of benzene is the sum of the rates of formation from the forward reaction, Equation (3-21), and the reverse reaction, Equation (3-22).

$$r_B \equiv r_{B,\text{net}} = r_{B,\text{forward}} + r_{B,\text{reverse}} \tag{3-23}$$

$$r_B = -k_1 C_B^2 + k_2 C_D C_H \tag{3-24}$$

Multiplying both sides of Equation (3-24) by -1, we obtain the kinetic expression for the rate of disappearance of benzene, $-r_B$:

$$-r_B = k_1 C_B^2 - k_2 C_D C_H = k_1 \left(C_B^2 - \frac{k_2}{k_1} C_D C_H \right)$$
$$= k_1 \left(C_B^2 - \frac{C_D C_H}{K_e} \right) \tag{3-25}$$

where

$$\frac{k_1}{k_2} = K_e = \text{concentration equilibrium constant}$$

An example of an elementary liquid phase reaction is the esterification of acetic acid and *n*-butyl alcohol to form *n*-butyl acetate and water:

$$CH_3COOH + C_4H_9OH \rightleftharpoons CH_3COOC_4H_9 + H_2O \tag{3-26}$$

Keeping in mind that the above reaction is reversible and elementary, write the rate law for the rate of disappearance of acetic acid.

(C19)

3.3C Nonelementary Reactions

It is interesting to note that although the reaction orders correspond to the stoichiometric coefficients for the reaction between hydrogen and iodine, the rate expression, for the reaction [shown in Equation (3-27)] between hydrogen and another halogen, bromine,

$$H_2 + Br_2 \longrightarrow 2HBr \tag{3-27}$$

is quite complex. This nonelementary reaction proceeds by a free-radical mechanism, and its *reaction rate law* is

$$r_{HBr} = \frac{k_1 C_{H_2} C_{Br_2}^{1/2}}{k_2 + \dfrac{C_{HBr}}{C_{Br_2}}} \tag{3-28}$$

Another reaction involving free radicals is the vapor-phase decomposition of acetaldehyde:

$$CH_3CHO \longrightarrow CH_4 + CO \tag{3-29}$$

At a temperature of about 500°C, the reaction is three-halves order w.r.t. acetaldehyde. Assuming that this reaction is carried out isothermally at a constant pressure of 10 atm and the feed consists of 80% acetaldehyde and 20% inerts, write the rate law in terms of a specific rate k and the conversion X and evaluate numerically all other terms.

(C20)

In many gas-solid catalyzed reactions it is sometimes preferable to write the rate law in terms of partial pressures rather than concentrations. One such example is the decomposition of cumene to form benzene, B, and propylene, P:

$$C_6H_5CH(CH_3)_2 \rightleftharpoons C_6H_6 + C_3H_6 \tag{3-30}$$

Equation (3-26) can be written symbolically as

$$C \rightleftharpoons B + P \tag{3-31}$$

The rate law is

$$-r'_C = \frac{k\left(P_C - \dfrac{P_B P_P}{K}\right)}{1 + K_A P_C + \dfrac{P_B}{K_D}} \tag{3-32}$$

To express the rate of decomposition of cumene $-r'_C$ as a function of conversion, replace the partial pressure with concentration, using the ideal gas law, e.g.,

$$P_C = C_C \cdot RT \tag{3-33}$$

and then express concentration in terms of conversion as we have done in the previous examples.

(C19) Let HAc = acetic Acid, BuOH = n-butyl alcohol, and BuAc = n-butyl acetate. Since the reaction is elementary the reaction order of each species corresponds to the stoichiometric coefficient of that species.

$$-r_{HAc} = k_1 C_{HAc} C_{BuOH} - k_2 C_{H_2O} C_{BuAc} \tag{C19-1}$$

$$= k_1 \left(C_{HAc} C_{BuOH} - \frac{C_{H_2O} C_{BuAc}}{K_e} \right); \quad K_e = \frac{k_1}{k_2} \tag{C19-2}$$

We could go one step further and write Equation (C19-2) in terms of the conversion of acetic acid. For liquid phase reactions, $V = V_0$ and $v = v_0$. By now we should be approaching the point where we will be able to form a mental picture of the stoichiometric table for this reaction without having to write it down on paper. By substituting for the concentration of each species in terms of the conversion of acetic acid, we obtain rate Equation (C19-3):

$$-r_{HAc} = k_1 (C_{HAc,0})^2 \left[(1-X)(\theta_{BuOH} - X) - \frac{(\theta_{H_2O} + X) \cdot (\theta_{BuAc} + X)}{K_e} \right] \tag{C19-3}$$

$$\theta_i = \frac{C_i}{C_{HAc}}$$

$$k_1(80°C) = 1.73 \times 10^{-6} \text{ gmole}/l/\text{sec}$$

(C20)

$$-r_{CH_3CHO} = k C_{CH_3CHO}^{3/2} \tag{C20-1}$$

Let A = acetaldehyde:

$$C_A = \frac{N_A}{V} = \frac{N_{A0}(1-X)}{V} = \frac{N_{A0}(1-X)}{V_0(1+\epsilon X)}$$

$$= C_{A0} \frac{(1-X)}{(1+\epsilon X)}, \quad f=1, \text{ isothermal} \tag{C20-2}$$

$$\delta = 1 + 1 - 1 = 1$$

$$\epsilon = y_{A0} \delta = (.8)(1) = +.8 \tag{C20-3}$$

$$C_{A0} = \frac{P_{A0}}{RT} = \frac{y_{A0} \Pi_0}{RT} = \frac{(.8)(10)}{(.082)(773)} \tag{C20-4}$$

$$= .126 \text{ gmole}/l$$

$$C_{A0}^{3/2} = .045 \, (\text{gmole}/l)^{3/2}$$

$$-r_A = .045 k \left(\frac{1-X}{1+.8X} \right)^{3/2} \tag{C20-5}$$

The rate of reaction per unit weight of catalyst, $-r'_A$, and the rate of reaction per unit volume, $-r_A$, are related through the bulk density ρ_B of the catalyst particles in the fluid media; i.e.,

$$-r_A = \rho_B[-r'_A]$$

$$\left[\frac{\text{moles}}{\text{time} \cdot \text{volume}}\right] = \left[\frac{\text{mass}}{\text{volume}}\right]\left[\frac{\text{moles}}{\text{time} \cdot \text{mass}}\right]$$

In fluidized catalytic beds the bulk density will normally be a function of the flow rate through the bed.

One of the major objectives of this chapter is to learn how to express any given rate law $-r_A$ as a function of conversion. The schematic diagram in Figure 3-3 helps to summarize our discussion on this point. The concentration of the key reactant, A (which is the basis of our calculation), is expressed as a function of conversion in both flow and batch systems, for various conditions of temperature, pressure, and volume.

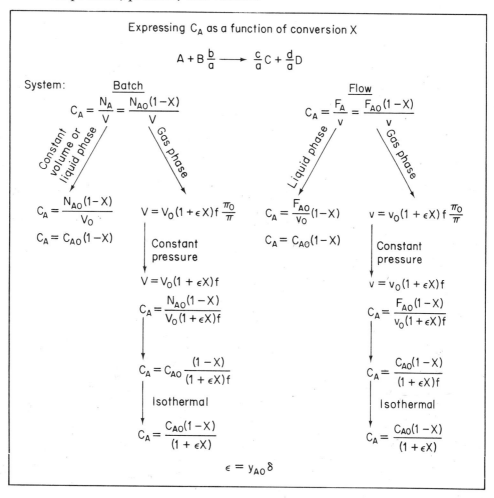

Figure 3-3

3.3D Reactions with Phase Change

When Equation (3-12) is used to evaluate ϵ, it should be remembered from the derivation of this equation that δ represents the change in the number of moles in the gas phase per mole of A reacted. As the last example in this chapter, we shall consider a gas phase reaction in which

one of the products, D, condenses immediately after being formed:

$$A(g) + 2B(g) \longrightarrow C(g) + D(l)$$

This reaction is first order w.r.t. A and first order w.r.t. B. The feed to a tubular reactor is equimolar in A and in B. Neglecting the volume of liquid formed in the reactor, express the rate law $-r_A$ solely as a function of the conversion of A, and evaluate numerically all the symbols possible.

(C21)

Additional information:

$$k = 10^4 \ l/\text{gmoles}/\text{sec}$$
$$C_{A0} = .02 \ \text{gmole}/l$$
$$T = 32°C$$
$$\text{Vapor pressure of D} = 1 \ \text{mm Hg}$$
$$R = .082 \ l \ \text{atm}/\text{gmole}/°K$$

SUMMARY

1. The reaction order is determined from experimental observation:

$$A + B \longrightarrow C$$
$$-r_A = kC_A^\alpha C_B^\beta \tag{S3-1}$$

The reaction in Equation (S3-1) is α order w.r.t. species A, β order w.r.t. species B, whereas the overall order is $\alpha + \beta$. The reaction order is determined from experimental observation. For example, if $\alpha = 1$ and $\beta = 2$, we would say that the reaction is first order w. r. t. A, second order w. r. t. B, and overall third order.

2. In addition to the reaction order, the following terms were defined:
   a. elementary reaction,
   b. exothermic and endothermic reactions, and
   c. homogeneous and heterogeneous reactions.

3. The temperature dependence of the specific reaction rate is given by the Arrhenius equation

$$k = Ae^{-E/RT} \tag{S3-2}$$

where A is the frequency factor and E the activation energy.

4. In the case of ideal gases, Equations (S3-3) through (S3-6) relate volume and volumetric flow rate to conversion.

$$\text{Batch:} \quad V = V_0 \left(\frac{\Pi_0}{\Pi}\right)(1 + \epsilon X)f \tag{S3-3}$$

$$\text{Flow:} \quad v = v_0 \left(\frac{\Pi_0}{\Pi}\right)(1 + \epsilon X)f \tag{S3-4}$$

where

$$\epsilon = y_{A0}\delta \tag{S3-5}$$

$$f = \frac{T}{T_0} \tag{S3-6}$$

5. For ideal gases, the concentration of A and C in the reaction

$$A + \frac{b}{a}B \longrightarrow \frac{c}{a}C + \frac{d}{a}D \tag{S3-7}$$

(C21)

$$A(g) + 2B(g) \longrightarrow C(g) + D(l)$$

Species	Entering	Change	Leaving
$A(g)$	F_{A0}	$-F_{A0}X$	$F_{A0}(1-X)$
$B(g)$	$F_{A0}\theta_B = F_{A0}$	$-2F_{A0}X$	$F_{A0}(1-2X)$
$C(g)$	0	$F_{A0}X$	$F_{A0}X$
$D(l)$	0	$F_{A0}X(\downarrow)$	$(F_{A0}X)$ liq†
	$F_{t0}(g) = 2F_{A0}$		$F_t(g) = F_{A0}(2-2X) = F_t$

$\theta_B = 1$

where $F_T(g)$ is the total molar gas flow rate $[F_t \equiv F_t(g)]$.

†The total pressure is 1 atm; therefore we shall neglect the vapor pressure of D. (See Problem P3-8 for the case when this assumption is not valid.)

$$v = F_t \frac{RT}{\Pi} = F_{A0}(2-2X)\frac{RT}{\Pi}$$

$$v_0 = F_{t0}\frac{RT_0}{\Pi_0} = 2F_{A0}\frac{RT_0}{\Pi_0}$$

$$\frac{v}{v_0} = \left(\frac{T}{T_0}\right)\left(\frac{\Pi_0}{\Pi}\right)(1-X)$$

If $T = T_0$ and $\Pi = \Pi_0$,

$$v = v_0(1-X)$$

We can also use Equation (3-12), remembering that δ is the change in the number of moles in the gas phase per mole of A reacted:

$$\delta = (0 \overset{(D(l) + C(g) - 2B(g) - A(g))}{+1 \quad -2 \quad -1}) = -2$$

$$\epsilon = y_{A0}\delta = (.5)(-2) = -1$$

$$v = v_0(1-X)$$

$$-r_A = k_1 C_A C_B = k_1 \frac{F_A}{v}\frac{F_B}{v}$$

$$= k_1 \frac{F_{A0}(1-X)F_{A0}(1-2X)}{[v_0(1-X)]^2}$$

$$= k_1 C_{A0}^2 \left(\frac{1-2X}{1-X}\right) = (10^4)(.02)^2\left(\frac{1-2X}{1-X}\right)$$

$$= 4\frac{(1-2X)}{(1-X)}$$

which is carried out at constant pressure, can be expressed in the form

$$C_A = C_{A0} \frac{(1-X)}{(1+\epsilon X)f}$$

$$C_C = C_{A0} \frac{\left(\theta_C + \frac{c}{a}X\right)}{(1+\epsilon X)f}$$

(S3-8)

For this reaction

$$\delta = \left(\frac{d}{a} + \frac{c}{a} - \frac{b}{a} - 1\right)$$

6. When the reactants and products are incompressible liquids, the concentrations of species A and C in the reaction given by Equation (S3-7) can be written as

$$C_A = C_{A0}(1-X)$$

$$C_C = C_{A0}\left(\theta_C + \frac{c}{a}X\right)$$

(S3-9)

Equations (S3-9) *also* hold for gas phase reactions carried out at constant volume in batch systems.

QUESTIONS AND PROBLEMS

P3-1. Table P3-1 gives the frequency of flashing of fireflies and the frequency of chirping of crickets as a function of temperature. (*J. Chem. Ed. 5*, (1972) 343).

TABLE P3–1(A)

Fireflies			
Temperature (°C)	21.0	25.0	30.0
Flashes per minute	9.0	12.16	16.7

TABLE P3–1(B)

Crickets			
Temperature (°C)	14.2	20.3	27.0
Chirps per minute	80	126	200

What do these two events have in common?

P3-2. A rule of thumb that is often used in predicting the increase in reaction rate with increase in temperature is that the rate doubles for every 10°C increase in temperature.

    a. If a reaction is carried out at room temperature (298°K), at what activation energy will the rule be valid?

    b. If the reaction is carried out at 500°K, at what activation energy will the rule be valid?

P3-3. Determine the activation energy and the frequency factor from the following data:

$k\,\text{min}^{-1}$	.001	.050
$T°C$	.0	100.0

using

    a. Equation (3-2).
    b. Equation (3-3) with $n = 1$.
    c. What is the percentage difference between the two activation energies E and E'?
    d. Calculate k at 1000°C using first Equation (3-2) and then Equation (3-3). What is the percent difference between the two ks at this temperature?

P3-4. For the reaction given in Equation (S3-7), write expressions similar to those in Equation (S3-8) for species B and D when the reaction is carried out at

    a. constant pressure
    b. constant volume.

P3-5. Calculate the percentage volume change for the following reactions, in which the conditions of constant pressure and constant temperature are maintained.

    a. 40% conversion of hydrogen to hydrogen iodide from an equimolar mixture of hydrogen and iodine.
    b. 90% conversion of acetaldehyde to form methane and carbon monoxide from an initial mixture of 60% acetaldehyde and 40% inerts.
    c. 50% conversion of sodium hydroxide in the liquid phase saponification of ethyl acetate. The feed is equimolar in sodium hydroxide and ethyl acetate.

P3-6. Express the rate of formation of hydrogen bromide in terms of the constants k_1 and k_2 and the conversion of bromine, X. Evaluate numerically all other quantities. The feed consists of 25% hydrogen, 25% bromine, and 50% inerts at a pressure of 10 atm and a temperature of 400°C.

P3-7. Write the rate of decomposition of cumene, $-r'_C$, in terms of the initial concentration of cumene, the conversion of cumene, and the specific rate and equilibrium constants for:

a) the case of constant volume.
b) the case of constant pressure.

The initial mixture consists of 75% cumene and 25% inerts.

P3-8. Does the concentration of reactant A in the reaction in Frame (C21) remain constant during the course of the reaction? Explain in detail! Rework this problem for the case when the vapor pressure of D is 120 mm Hg.

P3-9. In the homogeneous gas phase reaction

$$CH_4 + \tfrac{3}{2}O_2 \longrightarrow HCOOH + H_2O$$

what is the relationship between r_{CH_4} and r_{O_2}?

    a. $r_{CH_4} = r_{O_2}$ c. $r_{CH_4} = \tfrac{2}{3} r_{O_2}$
    b. Cannot tell without rate data.   d. $r_{CH_4} = \tfrac{3}{2} r_{O_2}$
                       e. None of the above.

P3-10. We said that if the reaction

$$aA + bB \longrightarrow cC + dD \tag{2-1}$$

was elementary *as written*, the rate expression was

$$-r_A = k C_A^a C_B^b$$

If we take A as our basis and divide through Equation (2-1) by the stoichiometric coefficient of A so that the equation can be expressed as:

$$A + \frac{b}{a} B \longrightarrow \frac{c}{a} C + \frac{d}{a} D \tag{2-2}$$

will the kinetic rate equation then become

$$-r_A = k C_A C_B^{b/a}$$

Explain.

P3-11. The elementary reaction
$$A + B \longrightarrow C$$
is taking place only in the gas phase of a square duct. The feed to the duct consists of a gas stream of pure A and a liquid stream of pure B. The flowing liquid B covers the bottom of the duct and evaporates into the gas phase, maintaining its equilibrium vapor pressure through the system. The gas phase flows in plug flow. Ignore the volume occupied by liquid B. (See Figure P3-11.)

    a. Express the rate law solely as a function of conversion, and evaluate numerically all possible symbols.

    b. What is the rate of reaction, $-r_A$, when the conversion is 50%?

Other information:

    Total pressure (considered constant) = 1 atm.
    Value of $k = 10^6$(cu ft/lb-mole/sec)
    Temperature within the reactor (considered constant) = 540°F
    Vapor pressure of B = 0.25 atm
    Inlet flow rate of A = 1.5 lb-mole/sec   [CU]

ANS. b. $0.174 \dfrac{\text{lb-mole}}{\text{cu ft} \cdot \text{sec}}$

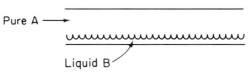
Figure P3–11

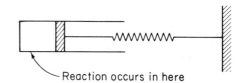
Figure P3–12

P3-12. Consider a cylindrical batch reactor that has one end fitted with a frictionless piston attached to a spring (Figure P3-12). The reaction
$$A + B \longrightarrow 8C$$
following the rate expression
$$-r_A = k_1 C_A^2 C_B$$
is taking place in this type of reactor.
    Write the rate law solely as a function of conversion, numerically evaluating all possible symbols.

Other information:

    Equal moles of A and B are present at $t = 0$.
    Initial volume = 0.15 cu ft.
    Value of $k_1 = 1.0$(cu ft)^2/(lb-mole)^2(sec).
    The relationship between the volume of the reactor and pressure within the reactor is
$$V = (0.1)(P) \quad (V \text{ in cu ft, } P \text{ in atm})$$
    Temperature of system (consider it isothermal) = 140°F.
    Gas constant = 0.73 (cu ft)(atm)/(lb-mole)(°R)   [CU]
    What is the rate of reaction when $V = .2$ cu ft?

SUPPLEMENTARY READING

1. Two references relating to the discussion of activation energy have already been given in this chapter. Activation energy is usually discussed in terms of either collision theory or transition state theory. A concise and readable account of these two theories can be found in Chapter 5 (pp. 44–52) of

> STEVENS, B., *Chemical Kinetics*. London: Chapman and Hall, Ltd., 1970.

An expanded but still elementary presentation can be found in Chapters 4 and 5 of

> FROST, A. A. and R. G. PEARSON, *Kinetics and Mechanism*, 2nd ed. New York: John Wiley and Sons, 1961.

Chapter 3 of

> PANNETIER, G. and P. SOUCHAY, *Chemical Kinetics*. New York: American Elsevier Publishing Co., 1967.

Chapter 2 of

> LAIDLER, K. J., *Reaction Kinetics*, vol. 1. New York: Pergamon Press, 1963.

and Chapters 4 and 5 of

> GARDINER, W. C., *Rates and Mechanism of Chemical Reactions*. New York: W. C. Benjamin, Inc., 1969.

A more advanced treatise of activation energies and collision and transition state theories can be found in

> BENSON, S. W., *The Foundations of Chemical Kinetics*. New York: McGraw-Hill Book Co., 1960.

2. The books listed above also give the rate laws and activation energies for a number of reactions; in addition, as mentioned earlier in this chapter, an extensive listing of rate laws and activation energies can be found in the NBS circulars,

> NATIONAL BUREAU OF STANDARDS, *Tables of Chemical Kinetics Homogeneous Reactions*. Circular 510, Sept. 28, 1951; Supplement 1, Nov. 14, 1956; Supplement 2, Aug. 5, 1960; Supplement 3, Sept. 15, 1961. Washington, D. C.

3. One might also consult the current chemistry literature for the appropriate algebraic form of the rate law for a given reaction. For example, check the Journal of Physical Chemistry in addition to the Journals listed in number 4 of the Supplementary Reading in Chapter 4.

4
Elementary Reactor Design

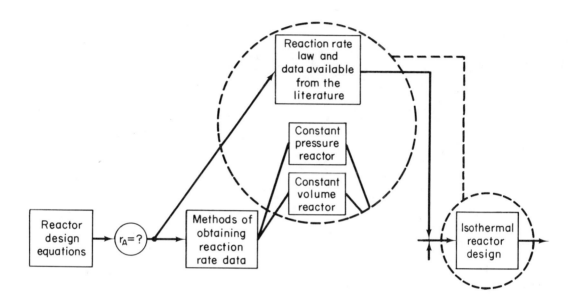

In this chapter we shall use the reaction rate data obtained from constant volume batch reactors to determine the specific reaction rate for gas phase reactions of known order. Using these values of k, we shall then determine the reaction volume needed to achieve a specified conversion for various flow systems operated at the same temperature at which the data were obtained.

Our studies on batch systems will center around obtaining data for constant volume reactors in which the total pressure will be monitored as a function of time. To interpret this data properly we must formulate the differential batch reactor design equation in terms of the total pressure. Finally, we shall briefly discuss equilibrium conversion and work through an example of the formation of smog from nitrogen oxide emission in automobile exhaust.

4.1–Total Pressure Variations in Batch Reactors

Where there is a net increase or decrease in the total number of moles in a gas phase reaction, it may be possible to determine the reaction order from experiments performed with a constant-volume batch reactor. One such reaction system is the decomposition of 1,1-dichloroethane (1,1-DCE).

$$1,1\text{-}C_2H_4Cl_2 \longrightarrow C_2H_3Cl + HCl \tag{4-1}$$

In this and similar systems, the total pressure is monitored as a function of time. For a constant volume reactor that is operated isothermally, rewrite Equation (3-14), $V = V_0(\Pi_0/\Pi)(1 + \epsilon X)f$, so that X is a function of ϵ, the total pressure Π at any time t, and the initial pressure Π_0.

(D1)

The differential equation resulting from a mole balance in a batch system was given in Equation (B10-4):

$$N_{A0} \frac{dX}{dt} = -r_A V \tag{B10-4}$$

We can combine Equations (B10-4) and (D1) to obtain the time rate of change of the total pressure:

$$\frac{N_{A0}}{V} \frac{d}{dt}\left[\frac{1}{\epsilon\Pi_0}(\Pi - \Pi_0)\right] = \left(\frac{1}{\epsilon}\frac{N_{A0}}{V\Pi_0}\right)\frac{d\Pi}{dt} = \left(\frac{1}{\delta RT}\right)\frac{d\Pi}{dt} = -r_A \tag{4-2}$$

where

$$\frac{N_{A0}}{\epsilon V\Pi_0} = \frac{N_{A0}}{y_{A0}\delta V\Pi_0} = \frac{N_{t0}}{\delta V\Pi_0} = \frac{1}{\delta RT_0}$$

The reaction rate $-r_A$ is a function of the temperature and the species concentrations. We must now express these concentrations in terms of the total pressure Π. To accomplish this in a general manner, we revert to the reaction given in Equation (2-1), i.e.,

$$aA + bB \longrightarrow cC + dD \tag{2-1}$$

taking A as our basis:

$$A + \frac{b}{a}B \longrightarrow \frac{c}{a}C + \frac{d}{a}D \tag{2-2}$$

For a constant volume isothermal reaction, Equation (D1-3) gives

$$X = \frac{1}{\epsilon}\left(\frac{\Pi}{\Pi_0} - 1\right) \tag{D1-3}$$

where

$$\epsilon = y_{A0}\delta$$

Remembering that under conditions of constant volume

$$\begin{aligned} C_A &= C_{A0}(1 - X) \\ C_B &= C_{A0}\left(\theta_B - \frac{b}{a}X\right) \\ C_C &= C_{A0}\left(\theta_C + \frac{c}{a}X\right) \\ C_D &= C_{A0}\left(\theta_D + \frac{d}{a}X\right), \end{aligned} \tag{4-3}$$

derive the following expressions for C_A and C_D in Frame (D2), starting from Equations (4-3) and using Equation (D1-3):

$$C_A = \frac{\left[P_{A0} - \dfrac{(\Pi - \Pi_0)}{\delta}\right]}{RT} \qquad C_B = \frac{\left[P_{B0} - \dfrac{b}{a}\dfrac{(\Pi - \Pi_0)}{\delta}\right]}{RT}$$

$$C_C = \frac{\left[P_{C0} + \dfrac{c}{a}\dfrac{(\Pi - \Pi_0)}{\delta}\right]}{RT} \qquad C_D = \frac{\left[P_{D0} + \dfrac{d}{a}\dfrac{(\Pi - \Pi_0)}{\delta}\right]}{RT}$$

(D1)

$$V = V_0, \quad f = \frac{T}{T_0} = \frac{T_0}{T_0} = 1 \tag{D1-1}$$

$$1 = \frac{\Pi_0}{\Pi}(1 + \epsilon X) \tag{D1-2}$$

$$X = \frac{1}{\epsilon}\left(\frac{\Pi}{\Pi_0} - 1\right) = \frac{1}{\epsilon \Pi_0}(\Pi - \Pi_0) \tag{D1-3}$$

The initial partial pressure can be obtained either from the product of the initial mole fraction and the initial total pressure (i.e., $P_{j0} = y_{j0}\Pi_0$), or from the ideal gas law ($P_{j0} = C_{j0}RT_0$).

(D2)

4.2–Decomposition of 1, 1 DCE
Example Problem

The constant volume apparatus described below in the Experimental Procedure (Figure 4-1) was used to study the kinetics of the gas phase decomposition of 1,1-dichloroethane (1,1-DCE), which corresponds to the overall reaction

$$1,1\text{-}C_2H_4Cl_2 \longrightarrow C_2H_3Cl + HCl \qquad (4\text{-}1)$$

This reaction is essentially irreversible and is thought to be first order with respect to 1,1-DCE.*

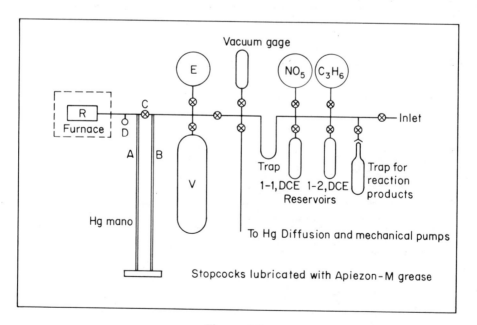

Figure 4-1

* G. R. DeMare et al., *J. Chem. Ed.* 46, (1969), 684.

(D2)

For species A:

$$C_A = C_{A0}(1-X) = \frac{P_{A0}}{RT_0}(1-X) = \frac{P_{A0} - P_{A0}X}{RT_0} \tag{D2-1}$$

From Equation (D1-3)

$$X = \frac{1}{\epsilon}\left(\frac{\Pi}{\Pi_0} - 1\right) = \frac{1}{y_{A0}\delta}\left(\frac{\Pi}{\Pi_0} - 1\right) \tag{D2-2}$$

$$P_{A0} = y_{A0}\Pi_0$$

$$P_{A0}X = y_{A0}\Pi_0 X = y_{A0}\Pi_0\left[\frac{1}{y_{A0}\delta}\left(\frac{\Pi}{\Pi_0} - 1\right)\right] = \frac{1}{\delta}(\Pi - \Pi_0) \tag{D2-3}$$

$$C_A = \frac{P_{A0} - \dfrac{(\Pi - \Pi_0)}{\delta}}{RT} \tag{D2-4}$$

For species D:

$$C_D = C_{A0}\left(\theta_D + \frac{d}{a}X\right) = \frac{P_{A0}\theta_D + \dfrac{d}{a}P_{A0}X}{RT} \tag{D2-5}$$

$$\theta_D = \frac{C_{D0}}{C_{A0}} = \frac{P_{D0}}{P_{A0}} \tag{D2-6}$$

Substituting for θ_D, [Equation (D2-6)] and $P_{A0}X$ [Equation (D2-3)], we obtain

$$C_D = \frac{P_{D0} + \dfrac{d}{a}\dfrac{(\Pi - \Pi_0)}{\delta}}{RT} \tag{D2-7}$$

In an analogous manner we can express C_B and C_C as functions of the total pressure. We usually express the concentrations and design equations in terms of total pressure when the measured variable in experiments carried out to obtain rate data is the *total pressure*.

Experimental Procedure. A 60 ml Pyrex reaction vessel (R), which is centered in a copper cylinder surrounded by asbestos insulation, is heated electrically. The temperature is controlled with the aid of a platinum resistance thermometer. Pressures are measured by means of mercury manometers. First, a sample of 1,1-DCE is transferred from the reservoir to the calibrated 400 ml volume (V). The pressure, from 30 to 200 torr (i.e., 30–200 mm Hg), is read on manometer B. Stopcock C is opened for a few seconds to allow 1,1-DCE to enter the reaction vessel; the timer is started and the pressure on manometer A is recorded at various times. At the end of the reaction, to avoid their passage through the diffusion and mechanical pumps, the reaction products are frozen in a removable trap for disposal.

4.2A Batch Data

For this constant-volume reaction system, the mole balance on 1,1-DCE is

$$-\frac{dC_A}{dt} = -r_A \quad \text{(let A = 1,1—DCE)} \tag{4-4}$$

The reaction rate is first order w.r.t. 1,1-DCE.

$$-r_A = kC_A \tag{4-5}$$

$$-\frac{dC_A}{dt} = kC_A \tag{4-6}$$

Starting with Equation (4-6), replace C_A with the appropriate expression, involving the total pressure, that will lead to a differential equation involving only Π, Π_0, and k for the 1,1-DCE decomposition. Under the conditions of this experiment only 1,1-DCE was initially present in the system.

(D3)

The pressure time data shown in Figure 4-2 was obtained from carrying out this reaction at 705°K and 716°K. With the aid of the integrated form of Equation (D3-1), explain how you would replot the data in Figure 4-2 so that the slope of the resulting linear plot is directly proportional to the specific reaction rate k.

(D3)

$$\frac{-dC_A}{dt} = kC_A \tag{4-6}$$

$$C_A = \frac{P_{A0} - \frac{(\Pi - \Pi_0)}{\delta}}{RT}$$

$$\delta = 1 + 1 - 1 = 1$$

Initially, only DCE is present; therefore

$$P_{A0} = \Pi_0$$

$$C_A = \frac{(2\Pi_0 - \Pi)}{RT}$$

Substituting for C_A in terms of Π in Equation (4–6) we obtain

$$\frac{-dC_A}{dt} = \left(\frac{d\Pi}{dt}\right)\frac{1}{RT} = k\frac{(2\Pi_0 - \Pi)}{RT}$$

$$\frac{d\Pi}{dt} = k(2\Pi_0 - \Pi) \tag{D3-1}$$

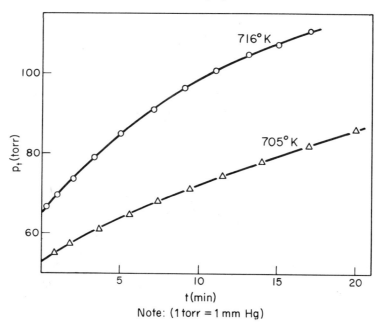

Figure 4-2. $P_t = \Pi$, $P_{t0} = \Pi_0$.

(D4)

4.2B CSTR Design

From the slope of the line in Figure 4–3, the specific reaction rate k was found to be .0414 min^{-1} at 705°K. What CSTR volume would be necessary to achieve 75% conversion of 1,1-DCE if pure 1,1-DCE is fed to the reactor at 705°K and 10 atm at a volumetric flow rate of 6 l/h?

$$R = .082 \frac{l \cdot \text{atm}}{\text{gmole-}°K}$$

(D4)

$$\frac{d\Pi}{dt} = k(2\Pi_0 - \Pi)$$ (D4-1)

$$\frac{d\Pi}{2\Pi_0 - \Pi} = kdt$$

Integrating:

$$\int_{\Pi_0}^{\Pi} \frac{d\Pi}{(2\Pi_0 - \Pi)} = \int_0^t k\,dt$$

$$\ln\left(\frac{\Pi_0}{2\Pi_0 - \Pi}\right) = kt$$ (D4-2)

Converting to base 10:

$$\log_{10}\left(\frac{\Pi_0}{2\Pi_0 - \Pi}\right) = \frac{k}{2.3}t$$

We see that when the data in Figure 4-2 are replotted in Figure 4-3 as $\log_{10}(\Pi_0/2\Pi_0 - \Pi)$ vs. t, the resulting plot is linear indicating the reaction is first order. The slope of this plot, which is equal to $k/2.3$ is 0.018 min^{-1}. Consequently $k = 2.3\,(.018)$ min^{-1} = .0414 min^{-1}.

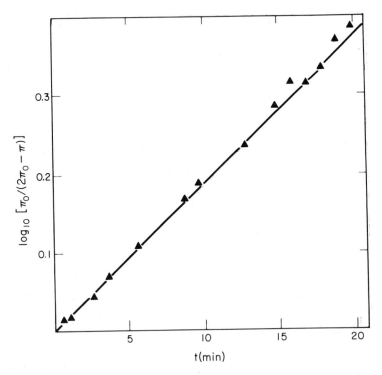

Figure 4-3

(D5)

4.3–Decomposition of Dihydrofuran
Example Problem

The decomposition of 2,5-dihydrofuran is a first-order irreversible reaction.* Studies on this reaction were carried out in a system similar to that shown in Figure 4-1. The stoichiometry of the reaction is

$$\text{(dihydrofuran)} \longrightarrow \text{(furan)} + H_2 \qquad (4\text{-}7)$$

4.3A Batch Data

Using the data in Table 4-1, we can determine the specific reaction rate constant for an initial mixture of 40% dihydrofuran and 60% inerts.

TABLE 4-1

	Pressure (mm Hg)	Time
	200	0
T = 685°K	208	2
	215	4
	226	8

First rewrite the differential form of the constant volume batch reactor design equation for this first-order reaction solely in terms of Π_0, Π, and k, numerically evaluating all possible symbols in Frame (D6).

(D6)

---

*Rubin, J. A. and S. V. Filseth, *J. Chem. Ed.* 46 (1969), 57.

(D5)

Let A = 1,1-DCE:

$$-r_A = kC_A$$

$$C_A = C_{A0}\frac{(1-X)}{(1+\epsilon X)f}$$

Isothermal operation; therefore $T = T_0$ and $f = 1$:

$\delta = 1 + 1 - 1 = 1$, pure 1,1-DCE initially ∴ $y_{A0} = 1$

$\epsilon = y_{A0}\delta = 1 \times 1 = 1$

$$V = \frac{F_{A0}X}{-r_A} = \frac{C_{A0}v_0 X}{kC_{A0}\frac{(1-X)}{(1+X)}}$$

$$V = v_0\frac{X(1+X)}{k(1-X)}$$

$$= 6\ l/hr \times \frac{1\ hr}{60\ min} \frac{(1+.75)(.75)}{\frac{(.0414)}{min}(1-.75)}$$

$$= (.1)\frac{(.75)(7)}{.0414} = 12.7\ \text{liters}$$

(D6)

The reaction is irreversible, is first order w.r.t. dihydrofuran (A), and is carried out in a constant-volume batch system.

$$\frac{-dC_A}{dt} = -r_A = kC_A$$

$$C_A = \frac{P_{A0} - \frac{(\Pi - \Pi_0)}{\delta}}{RT}, \quad \delta = 1 + 1 - 1 = 1$$

$$P_{A0} = y_{A0}\Pi_0 = .4\Pi_0$$

$$C_A = \frac{1.4\Pi_0 - \Pi}{RT} \tag{D6-1}$$

$$\frac{d\Pi}{dt} = k(1.4\Pi_0 - \Pi) \tag{D6-2}$$

Next, integrate Equation (D6-2), evaluating all constants of integration, and explain how you would plot the resulting expression to obtain the specific reaction rate constant from the slope of this plot.

(D7) _____

_____

The specific reaction rate constant for the first-order decomposition reaction of 2,5-dihydrofuran to form furan and hydrogen at 685°K is .05 min^{-1}.

4.3B Tubular Flow Design

Next, consider a flow system operated isothermally, in which furan is to be formed in a tubular reactor from a feed containing 80% 2,5-dihydrofuran and 20% inerts. If this mixture is fed to the reactor at 685°K, 8 atm, and a volumetric rate of 100 l/hr, calculate the tubular reactor volume necessary to achieve 75% conversion of 2,5-dihydrofuran. Remember that we are neglecting pressure drop in our continuous flow reactors and we are assuming that the pressure remains constant throughout the reactor. (*Hint:* Write $-r_A$ as a function of conversion and then use an integral table to evaluate the design integral.)

(D8) _____

_____

For a tubular reactor with a cross-sectional area of 100 cm² the concentrations of the reacting species along the reactor are as follows.

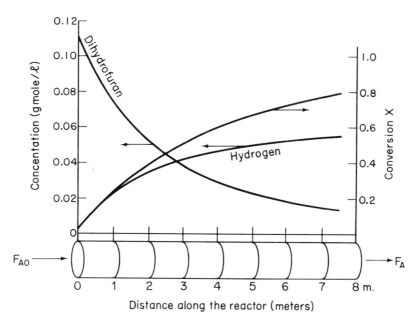

Figure 4-5

The concentration of furan will be equal to the concentration of hydrogen.

(D7)

Integrating, with limits $t = 0$ and $\Pi = \Pi_0$,

$$\ln\left(\frac{.4\Pi_0}{1.4\Pi_0 - \Pi}\right) = kt$$

t	Π	$\left(\frac{.4\Pi_0}{1.4\Pi_0 - \Pi}\right)$	$\ln\left(\frac{.4\Pi_0}{1.4\Pi_0 - \Pi}\right)$
0	200	1.00	0
2	208	1.11	.104
4	215	1.23	.201
8	226	1.48	.392

The specific reaction rate can be found from the slope of the plot of the natural log of $[.4\Pi_0/(1.4\Pi_0 - \Pi)]$ as a function of time shown in Figure 4-4.

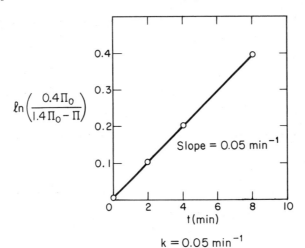

$k = 0.05 \text{ min}^{-1}$

Figure 4-4

(D8)

$$V = F_{A0} \int_0^{.75} \frac{dx}{-r_A}$$

$$V = C_{A0} v_0 \int_0^{.75} \frac{dX}{kC_{A0}\frac{(1-X)}{(1+\epsilon X)}} = v_0 \int_0^{.75} \frac{(1+\epsilon X)dX}{k(1-X)} \tag{D8-1}$$

Note: Since the reaction is carried out isothermally, the specific reaction rate is constant and *can* be taken outside of the integral sign.

$$\epsilon = y_{A0} \cdot \delta = (.8)(1) = .8$$

Using integral (a2-4) of Appendix A–2, we find

$$V = \frac{v_0}{k} \int_0^X \frac{(1+\epsilon X)dX}{(1-X)} = \frac{v_0}{k}\left\{[1+\epsilon]\ln\left(\frac{1}{1-X}\right) - \epsilon X\right\} \tag{D8-2}$$

$$V = \frac{v_0}{k}\left[[1.8]\ln\left(\frac{1}{1-X}\right) - .8X\right] \tag{D8-3}$$

$$X = .75$$

$$V = \frac{(100 \text{ l/hr})}{(.05/\text{min})} \times \frac{1 \text{ hr}}{60 \text{ min}} \cdot [1.90] = 63.4 \text{ l}$$

$$V = 63.4 \text{ l}$$

4.4–Nitric Oxide Example Problem

In the examples above we designed flow reactors for the decomposition of 1,1-dichloroethane and of dihydrofuran on the basis of information obtained from batch experiments performed in the laboratory. As one final example, let us consider a batch system in which there is a reversible reaction taking place. In particular, we shall consider the formation of nitric oxide in a constant volume reactor the size of an automobile engine cylinder. It is believed that nitric oxides play a major role in air pollution.* A brief and somewhat simplified summary of the role of formation and presence of nitric oxides in smog formation follows.

4.4A Role of Nitric Oxide in Smog Formation

The NO from automobile exhaust is oxidized to NO_2 in the presence of peroxide radicals. Nitrogen dioxide is then decomposed photochemically to give nascent oxygen,

$$NO_2 + h\nu \longrightarrow NO + O \qquad (4\text{-}8)$$

which reacts to form ozone.

$$O + O_2 \longrightarrow O_3 \qquad (4\text{-}9)$$

The ozone then becomes involved in a whole series of reactions. Of particular importance are its reactions with hydrocarbons in the atmosphere to form aldehydes, various free radicals, and other intermediates, which react further to produce undesirable products in air pollution.

$$\text{ozone} + \text{olefins} \longrightarrow \text{aldehydes} + \text{free radicals}$$

$$O_3 + RCH{=}CHR \longrightarrow RCHO + R\dot{O} + H\dot{C}O \qquad (4\text{-}10a)$$

$$\xrightarrow{h\nu} \dot{R} + H\dot{C}O \qquad (4\text{-}10b)$$

One specific example is the reaction of ozone with 1,3 butadiene to form acrolein and formaldehyde, which are *severe* eye irritants.

$$\tfrac{2}{3}O_3 + CH_2{=}CHCH{=}CH_2 \xrightarrow{h\nu} CH_2{=}CHCHO + HCHO \qquad (4\text{-}11)$$

By regenerating NO_2, more ozone can be formed and the cycle continued. One means by which this regeneration may be accomplished is through the reaction of NO with the free radicals in the atmosphere. For example, the free radical formed in Equation (4-10b) can react with O_2 to give the peroxy free radical,

$$\dot{R} + O_2 \longrightarrow R\dot{O}\dot{O} \qquad (4\text{-}12a)$$

which reacts with nitric oxide to form nitrogen dioxide.

$$R\dot{O}\dot{O} + NO \longrightarrow R\dot{O} + NO_2 \qquad (4\text{-}12b)$$

---

*M. S. Peters, *Important Chemical Reactions in Air Pollution Control*, CEP Symposium Series Vol. 67, No. 115, (New York: 1971) p. 1.

C. S. Tuesday, *Report GMR 332*, General Motors Research Laboratory, (Warren, Michigan: 1961).

Referring back to Equation (4-8), we see that the cycle has been completed and that, with a relatively small amount of nitrogen oxides, a large amount of pollutants can be produced. Of course there are many other reactions taking place, and one should not be misled by the brevity of the preceding discussion; however, it does serve to present, in rough outline the role of nitrogen oxides in air pollution.

The nitrogen oxides are formed from nitrogen and oxygen at temperatures of about 4400°F in the combustion cylinders of automobiles.

$$N_2 + O_2 \underset{k_{-1}}{\overset{k_1}{\rightleftharpoons}} 2NO \qquad (4\text{-}13)$$

The reaction is elementary, homogeneous, reversible, and is endothermic w.r.t. the formation of NO. Neglecting the effects of the other reactions occurring in the cylinder, write the rate law $-r_{N_2}$ for the above reaction in terms of the concentration of the various species, the forward specific rate constant k_1 and the equilibrium constant K_e.

(D9)

The formation of NO from N_2 and O_2 is to be carried out in a small batch reactor. As a first approximation, we shall consider that the reaction takes place isothermally at 2700°K in a constant volume reactor of .4 l under a pressure of 20 atm. Consider that the feed consists of 77% N_2, 15% O_2, and 8% other gases, which may be considered inerts. At this temperature the equilibrium constant K_e is .01 and the forward specific rate k_1 is 1.11×10^4 l/gmole sec.*

Calculate (1) the equilibrium conversion of N_2, X_e, and (2) the equilibrium concentration of NO.

(D10)

*E. O. Ermenc, *Chem. Engr.* 77, (June 1, 1970) 193.

(D9)
$$-r_{N_2} = k_1 C_{N_2} C_{O_2} - k_{-1} C_{NO}^2$$
$$= k_1 \left[C_{N_2} C_{O_2} - \frac{C_{NO}^2}{K_e} \right]$$

(D10) 77% N_2, 15% O_2, and 8% inerts:

$$-r_{N_2} = k_1 C_{N_2} C_{O_2} - k_{-1} C_{NO}^2 \tag{D10-1}$$

$$-r_{N_2} = k_1 \left(C_{N_2} C_{O_2} - \frac{C_{NO}^2}{K_e} \right) \tag{D10-2}$$

At equilibrium

$$-r_{N_2} = 0 = k_1 \left(C_{N_2} C_{O_2} - \frac{C_{NO}^2}{K_e} \right) \tag{D10-3}$$

At constant volume

$$C_{N_2} = C_{N_2,0}(1 - X)$$
$$C_{O_2} = C_{N_2,0}(\theta_{O_2} - X)$$
$$\theta_{O_2} = \frac{.15}{.77} = .195 \tag{D10-4}$$
$$C_{O_2} = C_{N_2,0}(1.95 - X)$$
$$C_{NO} = C_{N_2,0}(2X)$$

Rearranging Equation (D10-3) gives:

$$C_{NO}^2 = K_e C_{N_2} C_{O_2} \tag{D10-5}$$
$$4X_e^2 = K_e(1 - X_e)(.195 - X_e) \tag{D10-6}$$
$$399 X_e^2 + 1.195 X_e - .195 = 0 \tag{D10-7}$$

Solving for the equilibrium conversion

$$X_e = .02065$$
$$X_e \doteq .02000$$

The equilibrium concentration is

$$C_{NO,e} = C_{N_2,0}(2X) = \frac{y_{N_2,0} \Pi_0}{RT_0}(2X) = \frac{(.77)(20)}{(.082)(2700°K)}(.04) = 2.78 \times 10^{-3} \text{ gmole}/l$$

4.4B Batch Reactor

Calculate the time necessary to achieve 80% of the equilibrium conversion of N_2 for the reactor and conditions given above. If at first your numerical answer does not agree with the one given, try to work through the problem a second time before looking at the method of solution. (*Hint:* A table of integrals and/or Appendix A–2 may be useful.)

(D11)

Ans.: $t = 154$ μsec

Even for an engine operating at 6000 rpm, the time required for an 80% conversion of N_2 at 2700°K is still quite short compared to the compression cycle of 5000 μsec.

4.4C Tubular Reactor

Starting with the general mole balance equation [Equation (1-4)] derive the integral form of the steady-state tubular reactor design equation in terms of $F_{N_2,0}$, $-r_{N_2}$, and the conversion of

(D11)

For 80% of equilibrium conversion

$$X = (.8)(X_e) = .016$$

For constant volume

$$t = C_{N_2,0} \int_0^X \frac{dX}{-r_A}$$

From Equations (D10-2) and (D10-4)

$$-r_A = k_1 C_{N_2,0}^2 \left[(1-X)(.195-X) - \frac{4X^2}{K_e} \right]$$

$$C_{N_2,0} = \frac{(.77)(20)}{(.082)(2700°K)} = 6.95 \times 10^{-2} \text{ gmole}/l$$

$$K_e = .01$$

$$t = \frac{1}{k_1 C_{N_2,0}} \int_0^X \frac{dX}{[(1-X)(.195-X) - 400X^2]}$$

$$k_1 C_{N_2,0} = 7.7 \times 10^2 \text{ sec}^{-1}$$

$$t = 1.3 \times 10^{-3} \int_0^{.016} \frac{dX}{.195 - 1.195X - 399X^2} \text{ sec}$$

$$= (1.3 \times 10^{-3})(.118) = 1.54 \times 10^{-4} \text{ sec}$$

$$= 154 \ \mu\text{sec}$$

N_2, for the reaction of N_2 with O_2 given by Equation (4-13). The exit conversion is to be 80% of the equilibrium conversion at 2700°K. Review Frame (A19), letting $j = N_2$, and see Figure 4-6.

Figure 4-6

(D12)

Write the integral form of the tubular reactor design equation solely in terms of X. *Note:* The volumetric flow rate to the reactor will be 20 l/sec, and the pressure in the reactor will be 10 atm. Otherwise, all other conditions ($T = 2700°K$, 77% N_2, etc.) are the same as those applied to the batch reactor previously discussed. After you have substituted for $-r_{N_2}$ in the design integral, you should make as much use of the calculations delineated in Frame (D11) as possible to evaluate numerically all symbols other than X.

(D12)

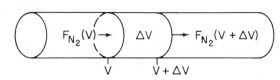

Figure 4-7

The general mole balance equation on a system volume V' is

$$F_{N_2,0} - F_{N_2} + \int_0^{V'} r_{N_2} \, dV' = \frac{dN_{N_2}}{dt} \quad (1\text{-}4)$$

Since the reactor is operated at steady state, $dN_{N_2}/dt = 0$,

$$F_{N_2,0} - F_{N_2} = \int_0^{V'} -r_{N_2} \, dV' \quad (D12\text{-}1)$$

Taking our system volume as ΔV, (i.e., $V' = \Delta V$) within which there is no spatial variation in reaction rate, the entering molar feed rate of N_2 is $F_{N_2}(V)$, i.e.,

$$F_{N_2,0} \longrightarrow F_{N_2}(V)$$

and the exiting molar feed rate of N_2 to our system volume ΔV is $F_{N_2}(V + \Delta V)$, i.e.,

$$F_{N_2} \longrightarrow F_{N_2}(V + \Delta V)$$

After integrating the right-hand side of Equation (D12-1) for the case of no spatial variations within ΔV and making the above substitutions, we obtain

$$F_{N_2}(V) - F_{N_2}(V + \Delta V) = -r_{N_2} \Delta V \quad (D12\text{-}2)$$

Dividing by ΔV and taking the limit as ΔV goes to zero, we get

$$-\frac{dF_{N_2}}{dV} = -r_{N_2} \quad (D12\text{-}3)$$

Rewriting F_{N_2} in terms of the conversion and the inlet molar flow rate of nitrogen

$$F_{N_2} = F_{N_2,0}(1 - X) \quad (D12\text{-}4)$$

and substituting in Equation (D12-3), we arrive at the differential form of the tubular design equation

$$F_{N_2,0} \frac{dX}{dV} = -r_{N_2} \quad (D12\text{-}5)$$

Integrating:

$$V = F_{N_2,0} \int_0^X \frac{dX}{-r_{N_2}} \quad (D12\text{-}6)$$

The alternate derivation of Equation (D12-6) is to substitute (D12-4) into (D12-1) to obtain

$$F_{N_2,0} X = \int_0^{V'} -r_{N_2} \, dV' \quad (D12\text{-}7)$$

Letting V' represent our reactor volume V, we can differentiate Equation (D12-8) w.r.t. V to obtain

$$F_{N_2,0} \frac{dX}{dV} = -r_{N_2} \quad (D12\text{-}8)$$

Rearranging and integrating, we obtain the integral form of the tubular reactor design equation:

$$F_{N_2,0} \int_0^X \frac{dX}{-r_{N_2}} = V \quad (D12\text{-}10)$$

(D13)

Again using the information obtained in Frame (D11), along with your calculations in Frame (D12), evaluate Equation (D13-5) to determine the tubular reactor volume needed to achieve 80% of equilibrium conversion at 2700°K.

(D14)

4.4D CSTR

Next, derive the CSTR design equation in terms of $F_{N_2,0}$, X, and $-r_{N_2}$, starting from the general mole balance Equation (1-4).

(D15)

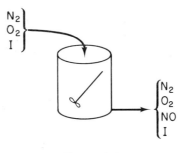

Figure 4-8

Calculate the CSTR volume needed to achieve 80% of equilibrium conversion of N_2 at 2700°K. The feed and operating conditions are identical with those previously delineated for the tubular reactor. Use any information from Frames (D14) and (D15) that might speed your calculation.

(D13)

80% of equilibrium conversion $= .8(.02) = .016$.

$$V = F_{N_2,0} \int_0^{.016} \frac{dX}{-r_{N_2}} = v_0 C_{N_2,0} \int_0^{.016} \frac{dX}{-r_{N_2}} \tag{D13-1}$$

$$-r_{N_2} = k_1 \left(C_{N_2} C_{O_2} - \frac{C_{NO}^2}{K_e} \right) \tag{D10-2}$$

$$= k_1 \left[C_{N_2,0}^2 (1-X)(\theta_{O_2} - X) - \frac{(2X)^2 C_{N_2,0}^2}{K_e} \right] \tag{D13-2}$$

$$-r_A = k_1 C_{N_2,0}^2 \left[(1-X)(.195 - X) - \frac{4X^2}{K_e} \right] \tag{D13-3}$$

$$V = \frac{v_0}{k_1 C_{N_2,0}} \int_0^{.016} \frac{dX}{(1-X)(.195-X) - \frac{4X^2}{K_e}} \tag{D13-4}$$

$$C_{N_2,0} = \frac{(.77)(10)}{(.082)(2700°K)} = 3.48 \times 10^{-2} \text{ gmole}/l$$

$$k_1 C_{N_2,0} = 3.86 \times 10^2 \text{ sec}, \quad v_0 = 20 \, l/\text{sec}$$

$$V = 5.2 \times 10^{-2} \int_0^{.016} \frac{dX}{.195 - 1.195X - 399X^2} \tag{D13-5}$$

(D14)

$$V = (5.2 \times 10^{-2})(.118) = 6.16 \times 10^{-3} \, l$$
$$= 6.16 \text{ cu cm}$$

(D15)

The CSTR is modeled as having no spatial variation within the vessel. Consequently the species concentrations of the outlet stream are identical with the species concentrations inside the tank. For no spatial variations in $-r_A$ and for steady state operation, the general mole balance equation

$$F_{N_2,0} - F_{N_2} + \int_0^V r_{N_2} \, dV = \frac{dN_{N_2}}{dt} \tag{1-4}$$

reduces to

$$V = \frac{F_{N_2,0} - F_{N_2}}{-r_{N_2}} \tag{D15-1}$$

Substituting for F_{N_2} in terms of X, we obtain

$$V = \frac{(F_{N_2,0})X}{-r_{N_2}} \tag{D15-2}$$

SUMMARY

1. For the gas phase reaction,

$$A + \frac{b}{a}B \longrightarrow \frac{c}{a}C + \frac{d}{a}D \tag{S4-1}$$

which is carried out in a constant volume reactor, the concentration of the individual species A, B, C, and D can be expressed in terms of the total pressure:

$$C_A = \frac{\left[P_{AO} - \frac{(\Pi - \Pi_0)}{\delta}\right]}{RT} \tag{S4-2}$$

$$C_B = \frac{\left[P_{BO} - \frac{b}{a}\frac{(\Pi - \Pi_0)}{\delta}\right]}{RT} \tag{S4-3}$$

$$C_C = \frac{\left[P_{CO} + \frac{c}{a}\frac{(\Pi - \Pi_0)}{\delta}\right]}{RT} \tag{S4-4}$$

$$C_D = \frac{\left[P_{DO} + \frac{d}{a}\frac{(\Pi - \Pi_0)}{\delta}\right]}{RT} \tag{S4-5}$$

2. The rate expression for the elementary reversible reaction

$$A + B \underset{k_{-1}}{\overset{k_1}{\rightleftharpoons}} C + D \tag{S4-6}$$

is

$$-r_A = k_1 C_A C_B - k_{-1} C_C C_D \tag{S4-7}$$

$$= k_1 \left(C_A C_B - \frac{C_C C_D}{K_e} \right) \tag{S4-8}$$

where K_e is the concentration equilibrium constant.

QUESTIONS AND PROBLEMS

P4-1. Write a paragraph discussing the role of nitrogen oxides in air pollution. Include in your discussion the balanced chemical reactions relating to air pollution.

P4-2. The irreversible gas phase reaction

$$A \longrightarrow 3B$$

will be carried out isothermally. The reaction is zero order, the initial concentration of A is 2 moles/l, and the system contains 40% inerts. The specific reaction rate constant is 0.01 moles/l/min. Calculate the time needed to achieve 80% conversion in

    a. a constant volume batch reactor,
    b. a constant pressure batch reactor.

(D16)

$$V = \frac{(F_{N_2,0})X}{-r_{N_2}} = \frac{(v_0 C_{N_2,0})X}{-r_{N_2}}$$

$$-r_{N_2} = k_1 C_{N_2,0}^2 [(1-X)(.195-X) - 400X^2]$$

The outlet conversion is identical with the conversion inside the tank:

$$V = \frac{v_0 X}{k_1 C_{N_2,0}(.195 - 1.195X - 399X^2)}$$

$$= (5.2 \times 10^{-2}) \frac{X}{.195 - 1.195X - 399X^2}$$

$X = .016$

$V = .0117 \; l = 11.7 \text{ cu cm}$

For a volumetric flow rate of 2 *l*/min, calculate the reactor volume, space time, and space velocity necessary to achieve 80% conversion in

    c. a CSTR
    d. a plug flow reactor

[Quiz, U. of M.]

P4-3. An ideal gas mixture is charged to a tubular (plug flow) reactor at the rate of 25 lb-moles/hr. The reactor is operated isothermally at 1000°F and the pressure is 6 atm (abs). The reactor is 6 inches in inner diameter.

The second-order irreversible reaction $A + B \xrightarrow{I} D$ that is taking place in the reactor has a specific reaction rate of 10^5 cu ft/lb-moles/hr at 1000 °F. The feed composition is: 40% A, 40% B, and 20% I (inert).

    a. Derive the design equation for this reactor.
    b. What reactor length is necessary for 80% conversion?

(Ans.: L = 236 ft.)   [Exam, U. of M.]

P4-4. The elementary reversible reaction $A \rightleftharpoons 2B$ is conducted at 540°F and 3 atm in a tubular flow reactor. The feed rate is 75 lb-moles/hr with 40% A in the feed stream. The specific reaction rate is $k = 1.6$ sec^{-1}, and the concentration equilibrium constant is $K_c = 0.0055$ moles/cu ft. To accomplish 75% equilibrium conversion, determine

    a. the volume, space time, and space velocity of the tubular reactor,
    b. the volume and space velocity of a CSTR reactor operated under the same conditions.

(Ans. $V_{CSTR} = 56$ ft^3)   [Exam, U. of M.]

P4-5. The elementary gaseous reaction $A \rightarrow B$ has a unimolecular reaction rate constant of 0.0015/min at 80°F. This reaction is to be carried out in parallel tubes 10 ft long and 1 inch in inside diameter under a pressure of 132 psig at 260°F. A production rate of 1000 lb/hr of B is required. Assuming an activation energy of 25,000 cal/gmole, how many tubes are needed if the conversion of B is to be 90%? Assume perfect gas laws. A and B each have molecular weights of 58.

(Calif. PECEE, April 1966)

P4-6. A biomolecular elementary second-order reaction, $A + B \rightarrow C + D$, takes place in a homogeneous liquid system. Reactants and products are mutually soluble, and the volume change as a result of reaction is negligible.

Feed to a tubular (plug flow) reactor that operates essentially isothermally at 260°F consists of 210 lb/hr of A and 260 lb/hr of B. Total volume of the reactor is 5.33 cu ft, and, with this feed rate, 50% of compound A in the feed is converted.

It is proposed that, in order to increase conversion, a stirred reactor of 100 gal capacity be installed in series with, and immediately upstream of, the tubular reactor. If the stirred reactor operates at the same temperature, estimate the conversion of A that can be expected in the revised system; neglect the reverse reaction. Other available data include:

	A	B
Density at 260°F, lb/cu ft	47.8	54.0
Molecular weight	139	172
Heat capacity, Btu/lb-°F	0.55	0.52
Viscosity, cp	0.32	0.45
Boiling point, °F	390	415

(Ans.: X = .775)   (Calif. PECEE, November 1966)

P4-7. The irreversible elementary reaction $2A \rightarrow B$ takes place in the gas phase in an isothermal tubular (plug flow) reactor. Reactant A and a diluent C are fed in equimolar ratio, and conversion of A is 80%.

If the molar feed rate of A is cut in half, what is the conversion of A assuming the feed rate of C is left unchanged?

Assume ideal behavior and that the reactor temperature remains unchanged.

(Calif. PECEE, August 1969)

P4-8. Compound A undergoes a reversible isomerization reaction, $A \rightleftharpoons B$, over a supported metal catalyst. Under pertinent conditions, A and B are liquid, miscible, and of nearly identical density; the equilibrium constant for the reaction (in concentration units) is 5.8.

In a fixed-bed isothermal flow reactor in which backmixing is negligible (i.e., plug flow), a feed of pure A undergoes a net conversion to B of 55%. The reaction is elementary.

If a second, identical flow reactor at the same temperature is placed downstream from the first, what *overall* conversion of A would you expect if

  a. the reactors are directly connected in series? (Ans. X = .74)
  b. the products from the first reactor are separated by appropriate processing and only the unconverted A is fed to the second reactor?

(Calif. PECEE, April 1968).

P4-9. 2500 gal/hr of metaxylene are being isomerized to a mixture of orthoxylene, metaxylene, and paraxylene in a reactor containing 1000 cu ft of catalyst. Reaction is being carried out at 750°F. and 300 psig. Under these conditions, 37% of the metaxylene fed to the reactor is isomerized. At a flow rate of 1667 gal/hr, 50% of the metaxylene is isomerized at the same temperature and pressure. Energy changes are negligible.

It is now proposed that a second plant be built to process 5500 gal/hr of metaxylene at the same temperature and pressure as above. What size reactor, i.e., what volume of catalyst, is required if conversion in the new plant is to be 46% instead of 37%? Justify any assumptions made for the scale-up calculation. (Ans. 2931 cu ft of cat.)

(Calif. PECEE, August 1971)

P4-10. A liquid organic substance, A, contains 0.1 mole% of an impurity, B, which can be hydrogenated to A:

$$B + H_2 \longrightarrow A$$

The material is purified by hydrogenation as a liquid in a continuous well-mixed reactor at 100°C. The feed rate of the liquid is constant at 730 lb/hr. The reactor holds 50 gal of liquid. Catalyst concentration and activity are kept constant. When the total pressure is held constant at 500 psig, the amount of B in the product levels out at .001 mole%.

What will be the concentration of B in the product if the hydrogen pressure is held at 300 psig?

Assume that the reaction behaves as though it were first order w.r.t. both B and H_2; i.e., in batch

$$-\frac{dB}{dt} = kBH_2$$

where

B = the concentration of B
H_2 = the concentration of H_2
t = time
k = constant

Assume perfect gas laws and Henry's law. Also assume the following properties:

	A	B
Vapor pressure at 100°C, mm	10	10
Molecular weight	196	194
Specific gravity at 100°C	0.980	0.960
Solubility of H_2 in liquid at 500 psig and 100°C, g/l	5.0	3.5

(Calif. PECEE, August 1967)

P4-11. An isothermal, constant-pressure plug flow reactor is designed to give a conversion of 63.2% of A to B for the first-order gas phase decomposition

$$A \xrightarrow{k} B$$

for a feed of pure A at a rate of 5 cu ft/hr. At the chosen operating temperature, the first-order rate constant $k = 5.0$ hr^{-1}. However, after the reactor is installed and in operation it is found that the conversion is 92.7% of the desired conversion. This is thought to be due to a flow disturbance in the reactor that gives rise to a zone of intense backmixing. Assuming that this zone behaves like a perfectly mixed, stirred-tank reactor in series with part of the plug flow reactor, what fraction of the total reactor volume is occupied by this zone? (Ans.: 57%)

P4-12. A semibatch reactor is sometimes utilized for a reaction that has a high exothermic heat of reaction or a particular selectivity problem. A schematic diagram of the reactor is shown in Figure P4-12.

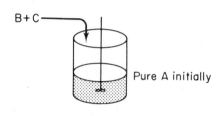

Initially the tank is charged with 10 cu ft of pure A. Compound A has a density of 52 lb/cu ft and a molecular weight of 130. A stream containing B and C enters at a rate of 1 cu ft/min. The density of this stream is 52 lb/cu ft and contains 10 wt % of C and 90 wt % of B. The molecular weight of B is 65.

The reaction A + 2B ⟶ C takes place within the tank, and the reaction is first order in A and in B.

In a separate batch experiment it was found that when the concentration of A was 0.2 lb-moles/cu ft and the concentration of B was 0.1 lb-moles/cu ft, compound B was being depleted from the system at a rate of 0.004 moles/min/cu ft.

Consider the reactor to be well mixed, so that the concentration of each component is uniform throughout the reactor.

  a. What are the molar flow rates of B and C into the tank?
  b. Determine the volume of liquid in the tank at any time t. (Neglect volume changes due to reaction.)
  c. From the data obtained in the separate experiment described above,
    (1) What is the rate of formation of C at the time the data were taken?
    (2) In terms of the concentrations of A and B, write the rate law for C (i.e., r_C) at any other time t.
  d. Derive a differential equation that expresses the time rate of conversion, X (i.e., moles of A reacted per mole of A initially present) solely in terms of the conversion X and the time t. Evaluate all constants, but *do not* attempt to solve this equation.
  e. Assuming the equation in part d that determines the conversion as a function of time, $X = f(t)$ is readily solved, write an expression for the concentration of C in the tank at any time t, evaluating all constants.

P4-13. You are designing a reactor system for carrying out the constant density liquid phase reaction

$$A \longrightarrow B + C$$

which has the rate law

$$r_B = \frac{k_1 \sqrt{C_A}}{1 + k_2 C_A}$$

  a. What system (i.e., type and arrangement) of flow reactors, either one alone or two in series, would you recommend for continuous processing of a feed of pure A in order to minimize the total reactor volume? (90% conversion of A is desired.)
  b. What reactor size(s) should be used?

   Data: $k_1 = 10.0$ [(lb-moles)/(cu ft)]$^{0.5}$ (hr)$^{-1}$
   $k_2 = 16.0$ cu ft/lb mole
   Feed = 100 lb-moles/hr of pure A
   $C_{A0} = .25$ lb-moles/cu ft

[CU]

P4-14. Pure liquid A is fed into a CSTR where it reacts in the liquid phase to form B, a liquid, and C, a gas. The reaction is elementary.

$$A \longrightarrow B + C$$

At the temperature of the reaction, A has a vapor pressure of .10 atm, B is a liquid with a negligible vapor pressure, and C is a gas. Gaseous A and C exit through the top of the CSTR while liquids A and B exit the reactor from the bottom. We want 70% conversion of a feed stream of 10 lb-moles/hr of pure A. If k has a value of 0.4 hr^{-1}, what is the liquid volume required in the reactor? The total pressure in the reactor is 1 atm.

$$\text{Molecular weights:} \quad A = 200, \quad B = 175, \quad C = 25$$
$$\text{Densities:} \quad A = 50 \text{ lb/cu ft}, \quad B = 50 \text{ lb/cu ft}$$

[CU]

P4-15. Consider a plug flow reactor with porous walls that you are operating under pressure. A first-order reaction is taking place:

$$A \longrightarrow B$$

The pressure drop along the length of the reactor is negligible. Therefore the gas within the reactor flows out through the walls at a constant rate per unit external area. Develop an equation that expresses the concentration of A as a function of the length of the reactor.

P4-16. Calculate the time necessary to achieve 90% of equilibrium conversion of the reaction discussed in Problem P4-4 when Pure A is charged to a 30 ft^3 constant volume isothermal batch reactor. All other conditions and parameters are the same as stated in Problem P4-4.

SUPPLEMENTARY READING

1. Further discussion of continuous-flow isothermal reactors, which includes a number of worked example problems, can be found in Chapters 4 and 5 of

   LEVENSPIEL, O., *Chemical Reaction Engineering*, 2nd ed. New York: John Wiley and Sons, 1972.

2. A number of worked example problems involving real reactions carried out in batch reactors can be found in Chapters 2 and 3 of

   STEVENS, B., *Chemical Kinetics* 2nd ed. London: Chapman and Hall, 1970.

3. Further discussion and illustrative examples on reactor design may be found in

   WALAS, S. M., *Reaction Kinetics for Chemical Engineers*. New York: McGraw-Hill, 1959.

   SMITH, J. M., *Chemical Engineering Kinetics*, 2nd ed. New York: McGraw-Hill, 1970.

   HOUGEN, O. A., and K. M. Watson, *Chemical Process Principles*, Part 3: *Kinetics and Catalysis*. New York: John Wiley and Sons, 1947.

4. Also, recent information on reactor design can usually be found in the following journals:

   Chemical Engineering Science
   Chemical Engineering Communications
   Industrial and Engineering Chemistry Quarterlies:
      I.E.C. Fundamentals
      I.E.C. Product Research and Development
      I.E.C. Process Design and Development
   Canadian Journal of Chemical Engineering
   A.I.Ch.E. Journal
   Chemical Engineering Progress

5
Analysis of Rate Data

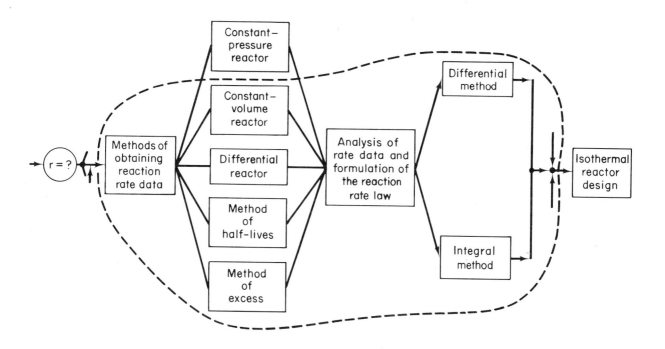

In this chapter we shall (1) discuss how two types of reaction systems (batch and differential) are used to collect rate data and (2) examine four methods of analyzing rate data: the differential, integral, and initial rate, and half-life methods. The method of half-lives and the method of initial rates require that a number of experiments at different initial conditions be carried out to determine the reaction order and specific reaction rate. On the other hand, with either the integral or differential method of analysis, it is possible to carry out only one experiment to find the specific reaction rate and the reaction order w.r.t. one of the reactants. In the batch reactor experiments, the concentration, pressure, and volume are usually measured and recorded at different times during the course of the reaction. Data is collected from the batch reactor during unsteady-state operation, while the measurements on the differential reactor are made during steady-state operation. The product flow rate from the differential reactor is monitored for a number of different feed conditions.

In addition to the discussion of these topics, we shall briefly present methods of plotting rate data and then close the chapter with a discussion of experimental planning and logic.

5.1–Differential Method

In many experiments it is possible to determine reaction order α and the specific reaction rate constant by numerically differentiating concentration vs. time data. This method is applicable when the reaction conditions are such that the rate is essentially a function of the concentration of only one reactant. E.g., if, for the decomposition reaction

$$A \longrightarrow \text{products}$$

the rate law is of the form

$$-r_A = kC_A^\alpha \tag{5-1}$$

the differential method may be used.

However, by utilizing the *method of excess*, it is also possible to determine the relationship between $-r_A$ and the concentration of other reactants. That is, for the reaction

$$A + B \longrightarrow \text{products}$$

with the rate law,

$$-r_A = kC_A^\alpha C_B^\beta \tag{5-2}$$

where α and β are both unknown, the reaction could first be run in an excess of B so that C_B remains essentially unchanged during the course of the reaction

$$-r_A = k'C_A^\alpha \tag{5-3}$$

where

$$k' = kC_B^\beta \simeq kC_{B0}^\beta \tag{5-4}$$

After determining α, the reaction is carried out in an excess of A, for which the rate law is approximated as

$$-r_A = k''C_B^\beta$$
$$[r_A] = [\text{gmole}/l/\text{min}] \tag{5-5}$$

To what is the pseudospecific reaction rate constant k'' equal, in terms of k and the concentration of A? Also, what are the units of k, k', and k''?

(E1)

Both α and β can be determined by using the experimental technique of the method of excess, coupled with differential analysis of data for batch systems.

To illustrate this differential method of analysis we shall work through the following data and determine the reaction order for the gas phase decomposition of di-*t*-butyl peroxide. This decomposition is described by the reaction

$$(CH_3)_3COOC(CH_3)_3 \longrightarrow C_2H_6 + 2CH_3\overset{\overset{O}{\|}}{C}CH_3. \tag{5-6}$$

This reaction was carried out in the laboratory in a constant-volume isothermal batch system in which the total pressure was recorded at various times during the reaction.[†] The following

---
[†] A. F. Trotman-Dickenson, *J. Chem. Ed.* 46, (1969), 396.

(E1) For the units of time we shall arbitrarily choose minutes.

$$[-r_A] = \text{gmole}/l/\text{min}$$

$$[k] = \frac{[-r_A]}{[C_A^\alpha C_B^\beta]} = (l/\text{gmole})^{\alpha+\beta-1}/\text{min}$$

$$[k'] = (l/\text{gmole})^{\alpha-1}/\text{min}$$

$$[k''] = (l/\text{gmole})^{\beta-1}/\text{min}$$

$$k'' = kC_A^\alpha \simeq kC_{A0}^\alpha$$

$$k' = kC_B^\beta \simeq kC_{B0}^\beta$$

data apply to this reaction. Only pure di-t-butyl peroxide was initially present in the reaction vessel. Let A represent di-t-butyl peroxide. Then

$$\frac{-dC_A}{dt} = -r_A = kC_A^\alpha \qquad (5\text{-}7)$$

where α and k are to be determined from the data listed in Table 5-1.

TABLE 5–1

Time (min)	Total Pressure (mm Hg)
0.0	7.5
2.5	10.5
5.0	12.5
10.0	15.8
15.0	17.9
20.0	19.4

First write the concentration of A solely in terms of the total pressure, Π and Π_0, and RT.

(E2) _____

_____

Transform Equation (5-7) into a differential equation involving Π_0, Π, RT, α, k, and t. After making this transformation, let $k' = k(2RT)^{1-\alpha}$, and then write the differential equation terms of k', Π, Π_0, t, and α.

(E3) _____

_____

Assuming that you would first calculate the derivative $(d\Pi/dt)$ from the data in Table 5-1, and then use the equal area method discussed in Appendix A3, outline how you would analyze and plot the data in Table 5-1 to determine α and k' for the decomposition of di-t-butyl peroxide.

(E4) _____

$$\text{HINT:} \quad \ln\left(\frac{d\Pi}{dt}\right) = \alpha \ln(3\Pi_0 - \Pi) + \ln k'$$

_____

After calculating $\Delta\Pi/\Delta t$, plot $\Delta\Pi/\Delta t$ vs. t in Figure 5-1 and, using the technique of equal area differentiation, determine $d\Pi/dt$ as a function of t.

(E2)

$$y_{A0} = 1$$
$$\delta = (1 + 2 - 1) = 2$$
$$\epsilon = y_{A0}\delta = 1 \cdot \delta = \delta = 2$$
$$P_{A0} = y_{A0}\Pi_0 = \Pi_0$$

$$C_A = \frac{P_{A0} - \frac{(\Pi - \Pi_0)}{\delta}}{RT} \tag{E2-1}$$

$$= \frac{\left(\Pi_0 - \frac{\Pi - \Pi_0}{2}\right)}{RT} = \frac{3\Pi_0 - \Pi}{2RT} \tag{E2-2}$$

(E3)

$$-\frac{dC_A}{dt} = kC_A^\alpha \tag{5-7}$$

Substituting for C_A and using Equation (E2-2),

$$\frac{1}{2RT}\frac{d\Pi}{dt} = k\left(\frac{3\Pi_0 - \Pi}{2RT}\right)^\alpha \tag{E3-1}$$

$$\frac{d\Pi}{dt} = k'(3\Pi_0 - \Pi)^\alpha \tag{E3-2}$$

$$k' = k(2RT)^{1-\alpha} \tag{E3-3}$$

(E4)

Differentiate the data either numerically or graphically (see Appendix A3) to obtain $d\Pi/dt$. The reaction order α can be determined from the slope of a plot of $\ln(d\Pi/dt)$ vs. $\ln(3\Pi_0 - \Pi)$. The constant k' may be calculated from the ratio

$$k' = \frac{(d\Pi/dt)}{(3\Pi_0 - \Pi)^\alpha}$$

and any pressure Π.

138

(E5)

t	Π	$\Delta\Pi/\Delta t$	$d\Pi/dt$
.0	7.5		
2.5	10.5		
5.0	12.5		
10.0	15.8		
15.0	17.9		
20.0	19.4		

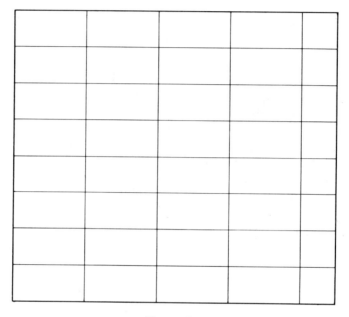

Figure 5–1

We shall determine the reaction order α from the slope of a log-log plot of $d\Pi/dt$ vs. the appropriate function of Π, which you determined in Frame (E3). First complete Table (E6).

(E6)

TABLE (E6)

t (min)	$d\Pi/dt$	$f(\Pi)$
.0		
2.5		
5.0		
10.0		
15.0		
20.0		

Now, from the slope of an appropriate plot of the "processed" data in Table E6, determine the reaction order for the decomposition of di-*t*-butyl peroxide. In addition, determine the specific reaction rate constants k and k'.

(E5)

t (min)	Π (mm Hg)	$\dfrac{\Delta\Pi}{\Delta t}$	$\dfrac{d\Pi}{dt}$
.0	7.5	—	1.44
2.5	10.5	1.20	0.95
5.0	12.5	0.80	0.74
10.0	15.8	0.66	0.53
15.0	17.9	0.42	0.34
20.0	19.4	0.30	0.25

Temperature = 170°C

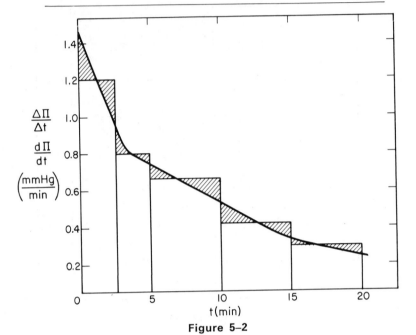

Figure 5–2

(E6)

t (min)	$d\Pi/dt$	$3\Pi_0 - \Pi$
.0	1.44	15.0
2.5	.95	12.0
5.0	.74	10.0
10.0	.53	6.7
15.0	.34	4.6
20.0	.25	3.1

(*Note*: If the reader is totally unfamiliar with any method of obtaining slopes from plots on log-log or semilog graphs, he should read Appendix E before proceeding further.)

(E7)

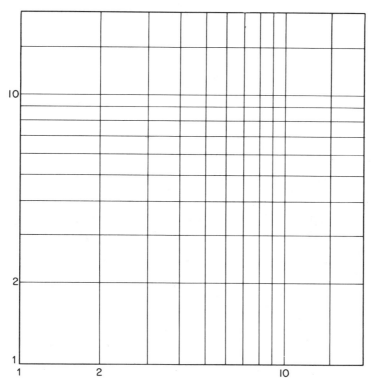

Figure 5–3

5.2–Integral Method

We shall again consider the data presented in Table 5-1 for the decomposition of di-*t*-butyl peroxide. To determine the reaction order by the integral method, we guess the reaction order and integrate the differential equation used to describe the batch system. If the order we assumed is correct, the appropriate plot (determined from this integration) of the pressure-time data should be linear. Recalling Equation (E3-2),

$$\frac{d\Pi}{dt} = k'(3\Pi_0 - \Pi)^\alpha \tag{E3-2}$$

As a first guess we might try zero order, $\alpha = 0$, for which Equation (E3) becomes

$$\frac{d\Pi}{dt} = k' \tag{5-8}$$

Integrating,

$$\Pi = \Pi_0 + k't \tag{5-9}$$

If this is the correct order, a plot of Π vs. t should be linear. From the plot in Figure 5-5, we see that Π is not a linear function of t; consequently we conclude the reaction is not zero order.

141

(E7) Since the length of the cycle on each axis is the same, we can determine the slope directly from the ratio of the distance Δy and Δx:

$$\alpha = \text{Slope} = \frac{\Delta y}{\Delta x} = \frac{2.1 \text{ cm}}{2.0 \text{ cm}} = 1.05$$

$$\alpha \simeq 1.0$$

$$-r_A = kC_A$$

$$k = .08 \text{ min}^{-1}, \qquad k' = k(2RT)^{1-\alpha} = k$$

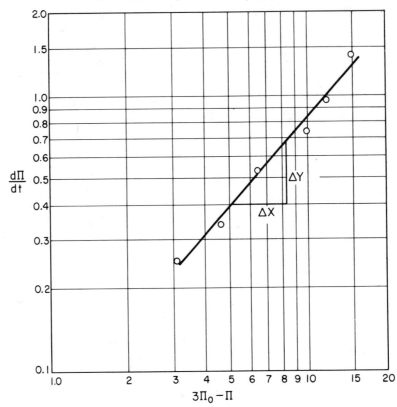

Figure 5–4

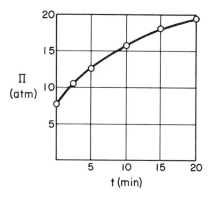

Figure 5-5

Try second order, $\alpha = 2$:
$$\frac{d\Pi}{dt} = k'(3\Pi_0 - \Pi)^2 \tag{5-10}$$

Supposing the reaction is second order, how would you plot the pressure-time data to verify this reaction order?

(E8) ───

───

Complete Table (E9); then plot $f(\Pi)$ vs. t and determine whether the reaction is second order.

(E9) ───

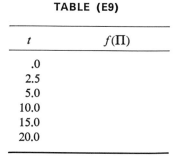

TABLE (E9)

t	$f(\Pi)$
.0	
2.5	
5.0	
10.0	
15.0	
20.0	

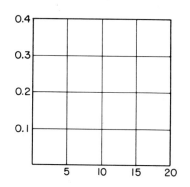

Figure 5-6

───

Finally, we try first order, i.e., $\alpha = 1$. If zero, first, or second order do not seem to describe the reaction rate equation, it is usually best to try some other method of determining the reaction order.

$$\frac{d\Pi}{dt} = k'(3\Pi_0 - \Pi) \tag{5-11}$$

(E8) Integrating,

$$\int_{\Pi_0}^{\Pi} \frac{d\Pi}{(3\Pi_0 - \Pi)^2} = k't$$

$$\frac{1}{(3\Pi_0 - \Pi)} - \frac{1}{2\Pi_0} = k't$$

If the reaction is second order, a plot of $1/(3\Pi_0 - \Pi)$ vs. t *should* be linear: $f(\Pi) = 1/(3\Pi_0 - \Pi)$

(E9)

TABLE (E9–a)

t (min)	Π (cm Hg)	$3\Pi_0 - \Pi$	$\dfrac{1}{3\Pi_0 - \Pi}$
.0	7.5	15.0	.0667
2.5	10.5	12.0	.0834
5.0	12.5	10.0	.1000
10.0	15.8	6.7	.1500
15.0	17.9	4.6	.2180
20.0	19.4	3.1	.3230

From the curvature of the plot shown in Figure 5-7, we conclude that the reaction is not second order.

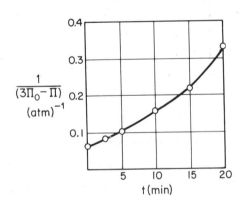

Figure 5–7

How would you plot the pressure-time data to verify that the reaction is first order?

(E10) _____

_____

Utilizing the graph in Figure 5-8, make the appropriate plot of $f(\Pi)$ vs. t and determine if the reaction is first order.

(E11)

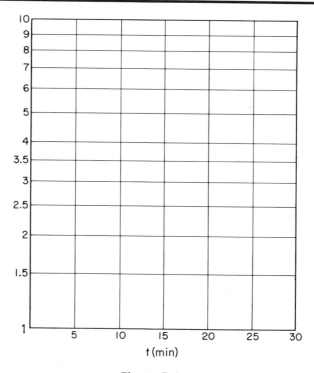

Figure 5–8

_____

In comparing the methods of analysis of the rate data presented above, we note that *the differential method tends to accentuate the uncertainties in the data, but the integral method tends to smooth the data*, thereby disguising the uncertainties in it. In most analyses it is imperative that the engineer know the limits and uncertainties in his data so that he may provide for a safety factor when he scales up or designs a process from the laboratory experiments to the pilot plant or full scale operation.

5.3–GRAPHICAL TECHNIQUES IN THE PLOTTING OF REACTION RATE DATA

In the integral method of analysis of rate data we were looking for the appropriate function of concentration corresponding to a particular rate law that, when plotted against time, would produce a straight line. One should be thoroughly familiar with the methods of obtaining these linear plots for reactions of zero, first, and second order.

First consider the reaction in an ultrasonic field between carbon tetrachloride and water that yields chlorine gas. If the rate of formation of chlorine is independent of the concentration of reactants, the reaction is said to be zero order. A mole balance on chlorine in the batch system

(E10)

$$\frac{d\Pi}{dt} = k'(3\Pi_0 - \Pi)$$

Integrating with limits $\Pi = \Pi_0$ when $t = 0$,

$$\ln\left(\frac{2\Pi_0}{3\Pi_0 - \Pi}\right) = k't$$

If the reaction is first order, a plot of $\ln[2\Pi_0/(3\Pi_0 - \Pi)]$ vs. t should be linear.

(E11)

t	Π	$2\Pi_0/(3\Pi_0 - \Pi)$
.0	7.5	1.00
2.5	10.5	1.25
5.0	12.5	1.50
10.0	15.8	2.25
15.0	17.9	3.27
20.0	19.4	4.85

From the semilog plot shown in Figure 5-9, we see that $\ln[2\Pi_0/(3\Pi_0 - \Pi)]$ is linear with time, and we therefore conclude that the decomposition of di-t-butyl peroxide follows first order kinetics. From the slope of the plot in Figure 5-9, we can determine the specific reaction rate, $k = .08$ min^{-1}.

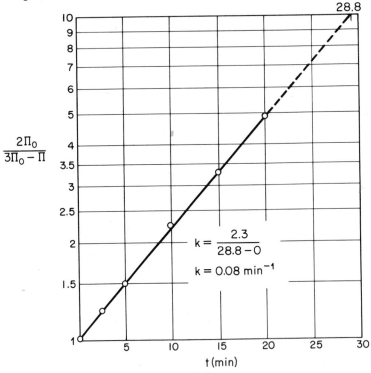

Figure 5-9

yields

$$\frac{1}{V}\frac{dN_{Cl_2}}{dt} = r_{Cl_2} \tag{5-12}$$

Let A = Cl_2. At constant volume

$$\frac{dC_A}{dt} = r_A \tag{5-13}$$

If the reaction is zero order, the reaction rate is independent of species concentrations; hence

$$\frac{dC_A}{dt} = r_A = k \tag{5-14}$$

Integrating with the initial conditions $t = 0$ and $C_A = 0$, we obtain

$$C_A = kt \tag{5-15}$$

The data in Table 5-2 was obtained for the reaction system described above.[†] The dependent variable is usually plotted on the y-axis, and the independent variable is usually plotted on the x-axis.

TABLE 5-2

Time (min)	Concentration of Cl_2 (meq/l)
0	.000
1	.090
2	.170
3	.265
4	.385

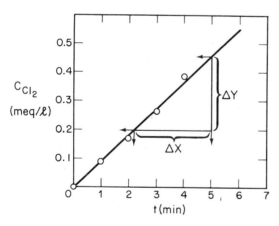

Figure 5-10

To determine the specific reaction rate k for this zero-order reaction, we simply measure the slope of the "best" straight line down through the data points. We choose two points, t and C_A, on this line, for example (2.25, .20) and (5, .46), to substitute into the equations

$$C_{A2} = kt_2$$
$$C_{A1} = kt_1$$

Subtracting the lower equation from the upper and then dividing by Δt, we find that

$$k = \frac{C_{A2} - C_{A1}}{t_2 - t_1} = \frac{.46 - .20}{5 - 2.25} = \frac{.26}{2.75}$$

$$= 0.0945 \text{ meq}/l/\text{min} \tag{5-16}$$

To summarize this reaction, we conclude that for a zero order reaction in which there is no volume change, a plot of the reactant (or product) concentration as a function of time will be linear with slope $-k$ (or slope k).

It is important to restate that, given a reaction rate law, one should quickly be able to choose the appropriate function of concentration or conversion that is plotted against time or *space time* to obtain a linear plot. In Frame (Ei2), only reactions taking place in constant-volume batch reactors are considered. Match the concentration- or conversion-time figures to their respective rate laws, and state how the specific reaction rate is determined in each case.

[†] P. K. Chendke and H. S. Fogler, *4th International UMA Symposium* (n.p., 1970).

(E12)
1. Zero order.   _____   _____
2. First order.   _____   _____
3. Second order.
   a. Type 1: $2A \rightarrow P$, pure A.   _____   _____
   b. Type 2: $A + B \rightarrow P$, $C_{A0} = C_{B0}$.   _____   _____

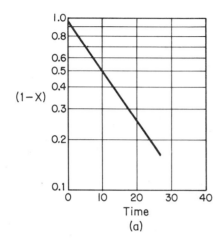
(a)

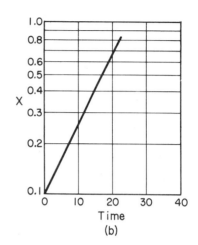
(b)

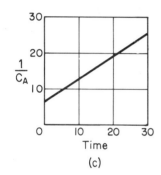
(c)

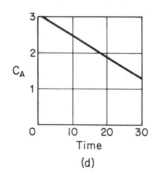
(d)

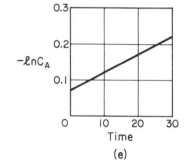
(e)

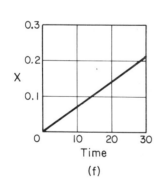
(f)

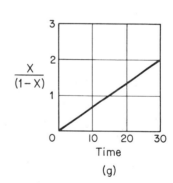
(g)

Figure 5–11(a-g)

In Frame (E13) we shall consider plots of rate data only for reactions taking place at constant pressure in a batch reactor. Again, match the concentration-or conversion-time figures to their respective rate laws. Assume that there are no inerts present.

149

(E12) Since all reactions are carried out isothermally at constant volume, the concentration can be written as

$$C_A = C_{A0}(1 - X) \tag{E12-1}$$

and the design equation as

$$\frac{-dC_A}{dt} = -r_A \tag{E12-2}$$

1. *Zero-order reaction*:

$$\frac{-dC_A}{dt} = k \tag{E12-3}$$

Integrating,

$$C_A = C_{A0} - kt \tag{E12-4}$$

For a zero-order reaction, a plot of C_A vs. t will be linear, with slope $-k$. Substituting Equation (E12-1) into (E12-4) and rearranging, we obtain

$$X = \frac{k}{C_{A0}} t \tag{E12-5}$$

A plot of X vs. t will also be linear. **Ans.: Figures (d) and (f)**

2. *First-order reaction*:

$$-\frac{dC_A}{dt} = kC_A \tag{E12-6}$$

$$\ln C_{A0} - \ln C_A = -\ln \frac{C_A}{C_{A0}} = kt \tag{E12-7}$$

For a first order reaction, a plot of $-\ln C_A$ vs. t will be linear, with the slope equal to the specific reaction rate. Substituting Equation (E12-1) into (E12-7),

$$\ln\left(\frac{1}{1-X}\right) = kt \tag{E12-8}$$

A plot of $\ln(1 - X)$ vs. t will also be a straight line with $-k$. **Ans.: Figures (a) and (e)**

3a. *Second-order reaction, type 1*:

$$-\frac{dC_A}{dt} = kC_A^2 \tag{E12-9}$$

Integrating,

$$\frac{1}{C_A} - \frac{1}{C_{A0}} = kt \tag{E12-10}$$

A plot of $1/C_A$ vs. t will be a straight line with the slope equal to the specific reaction rate. Substituting Equation (E12-1) into (E12-10) and rearranging,

$$\frac{X}{1-X} = kC_{A0}t \tag{E12-11}$$

we see that a plot of $X/(1 - X)$ vs. t will also be linear. **Ans.: Figures (c) and (g)**

3b. *Second-order reaction, type 2*:

Since $C_{A0} = C_{B0}$, then $\theta_B = 1$, and $C_A = C_B$:

$$-\frac{dC_A}{dt} = kC_A C_B = kC_{A0}^2(1 - X)(1 - X)$$

$$-\frac{dC_A}{dt} = kC_A^2$$

Same as 3a above. **Ans.: Figures (c) and (g)**

(E13)

1. Zero order with $\epsilon = -.5$
2. First order with $\epsilon = -.5$

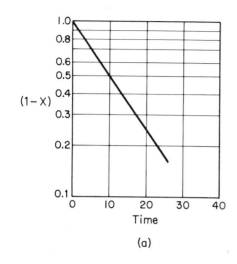
(a)

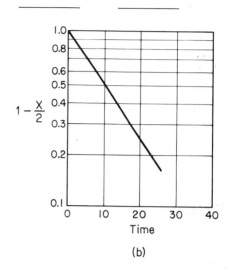
(b)

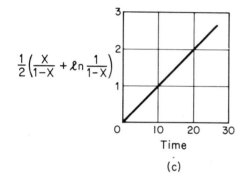
(c)

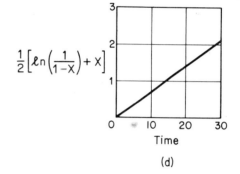
(d)

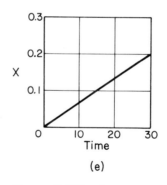
(e)

Figure 5–12(a-e)

In Frame (E13) we were concerned with the methods of plotting data for constant-pressure batch reactions. In our modeling of tubular reactors we assume that the pressure was constant along the length of the reactor. When there is *no* volume change with reaction, the concentration-time plots for the constant-pressure batch reactor are identical in form to their corresponding concentration-space time plots for the flow reactor! However, when $\epsilon \neq 0$ this is not the case. For example, for first-order reactions taking place in a tubular reactor, the design equation

(E13) All the reactions are carried out in a constant-pressure isothermal batch reactor, in which the concentration is

$$C_A = C_{A0}\left(\frac{1-X}{1+\epsilon X}\right)$$

and the mole balance on species A is

$$-\frac{1}{V}\frac{dN_A}{dt} = \frac{C_{A0}}{(1+\epsilon X)}\frac{dX}{dt} = -r_A$$

1. *Zero-order reaction:* $\epsilon = -.5$;

$$\frac{C_{A0}}{1-\frac{X}{2}}\frac{dX}{dt} = k$$

$$-\ln\left(1 - \frac{X}{2}\right) = \frac{\frac{1}{2}kt}{C_{A0}}$$

Ans.: Figure(b)

2. *First-order reaction:* $\epsilon = -.5$;

$$C_A = \frac{C_{A0}(1-X)}{\left(1 - \frac{X}{2}\right)}$$

$$\frac{C_A}{\left(1-\frac{X}{2}\right)}\frac{dX}{dt} = kC_A = kC_{A0}\frac{1-X}{\left(1-\frac{X}{2}\right)}$$

$$\frac{dX}{dt} = k(1-X)$$

$$-\ln(1-X) = kt$$

Ans.: Figure(a)

Although it was not requested in Frame (E13), let us derive the conversion time and relationship that would be plotted for the Type 1 second-order reaction given in Frame (E12).

Second-order reaction, type 1:
Pure A, $A \longrightarrow P/2$:

$$\epsilon = y_{A0}\delta = 1(\tfrac{1}{2} - 1) = -.5$$

$$C_A = C_{A0}\frac{(1-X)}{\left(1-\frac{X}{2}\right)}$$

$$\frac{C_{A0}}{\left(1-\frac{X}{2}\right)}\frac{dX}{dt} = kC_A^2 = \frac{kC_{A0}^2(1-X)^2}{\left(1-\frac{X}{2}\right)^2}$$

$$\frac{dX}{dt} = kC_{A0}\frac{(1-X)^2}{\left(1-\frac{X}{2}\right)}$$

Integrating,

$$\tfrac{1}{2}\left[\ln\left(\frac{1}{1-X}\right) + \frac{X}{1-X}\right] = C_{A0}kt$$

We see that the specific reaction rate will be equal to the slope of Figure (5-12c) divided by C_{A0}.

Ans.: Figure (c)

is

$$V = F_{A0} \int \frac{dX}{-r_A} = v_0 C_{A0} \int \frac{dX}{kC_A} = v_0 C_{A0} \int \frac{(1+\epsilon X)dX}{kC_{A0}(1-X)} \tag{5-17}$$

$$\tau = \frac{V}{v_0} = \frac{1}{k}\int_0^X \frac{(1+\epsilon X)dX}{(1-X)} = \frac{(1+\epsilon)}{k}\ln\left(\frac{1}{1-X}\right) - \frac{\epsilon X}{k} \tag{5-18}$$

If $\epsilon = -.5$ for this first order reaction that is carried out in a plug flow reactor under constant pressure, Equation (5-18) becomes

$$\tfrac{1}{2}\left[\ln\frac{1}{(1-X)} + X\right] = k\tau \tag{5-19}$$

and we see that a plot of $\tfrac{1}{2}[\ln(1/(1-X)) + X]$ vs. space time should be linear with slope k, as shown in Figure (5-12d). On the other hand, for a first-order reaction carried out in a constant-pressure batch reactor, a plot of $\ln(1/(1-X))$ vs. time will yield a straight line, as shown in Figure (E12-a). Further application of the various methods of plotting data can be found in the problems at the end of the chapter.

After working through Frames (E12) and (E13) in detail, see if you can go back and quickly determine how to plot your rate data for a given reaction order. *However, it is most important to know the appropriate concentration-time plots for zero-, first-, and second- order reactions that have no volume change; it is also important to be able to derive the appropriate plots for other reactions orders and for reactions with volume change.*

In the previous examples we determined the specific reaction rate k from concentration measurements taken at various times during the course of the reaction. We are now interested in studying how the specific reaction rate varies with temperature. The Arrhenius equation gives the dependence of k on temperature as

$$k = Ae^{-E/RT} \tag{3-2}$$

Taking the log of both sides of Equation (3-2), we obtain

$$\ln k = \ln A - \left(\frac{E}{R}\right)\frac{1}{T} \tag{5-20}$$

Converting to log base 10,

$$\log k = \log A - \left(\frac{E}{2.3R}\right)\frac{1}{T} \tag{5-21}$$

The activation energy can be determined from the slope of a plot of $\log k$ vs. $1/T$. Utilizing the decade method (Appendix E) determine the activation energy for the decomposition of benzene diazonium chloride to give chlorobenzene and nitrogen, i.e.,[†]

$$\text{C}_6\text{H}_5\text{–N=NCl} \longrightarrow \text{C}_6\text{H}_5\text{–Cl} + N_2 \tag{5-22}$$

from the plot shown in Figure 5-13.

[†]H. D. Gesser et al., *J. Chem. Ed.* 44 (1967), 387.

(E14)

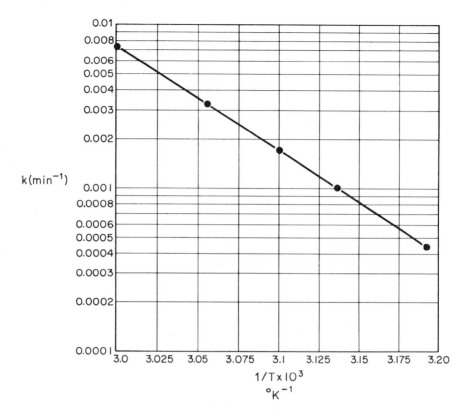

Figure 5–13

Before proceeding further, the reader may wish to work through some of the problems at the end of the chapter. Suggested problems on the material presented in Sections 5.1 through 5.3 might be chosen from Pb. 5-1 through Pb. 5-8.

5.4–METHOD OF INITIAL RATES

The use of the differential method of data analysis to determine reaction orders and specific reaction rates is clearly one of the easiest, since it requires only one experiment. However, the occurrence of other effects during the course of the reaction, such as the presence of a reverse reaction, could render the method ineffective. In these cases, one could use the method of initial rates to determine the reaction order and specific rate. Here, a series of experiments is carried out at different initial concentrations, C_{A0}, and the initial rate of reaction, r_{A0}, is determined for each run by differentiating the data, using the equal area (or other) technique and extrapolating to zero time. For example, in the di-*t*-butyl peroxide decomposition shown in Frame (E5), the initial rate was found to be 1.4 mm Hg/min. By various plotting or numerical analysis techniques relating $-r_{A0}$ to C_{A0}, we can obtain the appropriate rate law. If the rate law is in the form

$$-r_A = k C_A^\alpha \tag{5-1}$$

the slope of a plot of $\ln(-r_{A0})$ vs. $\ln C_{A0}$ will give the reaction order α. As an example of the initial rate method, consider the reaction between solid dolomite and hydrochloric acid

155

(E14)

$$\log k_1 = \log A - \left(\frac{E}{2.3R}\right)\frac{1}{T_1}$$

$$\log k_2 = \log A - \left(\frac{E}{2.3R}\right)\frac{1}{T_2}$$

Subtracting,

$$\log \frac{k_2}{k_1} = -\left(\frac{E}{2.3R}\right)\left(\frac{1}{T_2} - \frac{1}{T_1}\right)$$

$$E = -\frac{(2.3)(R)\left(\log \frac{k_2}{k_1}\right)}{\left(\frac{1}{T_2} - \frac{1}{T_1}\right)} = \frac{2.3R \log \frac{k_1}{k_2}}{\left(\frac{1}{T_2} - \frac{1}{T_1}\right)}$$

Choose $1/T_1$ and $1/T_2$ so that $\quad k_2 = .1k_1,\quad$ then $\quad \log\frac{k_1}{k_2} = 1$

when $k_1 = 0.005$, $1/T_1 = 0.003025$
when $k_2 = 0.0005$, $1/T_2 = 0.003185$

$$E = \frac{2.3R}{\left(\frac{1}{T_2} - \frac{1}{T_1}\right)} = \frac{(2.3)(1.98)}{(.003185 - .003025)}$$

$$= 28.4 \text{ kcal/gmole}$$

$$4HCl + CaMg(CO_3)_2 \longrightarrow Mg^{++} + Ca^{++} + 4Cl^- + 2CO_2 + 2H_2O$$

This reaction is of particular importance in the acidization of oil reservoirs. The concentration of HCl at various times was determined from atomic absorption spectrophotometer measurements of the calcium and magnesium ions. From the data presented in Frame (E15), in terms of concentration-time plots for this batch reaction, determine the reaction order w.r.t. HCl. Assume that the rate law is in the form given by Equation (5-1) and that the mole balance on HCl can be given by Equation (5-7).

(E15)

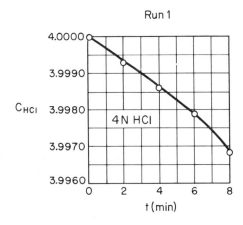
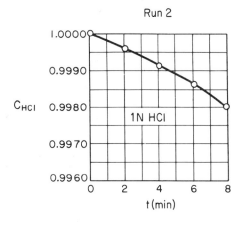
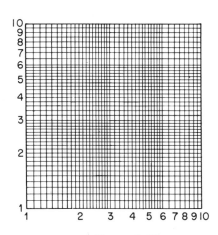

Figure 5–14

5.5–Differential Reactor

Data acquisition by the techniques of initial rate and differential reactor are similar in that the rate of reaction is determined for a specified number of predetermined initial or entering reactant concentrations. The differential reactor is normally used to determine the rate of reaction as a function of concentration for heterogeneous systems. It consists of a tube containing a very small amount of catalyst usually arranged in the form of a thin wafer or disk. A typical arrangement is shown schematically in Figure 5-16.

(E15)

Evaluating a mole balance on a constant volume batch reactor at time $t = 0$ gives

$$\left(-\frac{dC_{HCl}}{dt}\right)_0 = -(r_{HCl})_0 = kC_{HCl,0}^\alpha \tag{E15-1}$$

$$\ln\left[-\frac{dC_{HCl}}{dt}\right]_0 = \ln k + \alpha \ln C_{HCl,0} \tag{E15-2}$$

The derivative at time $t = 0$ can be found from the slope of the plot of concentration vs. time evaluated at $t = 0$.

4N HCl Solution	1N HCl Solution
$-r_{HCl,0} \doteq -\left[\dfrac{3.9982 - 4.0000}{5 - 0}\right]$	$-r_{HCl,0} = -\left[\dfrac{.9987 - 1.0000}{6 - 0}\right]$
$-r_{HCl,0} \doteq 3.6 \times 10^{-4}$ gmole/l/min	$-r_{HCl,0} = 2.165 \times 10^{-4}$ gmole/l/min

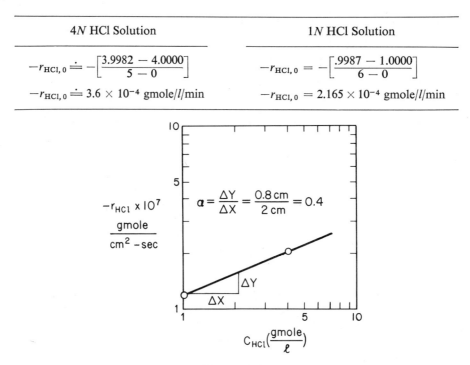

Figure 5–15

In each case the volume of the solution was 675 ml and the area of the dolomite surface was 20.2 sq cm. Basing the rate of reaction on the surface area of dolomite,

$$-r_{HCl,0} = \frac{3.60 \times 10^{-4} \text{ gmole}}{l \text{ min}} \times \frac{.675\, l}{20.2 \text{ cm}^2} \times \frac{1 \text{ min}}{60 \text{ sec}}$$

$$= \frac{2 \times 10^{-7} \text{ gmole}}{\text{cm}^2\text{-sec}} \quad \text{for } 4N \text{ HCl}$$

and

$$-r_{HCl,0} = \frac{1.2 \times 10^{-7} \text{ gmole}}{\text{cm}^2\text{-sec}} \quad \text{for } 1N \text{ HCl}$$

(*Note*: It was not necessary to base the rate on the surface area to determine α.) From the slope of a ln-ln plot of $-r_{HCl,0}$ vs. $C_{HCl,0}$ the rate law is $-r_{HCl} = kC_{HCl}^{.4}$. If more data points were included in this example we would find that $-r_{HCl} = kC_{HCl}^{.44}$ at 25°C.)[†]

[†]Lund, K., H. S. Fogler, and C. C. McCune, *Chem. Engr. Sci.* 28 (1973) 691.

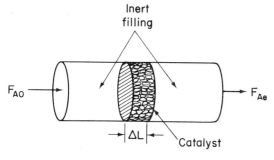

Figure 5-16

Because of the small amount of catalyst used, the conversion of the reactants in the bed is extremely small, as is the change in reactant concentration through the bed. As a result the reactant concentration through the reactor is essentially constant and approximately equal to the inlet concentration. That is, the reactor is considered to be *gradientless*, and the reaction rate is considered spatially uniform within the bed. The volumetric flow rate through the catalyst bed is monitored, as are the entering and exiting concentrations. If the weight of catalyst, W, is known, how is the rate of reaction per unit mass of catalyst, $-r'_A$, calculated from the preceding information?

(E16)

The rate of reaction (determined from Equation (E16-5)) can be obtained as a function of the reactant concentration in the catalyst bed, $C_{A,b}$, by varying the inlet concentration, i.e.,

$$-r'_A = -r_A(C_{A,b})$$

One approximation to the concentration of A within the bed, $C_{A,b}$ would be the arithmetic mean of the inlet and outlet concentrations.

$$C_{A,b} = \frac{C_{A0} + C_{Ae}}{2}$$

However, since very little reaction takes place within the bed, the bed concentration is essentially equal to the inlet concentration,

$$C_{A,b} = C_{A0}$$

and

$$-r'_A = -r_A(C_{A0}).$$

As with the method of initial rates, various numerical and graphical techniques can be used to determine the appropriate algebraic equation for the rate law.

The formation of methane from carbon monoxide and hydrogen using a nickel catalyst was studied by Pursley.[†] The reaction

$$3H_2 + CO \longrightarrow CH_4 + 2H_2O$$

was carried out at 500°F.

The reaction rate law is assumed to be the product of a function of the partial pressure of CO, f(CO), and a function of the partial pressure of H_2, $g(H_2)$, i.e.,

$$r'_{CH_4} = f(CO) \cdot g(H_2) \tag{5-23}$$

In Frame (E17), determine the reaction order w.r.t. carbon monoxide, using Equation (E16-4) to analyze the data in Table 5-3. Assume that the function dependence of r'_{CH_4} on P_{CO} is of the form

$$r'_{CH_4} \sim P_{CO}^\alpha \tag{5-24}$$

[†] J. A. Pursley, "An Investigation of the Reaction Between Carbon Monoxide and Hydrogen on a Nickel Catalyst Above One Atmosphere." (Ph.D. thesis, University of Michigan, 1951).

(E16) Since the differential reactor is assumed to be gradientless, the design equation will be similar to the CSTR design equation. A steady-state mole balance on reactant A gives

$$\text{In} - \text{Out} + \text{Generation} = 0$$

$$F_{A0} - F_{Ae} + \left(\frac{\text{rate of RXN}}{\text{mass of cat.}}\right)(\text{mass of cat.}) = 0$$

$$F_{A0} - F_{Ae} + (r'_A)(W) = 0 \tag{E16-1}$$

Solving for $-r'_A$,

$$-r'_A = \frac{F_{A0} - F_{Ae}}{W} \tag{E16-2}$$

The subscript e refers to the exit of the reactor. The mole balance equation can also be written in terms of the concentration:

$$-r'_A = \frac{v_0 C_{A0} - v C_{Ae}}{W} \tag{E16-3}$$

or in terms of the conversion:

$$-r'_A = \frac{F_{A0} X}{W} = \frac{F_P}{W} \tag{E16-4}$$

The term $F_{A0}X$ gives the rate of formation of the product, which is also the rate at which it leaves the reactor at steady state, F_p. For constant volumetric flow, Equation (E16-3) reduces to

$$-r'_A = \frac{v_0(C_{A0} - C_{Ae})}{W} \tag{E16-5}$$

By using very little catalyst and large volumetric flow rates, the concentration difference, $(C_{A0} - C_{Ae})$, can be made quite small.

The exit volumetric flow rate from a differential packed bed containing 10 grams of catalyst was maintained at 300 l/min for each run. The partial pressures of H_2 and CO were determined at the entrance to the reactor, and the methane concentration was measured at the reaction exit (see Table 5-3).

TABLE 5-3

Run	P_{CO} (atm)	P_{H_2} (atm)	C_{CH_4} gmole/l
1	1	1.0	2.44×10^{-4}
2	1.8	1.0	4.40×10^{-4}
3	4.08	1.0	10.0×10^{-4}
4	1.0	.1	1.65×10^{-4}
5	1.0	.5	2.47×10^{-4}
6	1.0	4.0	1.75×10^{-4}

(E17)

Run	P_{CO}	$-r'_{CO} = r'_{CH_4}$

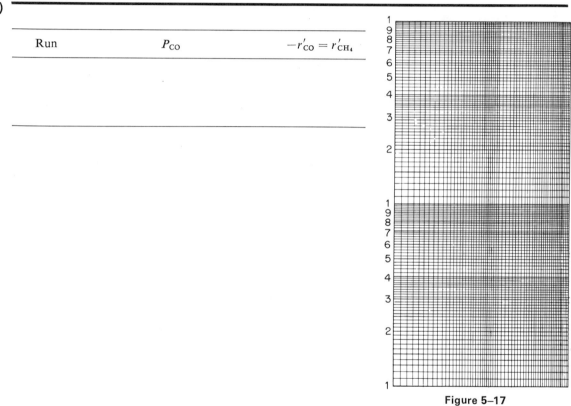

Figure 5–17

From the table in Frame (E17) it appears that the dependence of r'_{CH_4} on P_{H_2} cannot be represented by a power law. Comparing run 4 with run 5 and run 1 with run 6, we see that the reaction rate first increases with increasing partial pressure of hydrogen, and subsequently decreases with increasing P_{H_2}. That is, there appears to be a concentration of hydrogen at which the rate is maximum. One set of rate laws that is consistent with these observations is:

1. At low H_2 concentrations, where r'_{CH_4} increases as P_{H_2} increases, the rate law may be of the form,

$$r'_{CH_4} \sim P_{H_2}^{\beta_1}. \tag{5-25}$$

(E17) In this example the product composition, rather than the reactant concentration, is being monitored. Replacing $(F_{A0}X)$ by F_p in Equation (E16-4),

$$-r'_A = \frac{F_{A0}X}{W} \qquad (E16\text{-}4)$$

$-r'_A$ can be written in terms of the flow rate of methane from the reaction,

$$-r'_A = r'_{CH_4} = \frac{F_{CH_4}}{W} \qquad (E17\text{-}1)$$

Substituting for F_{CH_4} in terms of the volumetric flow rate and the concentration of methane,

$$-r'_A = \frac{v_0 C_{CH_4}}{W} \qquad (E17\text{-}2)$$

Since v_0, C_{CH_4}, and W are known for each run, we can calculate the rate of reaction.
For run 1:

$$-r'_A = \frac{300\ l}{min} \frac{2.44 \times 10^{-4}}{10\ g\ cat.} gmole/l = 7.33 \times 10^{-3} \frac{gmole\ CH_4}{g\ cat \times min}$$

Run	P_{CO} (atm)	P_{H_2} (atm)	C_{CH_4} gmole/l	$r'_{CH_4} \left(\frac{gmole\ CH_4}{g\ cat \times min}\right)$
1	1.0	1.0	2.44×10^{-4}	7.33×10^{-3}
2	1.8	1.0	4.40×10^{-4}	13.2×10^{-3}
3	4.08	1.0	10.0×10^{-4}	30.0×10^{-3}
4	1.0	.1	1.65×10^{-4}	4.95×10^{-3}
5	1.0	.5	2.47×10^{-4}	7.42×10^{-3}
6	1.0	4.0	1.75×10^{-4}	5.25×10^{-3}

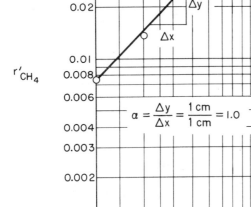

Figure 5-18

We now plot $\ln r'_{CH_4}$ vs. $\ln P_{CO}$ for runs 1, 2, and 3

$$r'_{CH_4} = k P_{CO}^\alpha \cdot g(P_{H_2})$$

For constant hydrogen concentration,

$$r'_{CH_4} = k' P_{CO}^\alpha$$

$$\ln r'_{CH_4} = \ln k' + \alpha \ln P_{CO}$$

From the slope of the plot in Figure 5-18, we find that $\alpha = 1$.

2. At high H_2 concentrations, where r_{CH_4} decreases as P_{H_2} increases,

$$r'_{CH_4} \sim \frac{1}{P_{H_2}^{\beta_2}}. \qquad (5\text{-}26)$$

We would like to find *one* rate law that is consistent with reaction rate data at both high and low hydrogen concentrations. When Equations (5-25) and (5-26) are combined into the form

$$r'_{CH_4} \sim \frac{P_{H_2}^{\beta_1}}{1 + b P_{H_2}^{\beta_2}} \qquad (5\text{-}27)$$

the resulting rate expression is consistent with the observed rate.

1. For condition 1: at low P_{H_2}, $b(P_{H_2})^{\beta_2} \ll 1$, and Equation (5-27) reduces to

$$r'_{CH_4} \sim P_{H_2}^{\beta_1}. \qquad (5\text{-}28)$$

2. For condition 2: at high P_{H_2}, $b(P_{H_2})^{\beta_2} \gg 1$, and Equation (5-27) becomes

$$r'_{CH_4} \sim \frac{(P_{H_2})^{\beta_1}}{(P_{H_2})^{\beta_2}} \sim \frac{1}{(P_{H_2})^{(\beta_2-\beta_1)}} \sim \frac{1}{(P_{H_2})^{\beta}} \qquad (5\text{-}29)$$

where $\beta > 0$.

In order for the proposed rate expression to agree with the data at high concentrations, β_2 must be greater than β_1. Theoretical considerations of the type to be discussed in Chapter 6 predict that if the rate-limiting step in the overall reaction is the reaction between atomic hydrogen absorbed on the nickel surface and CO in the gas phase, then the rate expression will be in the form

$$r'_{CH_4} = \frac{a P_{CO} P_{H_2}^{1/2}}{1 + b P_{H_2}}. \qquad (5\text{-}30)$$

This rate law is qualitatively consistent with experimental observations, and since it does indeed fit the data, a rearrangement of Equation (5-30) in the form

$$\frac{P_{CO} P_{H_2}^{1/2}}{r'_{CH_4}} = \frac{1}{a} + \frac{b}{a} P_{H_2} \qquad (5\text{-}31)$$

suggests that a plot of $P_{CO} P_{H_2}^{1/2}/r'_{CH_4}$ as a function of P_{H_2} should be a straight line with an intercept of $1/a$ and a slope of b/a.

From the plot in Figure 5-19 we see that the rate expression is indeed consistent with the

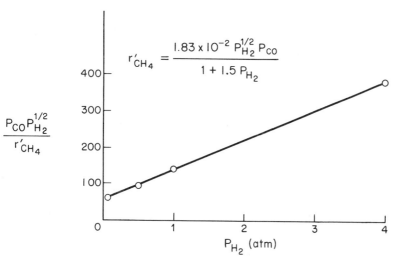

Figure 5-19

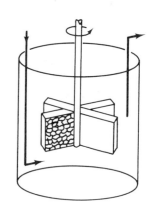

Figure 5-20. Spinning Basket Reactor

differential reactor data. A summary of some of the techniques used to obtain catalytic rate data is given by Anderson.[†] In addition to the differential reactor, three other types of reactors are currently used to obtain reaction rate data: the CSTR, the recycle reactor, and the spinning basket reactor. In the recycle reactor, the conversion per pass is low and the recycle rate is far in excess of the fresh feed rate, to assure negligible interphase and radial gradients. The spinning basket reactor consists of catalyst pellets held in wire cages in the form of impellers, which rotate in a cylindrical tank. The essential features of this reactor are shown in Figure 5-20. With this reactor, the bulk phase fluid is well mixed and in intimate contact with the catalyst particles so that it is essentially *gradientless*[‡]. This reactor is also known as the Carberry Reactor. (See Problem 5-12)

5.6–Method of Half-Lives

The half-life of a reaction, $t_{1/2}$, is defined as the time it takes for the concentration of the reactant to fall to half of its initial value. By determining the half-life of a reaction as a function of the initial concentration, the reaction order and the specific reaction rate can be determined. If there are two reactants involved in the chemical reaction, the experimenter will use the method of excess in conjunction with the method of half-lives in order to arrange the rate expression in the form

$$-r_A = k C_A^\alpha. \qquad (5\text{-}1)$$

For the reaction A $\to$ products, a mole balance on species A in a constant volume batch reaction system results in the following expression:

$$\frac{-dC_A}{dt} = -r_A = k C_A^\alpha. \qquad (5\text{-}7)$$

Integrating with the initial condition $C_A = C_{A0}$ when $t = 0$, we find that

$$t = \frac{1}{k(\alpha-1)} \left(\frac{1}{C_A^{\alpha-1}} - \frac{1}{C_{A0}^{\alpha-1}} \right) \qquad (5\text{-}32)$$

$$= \left[\frac{1}{k C_{A0}^{\alpha-1} [\alpha-1]} \right] \circ \left[\left(\frac{C_{A0}}{C_A} \right)^{\alpha-1} - 1 \right] \qquad (5\text{-}33)$$

Simplify Equation (5-33) so that the condition $t = t_{1/2}$ expresses $t_{1/2}$ as a function of k, C_{A0}, and α.

(E18)

Remembering that $t_{1/2}$ varies with C_{A0}, explain how the reaction order could be determined utilizing the method of half-lives, placing particular emphasis on how your data might be plotted.

(E19)

[†] R. B. Anderson, ed, *Experimental Methods in Catalytic Research* New York: Academic Press, 1968.
[‡] *Canad. J. Chem Eng* 47 (1964) 154; *IEC Process Design* and *Develop.* 8, No. 3 (1969) 364.

(E18)

Half-life is defined as the time required for concentration to drop to half of its initial value, i.e.,

$$t_{1/2} \quad \text{when } C_A = \tfrac{1}{2}C_{A0} \tag{E18-1}$$

Substituting for C_A in Equation (5-33),

$$t_{1/2} = \left(\frac{2^{\alpha-1}-1}{k(\alpha-1)}\right)\left(\frac{1}{C_{A0}^{\alpha-1}}\right) \tag{E18-2}$$

(E19)

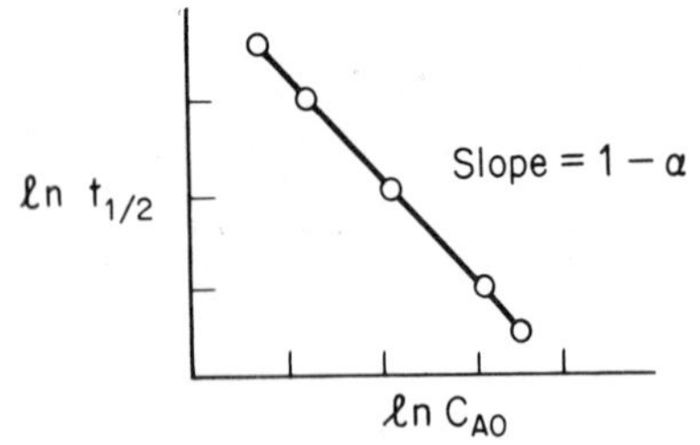

Taking the log of both sides of Equation (E18-2),

$$\ln t_{1/2} = \ln\left[\frac{2^{\alpha-1}-1}{(\alpha-1)k}\right] + (1-\alpha)\ln C_{A0} \tag{E19-1}$$

we see that the slope of a plot of $\ln t_{1/2}$, as a function of $\ln C_{A0}$, is equal to 1 minus the reaction order:

$$\alpha = 1 - \text{slope}$$

For the plot shown in Figure 5-21 the slope is -1:

$$\alpha = 1 - (-1) = 2$$

The corresponding rate law is

$$-r_A = kC_A^2$$

Figure 5-21 (plot of $\ln t_{1/2}$ vs $\ln C_{A0}$, Slope $= 1 - \alpha$)

5.7–EXPERIMENTAL DESIGN

In the preceding pages of this chapter various methods of analyzing rate data have been presented. It is just as important to know in which circumstances to use each method as it is to know the mechanics of these methods.

5.7A–Finding the Rate Law

A chart that may aid you in deciding the type of experimental plan to follow is presented in Figure 5-22.† This chart applies primarily to isothermal batch reaction systems in which only one chemical reaction is taking place.

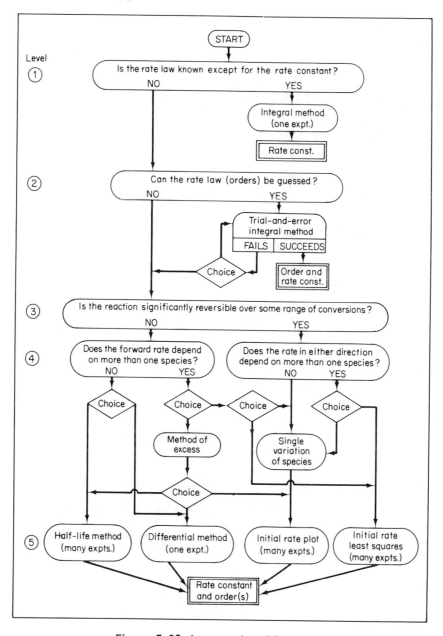

Figure 5–22. Interpretation of Rate Data

†R. L. Curl, Unpublished Notes, University of Michigan, 1968.

As level 1 on the chart shows, when the reaction rate law is known, the integral method of analysis may be used after performing an experiment to determine the specific reaction rate. This procedure is useful when the reaction order and specific rate at one temperature are already known from previous experiments, and the specific reaction rate at some other temperature is being sought. Knowing the value of k at two or more different temperatures makes it possible to calculate the activation energy and the frequency factor. In the trial and error method of level 2, it is probably best to try some other method of analysis if, after trying zero, first, second, and finally third order, you find that none of these orders fit the data.

When the reaction order is unknown and cannot be guessed and the reaction is irreversible, proceed down the left-hand side of the chart to level 4. If the reaction rate depends on only one species as it might in isomerization or decomposition reactions, use the differential method, which requires only one experiment, or the method of half-lives, which requires many experiments, to determine the reaction order and specific reaction rate. If certain constraints imposed by a given reaction prevent meaningful experimental measurements from being made other than in the initial rate period, neither the differential nor the half-life method may succeed. This is particularly true in solid-liquid reactions where flaking or crumbling of the solid occurs and in certain autocatalytic and simultaneous reactions. In such cases, it is usually best to use the initial rate plot method given in level 5. In this technique the initial rate is measured at various initial concentrations of the reacting species. The reaction order can be determined from a plot of the log initial reaction rate, $\ln(-r_{A0})$, vs. the log of the initial concentration, $\ln C_{A0}$. Which techniques or methods in the chart require only one experiment to determine the desired information?

(E20)

If the reaction rate depends on more than one species, the experimenter has a choice; he can use the method of excess coupled either with the half-life method or the differential method or he can use the single variation of species coupled with the initial rate plot technique. In the latter method, which is used when the method of excess is not feasible, the concentration of one reactant is varied while the concentrations of the others are held constant. After determining the reaction order for this species by the initial rate plots discussed above, the concentration of one of the other species is varied while the remaining concentrations are held constant. This process is repeated until the orders of reaction of each species and the specific reaction rate are evaluated. At level 5 one could also use the least squares analysis. (See Problem 5-12.)

The chart in Figure 5-22 can be of great help in the logical planning of a series of kinetic experiments to determine reaction orders and specific rate constants. However, the overall goals and design of the entire experimental program should always be kept in mind. Some guidelines on *experimental design and planning* are presented below.

(E20) One-experiment analysis method:
Level 1: Known order—use the integral method to determine k.
Level 2: If reaction is zero, first, second, or third order, you should succeed in the trial-and-error procedure, using the integral method to determine the reaction order and specific reaction rate.
Level 5: The differential method of analysis is a severe test of experimental data and can be used to determine the reaction order and specific reaction rate from one experiment.

5.7B—Experimental Planning

Experimentation design can best be discussed by asking the following questions about the experiment you wish to perform.

1. *Why perform the experiment?* Is the experiment necessary or can the information be obtained by some other means, such as researching literature and books or by calculations? A need for the experiment must be established.
2. *What are the constraints on the experimental program?* How much time and money has been budgeted (or should be budgeted) for the program? Are you restricted to specific equipment, materials, or systems? Is the safety of the project investigators endangered to such a degree that the experiment should not be carried out?
3. *What are the important measurements to make?* Define experimental objectives.
    a. How can the experiment be satisfactorily designed to achieve the experimental objectives in the simplest manner with the *minimum* number of measurements and *minimum* expense? If this experimental project is one of many subprojects of an overall program, are you sure that you are not losing sight of the goals and possible alternative solutions for the overall program?
    b. What are the controlled or independent variables? (1) What are the ranges of these variables? (2) How many settings of the controlled variables should be made in the various ranges to insure good sampling of the data? (3) Instead of varying each of the independent variables separately, can dimensionless groups or ratios (e.g., Reynolds number) of the controlled parameters be formed and varied so as to produce the same end results with fewer measurements?
    c. What are the dependent variables? Will variation of the independent variables over their entire range produce significant or measurable changes in the dependent variables?
4. *Is an absolute reference point or standard necessary for the experiment?* Do the instruments that are necessary to make proper measurements need to be calibrated, and if so, have they been calibrated?
5. *How good are the measurements in terms of accuracy and precision?* Is the uncertainty on the last significant digit small enough to yield any meaningful determinations of the value of some property (e.g., heat capacity)? Are the data precise and accurate enough to distinguish between mechanisms, theories, or possible outcomes?
    a. What modifications, if any, of the existing equipment are necessary to increase the precision or accuracy of the measurements or to better achieve the overall experimental objectives?
    b. In addition to recording the data, are there any descriptions of the phenomena or unusual observations that should be included in the data book?
6. *How are the data to be processed?* Are computer programs available to perform least-squares analysis, set confidence limits, or other statistical analyses? Are there any inaccuracies in the computational procedure that could be avoided?
7. *How are the processed data to be analyzed?*
    a. Are there any theories or mathematical models available that suggest the manner in which the data should be plotted or correlated? What information or generalizations can be obtained from the data?
    b. Are there any new experiments, measurements, or equipment modifications that should be made to extend the data into different regions or to better delineate the proposed model or theory?

c. Has an error analysis been performed on the experiment? Have all the important sources of error been listed and discussed in relation to how significantly they affect the final result (i.e., by what magnitude and in what direction)?

8. *What concrete conclusions can be drawn from the experimental results?* Were all the experimental objectives fulfilled? The conclusions should include all the significant and pertinent information that can be drawn from the data within the report.

9. *Have you clearly communicated your conclusions along with a description of your experiment to others?* This is usually accomplished by means of a technical report. Guidelines for writing such a report can be found in books on technical report writing, such as *Technical Reporting*, rev. ed. by J. N. Ulmann and J. R. Gould, (New York: Holt, Rinehart and Winston, 1959).

5.8–METHODS OF CHEMICAL ANALYSIS FOR REACTING SPECIES

During the planning of an experimental program to obtain information about reaction mechanisms and rate laws, one of the more difficult decisions will arise when you arrive at step 3 in the experimental design procedure; that is, what type of measurement should be made to monitor the concentration of the various species in the reacting system. Table 5-4, which lists

TABLE 5–4

Methods of Analysis	Applications	References
1. Physical methods		
a. Total pressure change	Gas phase reactions carried out under constant volume in which there is a change in the total number of moles in the gas phase.	*J. Chem. Ed.* 46 (1969), 684. A. P. Frost and R. G. Pearson, *Kinetics and Mechanism*, 2nd ed. (New York: John Wiley and Sons, 1968), Chap. 3.
b. Temperature change	Endothermic or exothermic reactions that are carried out adiabatically.	*Chem. Engr. Sci.*, 21 (1966), 397.
c. Volume change	(1) Gas phase reactions carried out under constant pressure and involving a change in the total number of moles.	(1) O. Levenspiel, *Chemical Reaction Engineering*, 2nd ed. p. 91, (New York: John Wiley and Sons, 1972).
	(2) Solid or liquid phase reactions in which there is a density change on reaction.	(2) *J. Am. Chem. Soc.*, 70, (1948), 639. *I.E.C. Process Design & Development*, 8, No. 1, (1969), 120.
2. Optical methods		
a. Absorption spectrometry	Transitions within molecules, which can be studied by the selective absorption of electromagnetic radiation. Transitions between electronic levels are found in the UV and visible regions; those between vibrational levels, within the same electronic level, are in the IR region.	
(1) Visible	Reactions involving one (or at most two) colored compound(s).	*J. Chem. Ed.*, 49, No. 8, (1972), 539. *I.E.C. Process Design & Development*, 8, No. 1, (1969), 120.

TABLE 5-4 (cont.)

Methods	Applications	References
(2) Ultraviolet	The determination of organic compounds, especially aromatic and heterocyclic substances or compounds with conjugated bonds.	
(3) Infrared	Organic compounds only.	*I.E.C. Product Research and Development*, 7, No. 1, (1968), 12.
(4) Atomic Absorption	Metallic ions.	K. Lund, H. S. Fogler, and C. C. McCune, *Chem. Engr. Sci.* 28, (1973), 691.
b. Polarimetry	Liquid phase reactions involving optically active species.	*Nature* 175, (London: 1955), 593.
c. Refractometry	Reaction in which there is a measurable difference between the refractive index of the reactants and that of the products.	
3. Electrochemical methods		
a. Potentiometry	Used in the measurement of the potentials of nonpolarized electrodes under conditions of zero current. Seldom used in organic reactions.	*J. Am. Chem. Soc.* 68, (1946), 658. *Ibid*, 69, (1947), 1325.
b. Voltammetry and polarography	Used for dilute electrolytic solutions. Not applicable to reactions that are catalyzed by mercury.	*J. Am. Chem. Soc.* 71, (1949), 3731.
c. Conductimetry	Reactions involving a change in the number or kind of ions present, thereby changing the electrical conductivity. Suitable for both organic and inorganic reactions.	*J. Chem. Soc.* 97, (London: 1910), 732. Frost & Pearson, *Kinetics and Mechanism*, Chap. 3.
4. Nuclear Methods		
a. Magnetic resonance spectrometry		
(1) Nuclear magnetic resonance	Primarily compounds containing Hydrogen	*J. Chem. Ed.* 49, No. 8, (1972), 560. Pople, Schneider & Bernstein, *High-Resolution Nuclear Magnetic Resonance*, (New York: McGraw-Hill, 1959).
(2) Electron spin resonance	Free radical studies and for estimation of trace amounts of paramagnetic ions	*J. Chem. Phys.* 46, No. 2, (1967), 490. *Ibid.* 48, No. 10, (1968), 4405.
b. Mass spectrometry		*Proc. Royal Soc.* A199, (London: 1949), 394. *I.E.C. Process Design & Development*, 8, No. 4, (1969), 450, 456.
c. Nuclear radiation (radioisotopes)	Elucidation of reaction mechanisms.	*Trans. Faraday Soc.* 30, (1934), 508.
5. Methods of interphase separations		
a. Gas-liquid chromatography	Gas phase reactions; also, liquid phase reactions involving only volatile substances.	*I.E.C. Product Research and Development* 8, No. 3, (1969), 319. *Ibid.* 10, No. 2, (1971), 138

*For detailed descriptions of the various methods, the reader is referred to *Instrumental Methods of Chemical Analysis* by G.W. Ewing, New York: McGraw-Hill, 1969).

some of the current methods used to obtain either direct or indirect measurements of a species concentration, will be useful in deciding this point. *Most of the references listed at the right of the table refer to a specific experiment in chemical kinetics in which the corresponding method analysis was used to obtain rate data.*

SUMMARY

1. Differential method for *constant-volume* systems:

$$-\frac{dC_A}{dt} = kC_A^\alpha \tag{S5-1}$$

   a. Plot $\Delta C_A/\Delta t$ as a function of t.
   b. Determine dC_A/dt from this plot.
   c. $\ln\left(-\dfrac{dC_A}{dt}\right) = \ln k + \alpha \ln C_A.$ \hfill (S5-2)

   Plot $\ln\left(-\dfrac{dC_A}{dt}\right)$ vs. $\ln C_A$. The slope should be the reaction order α.

2. Integral method:
   a. Guess the reaction order and integrate the mole balance equation.
   b. Calculate the resulting function of concentration for the data and plot this as a function of time. If the resulting plot is linear, you have probably guessed the correct reaction order.
   c. If the plot is not linear, guess another order and repeat the procedure. You probably should try another method of interpretation of the rate data if zero, first, or second order does not fit the experimental data.
   d. The integral method is most useful when you know the reaction order and want to determine the specific reaction rate k.

3. *Method of initial rates*: In this method of analysis of rate data, the initial reaction rate is measured for a number of initial concentrations. If the rate law is in the form $-r_A = kC_A^\alpha$, the slope of a plot of $\ln(-r_{A0})$ vs. $\ln C_{A0}$ will be the reaction order plots and semilog plots.

4. *Method of half-lives*: Plot $\ln t_{1/2}$ as a function of $\ln C_{A0}$. The reaction order α will be minus the slope of this plot; i.e.,

$$\alpha = 1 - \text{slope}. \tag{S5-3}$$

5. To determine the activation energy, we plot the natural log of the specific rate constant as a function of the reciprocal of the absolute temperature. We see from the natural log of Equation (3-1) that

$$\ln k = \left(-\frac{E}{R}\right) \cdot \frac{1}{T} + \ln A \tag{S5-4}$$

The activation energy can be found from the slope of the plot on semilog graph paper of k vs. $\frac{1}{T}$,

$$E = -2.3(R)(\text{slope})$$

where 2.3 represents the conversion factor from the natural log to the log base 10.

6. In modeling the differential reactor, we neglect any spatial variations in the species concentration. The rate of reaction is calculated from the equation

$$-r'_A = \frac{F_{A0}X}{W} = \frac{F_P}{W} = \frac{v_0(C_{A0} - C_{Ae})}{W}$$

In calculating the reaction order

$$r'_A = kC_A^\alpha,$$

the concentration of A is evaluated either at the entrance conditions or at a mean value between C_{A0} and C_{Ae}.

QUESTIONS AND PROBLEMS

P5-1. Tests have been run on a small experimental reactor for decomposing nitrogen oxides in an automobile exhaust stream. In one series of tests, a nitrogen stream containing various concentrations of NO_2 was fed to a reactor and the kinetic data obtained are shown in Figure P5-1. Each point represents one complete run.

The reactor essentially operates as an isothermal backmix reactor (CSTR); what can you deduce about the apparent order of the reaction over the temperature range studied?

The plot gives the fractional decomposition of NO_2 fed vs. the ratio of reactor volume V (in cu cm) to the NO_2 feed rate, $F_{NO_2,0}$ (gmoles/h), at different feed concentrations of NO_2 (in parts per million by weight).

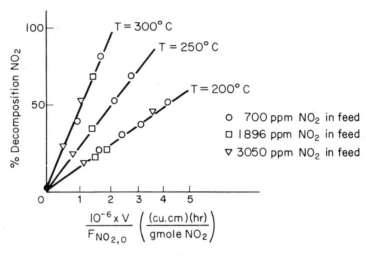

Figure P5–1

P5-2. When arterial blood enters a tissue capillary, it exchanges oxygen and carbon dioxide with its environment. The kinetics of this deoxygenation of hemoglobin in blood was studied with the aid of a tubular reactor by Nakamura and Staub (*J. Physiol.* 173, (1964), 161).

$$HbO_2 \underset{k_{-1}}{\overset{k_1}{\rightleftharpoons}} Hb + O_2$$

Although this is a reversible reaction, measurements were made in the initial phases of the decomposition so that the reverse reaction could be neglected. Consider a system similar to the one used by Nakamura and Staub: the solution enters a tubular reactor (0.158 cm in diameter) that has oxygen electrodes placed at 5 cm intervals down the tube. The solution flow rate into the reactor is 19.6 cu cm/sec.

Electrode position	1	2	3	4	5	6	7
% decomposition of HbO_2	.00	1.93	3.82	5.68	7.48	9.25	11.00

Using the method of differential analysis of rate data, determine the reaction order and the forward specific reaction rate constant k for the deoxygenation of hemoglobin.

P5-3. The following data[†] has been reported for the gas phase constant-volume decomposition of dimethylether at 504°C. Initially only $(CH_3)_2 O$ was present.

[†]C. N. Hinshelwood and P. J. Ackey, *Proc. Roy. Soc.* (*London*), A115, (1927), 215.

Time (sec)	390	777	1195	3155	∞
Total pressure (mm Hg)	408	488	562	799	931

Assuming that the reaction

$$(CH_3)_2O \longrightarrow CH_4 + H_2 + CO$$

goes to completion, determine the reaction order.

P5-4. The following reaction is carried out isothermally in a constant volume batch reactor:

$$2A \longrightarrow B + C + D$$

Initially there is pure A in the reactor at STP. The following pressures were recorded at subsequent times during the reaction. The reaction takes place isothermally.

Time (min)	0	1.20	1.95	2.90	4.14	5.70	8.10
Total pressure (atm)	1	1.10	1.15	1.20	1.25	1.30	1.35

Determine the order of the reaction and the specific reaction rate constant.

P5-5. For the irreversible gas phase dissociation of the dimer A_2,

$$A_2 \longrightarrow 2A,$$

determine the CSTR volume necessary to achieve 80% conversion and produce 1000 gmoles of A per minute. The feed stream consists of 60% A_2 and 40% inerts at a pressure of 10 atm and a temperature of 40°C.

The following data was obtained in the laboratory in a well-mixed constant-pressure batch reactor, which had an initial charge consisting of 85% A_2 and 15% inerts:

Temperature = 40°C.
Pressure = 3 atm.

Time (sec)	Volume (ml)
0	200
30	251
60	276
120	302
240	322

Process your data in terms of the measured variables (i.e., time and volume).

P5-6. A homogeneous irreversible gas phase cracking reaction whose stoichiometry can be represented by

$$A \longrightarrow B + 2C$$

is carried out in the laboratory at 300°C and 1 atm in a 1-liter continuous-flow stirred vessel. For a feed of pure reactant A, a graph of volumetric feed rate v_0 vs. the quantity $(1 - X)/((1 + 2X)X)$, where X is fractional conversion of A, gives a straight line with slope equal to 35 l/sec.

a. What is the apparent order of the reaction w.r.t. reactant A, and what is the value of the rate constant at 300°C?

It is desired to carry out this same reaction on a commercial scale, at 300°C and 5 atm pressure, to produce 100 lb-moles/sec of product C, with a 90% conversion of reactant A. The available feed consists of 80 mole % A and 20 mole % inert.

b. What volume tubular flow reactor will be necessary in the proposed commercial process?

(Exam U. of M.)

P5-7. It is desired to carry out the gaseous reaction A $\longrightarrow$ B in an existing tubular reactor consisting of 50 parallel tubes 40 ft long with 0.75 in. inside diameter.

Bench scale experiments have given the reaction rate constant for this first-order reaction as .00152 sec^{-1} at 200°F and .0740 sec^{-1} at 300°F.

At what temperature should the reactor be operated to give a conversion of A of 80%, with a feed rate of 500 lb/hr of pure A and an operating pressure of 110 psig? A has a molecular weight of 73. Departures from perfect gas behavior may be neglected, and the reverse reaction is insignificant at these condition. (*Ans.* T = 278°F)

(Calif. PECEE, November 1966).

P5-8. In order to study the photochemical decay of aqueous bromine in bright sunlight, a small quantity of liquid bromine was dissolved in water contained in a glass battery jar and placed in direct sunlignt. The following data were obtained:

Time (min)	10	20	30	40	50	60
Ppm Br$_2$	2.45	1.74	1.23	.88	.62	.44
			Temperature = 25°C			

a. Determine whether the reaction rate is zero, first, or second order in bromine, and calculate the reaction rate constant in units of your choice.

b. Assuming identical exposure conditions, calculate the required hourly rate of injection of bromine (in pounds) into a sunlit body of water, 25,000 gallons in volume, in order to maintain a sterilizing level of bromine of 1.0 ppm.

NOTE: Ppm = parts of bromine per million parts of water by weight. In dilute aqueous solutions, 1 ppm $\equiv$ 1 milligram per liter.

(Calif. PECEE, April 1966)

P5-9. Penicillin G reacts with hydroxylamine (NH$_2$OH) to form hydroxamic acid, which gives a colored complex with iron (III). To determine the overall reaction order, equal concentrations of penicillin and NH$_2$OH were mixed together in a 250 ml flask.† Samples were withdrawn every 10 min and added to a solution containing iron (III) chloride. With the aid of a colorimeter, the concentration of the colored complex, and hence the concentration of hydroxamic acid, was obtained as a function of time. The absorbence shown in the table below is directly proportional to the hydroxamic acid concentration.

Time (min)	.0	10	20	30	40	50	∞
Absorbency	.00	.337	.433	.495	.539	.561	.685

Determine the overall order of reaction assuming the reaction is irreversible.

P5-10. Nitrogen oxide is one of the pollutants in automobile exhaust and can react with oxygen to form nitrogen dioxide according to

$$2NO + O_2 \xrightarrow{k} 2NO_2$$

At 298°K the specific reaction rate is

$$k = 14.8 \times 10^3 \, (l)^2/(mole)^2/sec$$

or in parts per million

$$k = 1.4 \times 10^{-9} \, (ppm)^{-2}(min)^{-1}$$

†*J. Chem. Ed.* 49, (1972), 539.

a. What is the half-life of 3000 ppm NO (a typical precontrol auto exhaust value) in air?

Ans.: 1.2 min

b. What is the half-life of 1 ppm NO (a typical polluted atmospheric value)?

P5-11. The gas phase decomposition

$$A \longrightarrow B + 2C$$

is carried out in a constant-volume batch reactor. Runs 1 through 5 were carried out at 100°C while run 6 was carried out at 110°C.

a. From the data below determine the reaction order and specific reaction rate.

b. What is the activation energy for this reaction?

Run	Initial Concentration, C_{A0} (gmole/l)	Half-life $t_{1/2}$ (min)
1	.0250	4.1
2	.0133	7.7
3	.010	9.8
4	.05	1.96
5	.075	1.3
6	.025	2.0

P5-12. The ethane hydrogenolysis over a commercial nickel catalyst was studied in a Carberry reactor.

$$H_2 + C_2H_6 \longrightarrow 2CH_4$$

a) From the data below determine the rate law.

Total molar Feed Rate to Reactor gmole/hour	Partial Pressure (atm) in Feed		Mole Fraction CH_4 in Exit Stream
	Ethane P_{A0}	Hydrogen P_{B0}	
1.7	.5	.5	.05
1.2	.5	.5	.07
0.6	.5	.5	.16
0.3	.4	.6	.16
0.75	.6	.6	.10
2.75	.6	.4	.06

There are 4 spinning baskets, each with 10 grams of catalyst. Only hydrogen and ethane are fed to the reactor at 300°C.

b) Rework this problem using multiple regression[†] assuming $-r_A = kP_A^\alpha P_B^\beta$ and letting $Y = \ln(-r_A)$, $X_1 = \ln P_A$, $X_2 = \ln P_B$, $a \equiv \ln k$. Then $Y = a + \alpha X_1 + \beta X_2$.

Use the data in the above table in conjunction with the above transformations to form the following sums from which we can solve the following equations for the three unknowns, a, α and β. For N runs ($N = 6$)

$$Na + \alpha \sum_1^N X_1 + \beta \sum_1^N X_2 = \sum_1^N Y$$

$$a \sum_1^N X_1 + \alpha \sum_1^N X_1^2 + \beta \sum_1^N X_1 X_2 = \sum_1^N X_1 Y$$

$$a \sum_1^N X_2 + \alpha \sum_1^N X_1 X_2 + \beta \sum X_2^2 = \sum_1^N X_2 Y$$

[†]O. Levenspiel, et al., *Ind. Eng. Chem.*, 48, (1956), 324.

SUPPLEMENTARY READING

1. A wide variety of techniques for measuring the concentrations of the reacting species may be found in

    H. H. WILLIARD, L. L. MERRITT, and J. A. DEAN, *Instrumental Methods of Analysis*, 4th ed., Van Nostrand Reinhold, New York, 1965

   or

    G. W. EWING, *Instrumental Methods of Chemical Analysis*, McGraw-Hill, New York, 1969

   or Chapter 1 of

    G. PANNETIER and P. SOUCHAY, *Chemical Kinetics*, American Elsevier, New York, 1967.

2. A number of techniques in current use for following chemical reactions which take place very rapidly are discussed in

    KENNETH KUSTIN, ed., *Methods in Enzymology, Vol. XVI—Fast Reactions*, Academic Press, New York, 1969.

   The methods discussed in this book are not at all limited to enzymatic reactions. Also see Chapter 9 of

    W. C. GARDINER, *Rates and Mechanisms of Chemical Reaction*, Benjamin, Reading, Mass., 1969

   and

    D. N. HAGUE, *Fast Reactions*, Wiley-Interscience, New York, 1971.

3. A discussion on the methods of interpretation of batch reaction data can be found in Chapter 3 of

    O. LEVENSPIEL, *Chemical Reaction Engineering*, 2nd ed., Wiley, New York, 1972

   and in Chapter 3 of

    A. A. FROST and R. G. PEARSON, *Kinetics and Mechanism*, 2nd ed., Wiley, New York, 1961.

4. The interpretation of data obtained from flow reactors is also discussed in Chapter 4 of

    J. M. SMITH, *Chemical Engineering Kinetics*, 2nd ed., McGraw-Hill, New York, 1970.

   A number of example problems on the interpretation of rate data for both heterogeneous and homogeneous systems can be found in

    O. A. HOUGEN and K. M. WATSON, *Chemical Process Principles, Part 3—Kinetics and Catalysis*, Wiley, New York, 1947.

5. The design of laboratory catalytic reactors for obtaining rate data is presented in

    J. J. CARBERRY, *I.E.C.* **56**, No. 11 (1964), 39.

   The reactors discussed in this article are the differential reactor, the microcatalytic reactor, the "ideal" reactor, the recycle reactor, and the stirred basket reactor. Most of these types of reactors are also discussed in

    R. B. ANDERSON, ed., *Experimental Methods in Catalytic Research*, Academic Press, New York, 1968.

6. Current statistical methods applied to interpretation of rate data are presented in

    J. R. KITTRELL, *Advances in Chemical Engineering*, Vol. 8 (T. B. DRAW et al. eds.,) Academic Press, New York, 1970, pp. 97–183.

7. A review of current experimental and industrial work in chemical kinetics is given by

    V. W. WEEKMAN, *I.E.C.* **62**, No. 5 (1970), 53.

   An annual review of this type may be found in preceding yearly issues of *Industrial and Engineering Chemistry (I.E.C.)*.

6

Heterogeneous Reactions

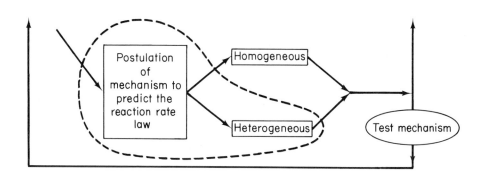

6.1–Fundamentals

There has been a constant search for new ways of increasing product *selectivity* and *yield* in chemical reactions. *Selectivity* is related to the number of moles of desired product formed per mole of undesired product formed, and the term *yield* refers to the number of moles of a specific product formed per mole of reactant consumed. Since the use of a catalyst makes it possible to obtain the end product by a different mechanism, it could affect both the yield and the selectivity. *A catalyst can either slow or accelerate the rate of formation of a particular product species; however, it does not affect the equilibrium.* In the absence of catalyst poisoning, the catalyst remains unaffected by the reaction. Homogeneous catalysis concerns reactions carried out in a single phase, such as the acid-catalyzed aldol condensation of acetone to form mesityl oxide:

$$2CH_3\overset{\overset{O}{\|}}{C}CH_3 \xrightarrow{H^+} CH_3\overset{\overset{O}{\|}}{C}CH=\underset{\underset{CH_3}{|}}{C}CH_3 + H_2O \qquad (6\text{-}1)$$

The cyclization of hexane to form cyclohexane is an example of a heterogeneous catalytic reaction, i.e., a reaction in which two or more phases are involved. One catalyst that can be used in this reaction is platinum supported on an Al_2O_3 base

$$CH_3(CH_2)_4 CH_3 \xrightarrow[\text{Catalyst}]{\substack{500° C \\ 300-700 \text{ psi}}} \boxed{S} + H_2 \qquad (6\text{-}2)$$

Define the term catalyst and describe how it may affect a chemical reaction.

(F1) _____

_____

6.1A Definitions

We shall focus our attention on gas phase reactions catalyzed by solid surfaces, i.e., *heterogeneous catalysis*. When gas molecules contact a solid surface, they can be adsorbed onto the surface by two different mechanisms: physical adsorption and chemisorption.

Physical adsorption is similar to condensation. The heat of adsorption, which is exothermic, is relatively small, on the order of 1–15 Kcal/gmole. The forces of attraction between the gas molecules and the solid surface are relatively weak. These forces, which are of van der Waal's type, consist of interaction between permanent dipoles, between a permanent dipole and an induced dipole, and/or between neutral atoms or molecules. The amount of gas physically adsorbed decreases rapidly with increasing temperature, and above the critical temperature only very small amounts of the various components are physically adsorbed.

The type of adsorption that is responsible for affecting the rate of chemical reaction is *chemisorption*. Here the adsorbed molecules are held to the surface by valence forces of the same type occurring between bonded atoms and molecules. These forces extend only a short distance beyond the adsorbed monomolecular layer. Conceptually, we consider that a chemical reaction has taken place when a gas molecule is chemically adsorbed onto a solid surface. The heat of adsorption is exothermic and of the same magnitude as the heat of chemical reaction, 10–100 kcal/gmole. For a catalyst to be effective, it must be used in the temperature range in which the chemisorption of one or more of the reactants is appreciable. (When referring to adsorption in the remainder of this chapter, we shall be speaking only of chemisorption.)

What are two similarities between chemisorption and physical adsorption?

(F2) _____

_____

(F1)
Although a catalysis does not affect the chemical equilibrium of a chemical reaction, it can cause the reaction to proceed at an accelerated rate. For example, the reaction
$$A \longrightarrow B$$
may be accelerated by utilizing a catalyst to proceed along a different reaction path:
$$A + \text{cat.} \longrightarrow A*\text{cat.} \longrightarrow B + \text{cat.}$$
In the absence of catalyst poisoning, the catalyst is chemically unchanged at the end of the reaction.

(F2)
In both chemisorption and physical adsorption,
    1. bonds are formed between the gas molecule and the solid surface, and
    2. the heat of adsorption is exothermic.

What is the major difference between physical adsorption and chemisorption?

(F3) _____

_____

Taylor[†] has suggested that the reaction is not catalyzed over the entire solid surface, but only at certain *active sites* or centers. He visualized these sites as unsaturated atoms in the solids that had resulted from surface irregularities, dislocations, and edges of crystals and cracks along grain boundaries. Other investigators in the field have taken exception to this definition, pointing out that other properties of the solid surface are more important in describing catalyst activity. In an attempt to reconcile these points, Boudart[‡] suggests that it is essential to characterize a catalytic surface only in connection with the reaction being catalyzed. Further discussion on this point is beyond the scope of this text,[§] and it will suffice for our purposes to say that *active sites are points on the catalystic surface that can form strong chemical bonds with the adsorbed molecules.*

If we divide the number of active sites per unit mass of catalyst by Avogadro's number, we obtain a molar concentration of the total number of the active sites per unit mass of catalyst, C_t. In an analogous manner, the molar concentration of vacant sites, C_v, is equal to the number of vacant sites per unit mass of catalyst divided by Avogadro's number. In the absence of catalyst poisoning the total concentration of sites remains constant. Further definitions are as follows:

P_i = partial pressure of species i in the gas phase, atm
C_v = surface concentration of vacant sites, gmoles/g of cat.
$C_{i \cdot S}$ = surface concentration of sites occupied by species i, gmoles i/g of cat.
S_a = surface area per unit mass of catalyst, sq cm/g of cat.
$C'_{i \cdot S}$ = surface concentration of species i based on the surface area, gmoles i/sq cm

$$C'_{i \cdot S} = \frac{C_{i \cdot S}}{S_a}$$

A conceptual model depicting species A and B adsorbed on two sites is shown in Figure 6-1.

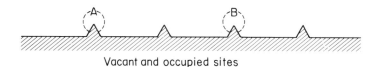

Vacant and occupied sites

Figure 6-1

Write an algebraic equation relating the total concentration of sites C_t to C_v and the adsorbed species shown in Figure 6-1.

(F4) _____

_____

[†]H. S. Taylor, *Proc. Roy. Soc.* 108A (1925), 105.
[‡]Boudart, M., *Am. Scientist* 57 (1969), 97.
[§]For further discussion see E. E. Petersen, *Chemical Reaction Analysis*, (Englewood Cliffs: Prentice-Hall, Inc., 1965), Chap. 3; M. Boudart, *Kinetics of Chemical Processes*, (Englewood Cliffs: Prentice-Hall, Inc., 1968).

(F3)

In physical adsorption the bonds formed between the surface and the adsorbed gas molecule are weak; consequently the rates of chemical reactions are not influenced to any reasonable extent. On the other hand, in chemisorption, chemical-type bonds are formed between the adsorbed gas molecule and the chemical surface.

(F4)

$$\text{Total sites} = \text{vacant sites} + \text{occupied sites}$$

Since species A and B are adsorbed, the concentration of occupied sites is $C_{A \cdot S} + C_{B \cdot S}$

$$C_t = C_V + C_{A \cdot S} + C_{B \cdot S}$$

6.1B Steps in a Catalytic Reaction

A schematic diagram of a tubular reactor packed with catalytic pellets is shown in Figure 6-2.

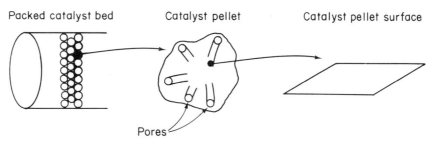

Figure 6–2

The overall process by which heterogeneous catalytic reactions proceed can be broken down into the sequence of individual steps shown in Table 6-1 and pictured in Figure 6-3. For this sequence we shall consider the isomerization

$$A \longrightarrow B. \tag{6-3}$$

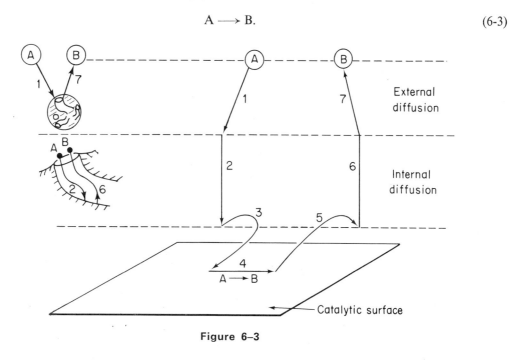

Figure 6–3

TABLE 6–1

Steps in a Catalytic Reaction

1. Mass transfer (diffusion) of the reactant(s) (e.g., species A) from the bulk fluid to the external surface of the catalyst pellet.
2. Diffusion of the reactant from the pore mouth through the catalyst pores to the immediate vicinity of the internal catalytic surface.
3. Adsorption of the reactant A onto the catalyst surface.
4. Reaction on the surface of the catalyst (e.g., A $\longrightarrow$ B).
5. Desorption of the products (e.g., B) from the surface.
6. Diffusion of the products from the interior of the pellet to the pore mouth at the external surface.
7. Mass transfer of the products from the external pellet surface to the bulk fluid.

In systems where diffusion from the bulk gas (or liquid) to the mouth of the catalyst pore or to the external catalyst surface limits the overall rate of reaction, the rate may be affected by the flow conditions through the catalyst bed. If the fluid velocity past the catalyst particles is increased, the concentration-boundary layer thickness should decrease, causing the rate of mass transfer to the pore mouth or external surface to increase. Consequently, the overall rate of reaction is increased. When internal diffusion within the catalyst pores is rate limiting, the overall rate will be unaffected by external flow conditions; in this case, catalyst effectiveness factors are useful in describing the catalyst's activity. A thorough discussion of the effects involving diffusion limitations can be found in Chapter 10.

In the following material we shall discuss only the steps of *adsorption, surface reaction,* and *desorption*.

6.1C Adsorption

First we shall consider the adsorption of a nonreacting gas on the surface of a catalyst. Since species A does not react further after being adsorbed, we need consider only the adsorption process:

$$A + S \rightleftharpoons A \cdot S \tag{6-4}$$

The *net* rate of adsorption is equal to the rate at which the reacting species "jumps" onto the surface, $k_A P_A C_v$, minus the rate at which A is desorbed from the surface, $k_{-A} C_{A \cdot S}$ (i.e., the rate at which molecules "jump" off the surface and return to the gas phase):

$$r_{ADS} = k_A P_A C_v - k_{-A} C_{A \cdot S} \tag{6-5}$$

The ratio $K_A = k_A/k_{-A}$ is the adsorption equilibrium constant. We now rearrange Equation (6-5):

$$\boxed{r_{ADS} = k_A \left(P_A C_v - \frac{C_{A \cdot S}}{K_A} \right)} \tag{6-6}$$

Recalling either Le Chatelier's principle or Van't Hoff's equation (Appendix C) along with the previous discussion on chemisorption, how would you expect k_A and K_A to vary with temperature?

(F5)

(F5) Since the heat of adsorption is exothermic, the adsorption-equilibrium constant K_A decreases with increasing temperature. The rate of chemisorption is analogous to the rate of chemical reaction. Consequently, one expects an Arrhenius temperature dependence for k_A. In other words, k_A increases with increasing temperature.

We shall develop Langmuir's adsorption isotherm equation by considering a system in which gaseous A is chemisorbed onto a catalytic surface. At equilibrium the rate of adsorption of A is equal to the rate of desorption of A. Solve Equation (6-6) for the equilibrium concentration of A adsorbed onto the surface and then list the symbols which you feel could readily be measured.

(F6)
___

___

The total concentration of active sites is constant and is equal to the sum of the concentration of vacant sites and the concentration of adsorbed species A.

$$C_t = C_{A \cdot s} + C_v \qquad (6\text{-}7)$$

Combining Equations (F6-1) and (6-7) to eliminate C_v, we obtain

$$C_{A \cdot s} = \frac{K_A P_A C_t}{1 + K_A P_A} \qquad (6\text{-}8)$$

The adsorption isotherm is usually represented by a plot of the volume of A that is adsorbed (which is directly proportional to $C_{A \cdot s}$) as a function of the total pressure. Since only pure A is present in the gas phase, the partial pressure equals total pressure. Sketch this isotherm in the figure in Frame (F7).

(F7)
___

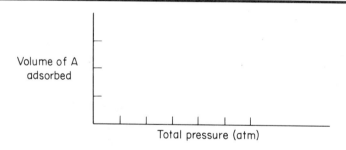

Figure 6–4
___

6.1D Surface Reaction

The rate of adsorption of species A onto a solid surface is given by Equation (6-6). Once a reactant has been adsorbed onto the surface, it is capable of reacting in a number of ways to form the reaction product. Three of these ways are:

1. The surface reaction may be a *single-site mechanism* in which only the site upon which the reactant is adsorbed is involved in the reaction. For example, an adsorbed molecule

(F6)

At equilibrium,
$$r_{ADS} = 0 = k_A \left(P_A C_v - \frac{C_{A \cdot S}}{K_A} \right) \qquad \text{(F6-1)}$$

The concentration of A adsorbed on the surface (gmole A/g of cat.) is
$$C_{A \cdot S} = K_A P_A C_v \qquad \text{(F6-2)}$$

Only the partial pressure of A in the gas phase can readily be measured.

(F7)

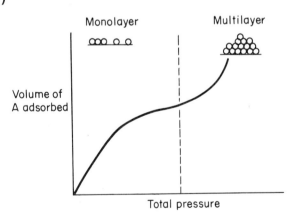

Figure 6-5

In the case of monomolecular adsorption, the volume adsorbed is given by Langmuir's isotherm:
$$\text{Volume adsorbed} \approx \frac{kP_A}{1 + k'P_A}$$

Langmuir's isothem, of course, would not predict the volume of gas adsorbed in multilayers as shown on the right side of Figure 6-5. In the material presented here we shall only consider monomolecular adsorption as shown on the left side of Figure 6-5.

of A may isomerize (or perhaps decompose) directly upon the site to which it is attached, i.e.,

$$A \cdot S \longrightarrow B \cdot S$$

2. The surface reaction may be a *dual-site mechanism* in which the adsorbed reactant interacts with another site (either unoccupied or occupied) to form the product. For example, adsorbed A may react with an adjacent vacant site to yield a vacant site and a site on which the product is adsorbed, i.e.,

$$A \cdot S + S \rightleftharpoons B \cdot S + S$$

Another example of a dual site mechanism is the reaction between two adsorbed species:

$$A \cdot S + B \cdot S \longrightarrow C \cdot S + D \cdot S$$

3. A third mechanism is the reaction between a molecule adsorbed on the surface and a molecule in the gas phase, e.g.,

$$A \cdot S + B(g) \longrightarrow C \cdot S + D(g)$$

In each of the above cases, the products of the surface reaction adsorbed on the surface are subsequently desorbed into the gas phase. Recalling Equations (6-4), (6-5), and (6-6), we see that the rate of desorption of a species A,

$$A \cdot S \rightleftharpoons A + S$$

is opposite in sign to the rate of adsorption of species A.

When modeling heterogeneous reactions, it is usually assumed that one of these steps (adsorption, surface reaction, or desorption) is the slowest step (i.e., rate limiting or rate controlling) in the catalytic reaction sequence.

6.2–Cumene Decomposition Example

We now wish to develop rate laws for catalytic reactions that are not diffusion limited. In particular we shall discuss the catalytic decomposition of cumene to form benzene and propylene. The overall reaction is:

$$C_6H_5(CH_3)_2 \rightleftharpoons C_6H_6 + C_3H_6 \tag{6-9}$$

A conceptual model depicting the sequences of steps in this platinum catalyzed reaction is in Figure 6-6.

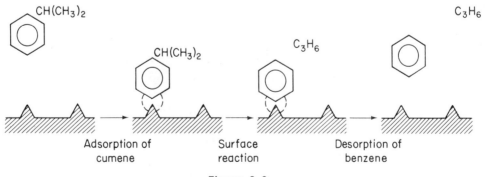

Figure 6–6

Figure 6-6 is only a schematic representation of the adsorption of cumene; probably a more realistic model is the formation of a complex of the pi-orbitals of benzene with the catalytic surface, as shown in Figure 6-7.

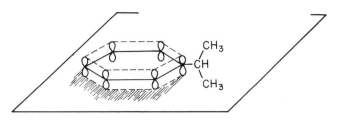

Figure 6-7

The following nomenclature will be used to denote the various species in this reaction:

C = Cumene, B = Benzene, P = Propylene, I = inhibitor.

The reaction sequence for this decomposition is

$$C + S \underset{k_{-A}}{\overset{k_A}{\rightleftharpoons}} C \cdot S \quad \text{Adsorption of cumene on the surface} \tag{6-10}$$

$$C \cdot S \underset{k_{-S}}{\overset{k_S}{\rightleftharpoons}} B \cdot S + P \quad \text{Surface reaction to form adsorbed benzene and propylene in the gas phase} \tag{6-11}$$

$$B \cdot S \underset{k_{-D}}{\overset{k_D}{\rightleftharpoons}} B + S \quad \text{Desorption of benzene from the surface.} \tag{6-12}$$

We may also have an inhibitor present which does not participate in the reaction, but does occupy active sites on the surface:

$$I + S \rightleftharpoons I \cdot S \quad \text{Adsorption of inhibitors on the surface.} \tag{6-13}$$

When writing rate laws for these steps, we treat each step as an elementary reaction; the only difference is that the species concentrations in the gas phase are replaced by their respective partial pressures:

$$C_A \Longrightarrow P_A$$

The rate expression for the adsorption of cumene as given in Equation (6-10) is

$$r_{ADS} = k_A P_C C_v - k_{-A} C_{C \cdot S} \tag{6-14}$$

$$\boxed{r_{ADS} = k_A \left(P_C C_v - \frac{C_{C \cdot S}}{K_A} \right)} \tag{6-15}$$

What are the units of k_A, k_{-A}, and K_A, if r_{ADS} has units of gmole/g of cat./hr?

(F8)

Write the rate expression for the rate of surface reaction, Equation (6-11), in terms of k_S, k_{-S}, C_v, $C_{C \cdot S}$, $C_{P \cdot S}$, $C_{B \cdot S}$, P_i, or any other appropriate variables.

(F8)

$$r_{ADS} = k_A P_C C_v - k_{-A} C_{C \cdot S}$$

$$r_{ADS} = k_A [\text{atm}] \left[\frac{\text{gmole}}{\text{g of cat.}} \right] - k_{-A} \left[\frac{\text{gmole}}{\text{g of cat.}} \right]$$

$$[r_{ADS}] = \frac{\text{gmole A}}{\text{g of cat. hr}^*}$$

$$[k_A] = 1/(\text{atm} \cdot \text{hr})$$

$$k_{-A} = \text{hr}^{-1}$$

$$K_A = \frac{k_{-A}}{k_A} = \text{atm}^{-1}$$

(*hr, sec, min, etc.)

The rate of benzene desorption is

$$r_D = k_D C_{B \cdot S} - k_{-D} P_B C_v \tag{6-16}$$

$$\boxed{r_D = k_D \left(C_{B \cdot S} - \frac{P_B C_v}{K_D} \right)} \tag{6-17}$$

Since the reaction is carried out under steady-state operation, the rates of each of the steps in the sequence are equal:

$$r_{ADS} = r_S = r_D \tag{6-18}$$

The rate of adsorption of the inhibitor is

$$\boxed{r_I = k_I \left(P_I C_v - \frac{C_{I \cdot S}}{K_I} \right)} \tag{6-19}$$

At steady-state operation the net rate of adsorption of the inhibitor is zero, i.e., $r_I = 0$, and the equilibrium surface concentration of the inhibitor is

$$C_{I \cdot S} = K_I P_I C_v \tag{6-20}$$

We note that the system may contain inert gases which do not act as inhibitors.

For the mechanism postulated in the sequence given by Equations (6-10) through (6-13), we wish to determine which step is rate determining. We first assume one of the steps to be rate limiting (rate controlling) and then formulate the reaction rate law in terms of the partial pressures of the species present. From this expression we can determine the variation of the initial reaction rate with the initial total pressure. If the predicted rate varies with pressure in the same manner as the rate observed experimentally, this implies that the assumed rate-limiting step may be the correct one.

6.2A Is the Adsorption of Cumene Rate Limiting?

To answer the above question we shall assume that the adsorption of cumene is rate limiting and then derive the corresponding rate law checking to see if it is consistent with experimental observation. One way this check could be made is by constructing a plot of the initial rate as a function of the total pressure and then comparing it with a corresponding plot of the experimental data. By assuming that this (or any other) step is rate limiting, we are considering that the specific reaction rate (in this case k_A) of this step is small w.r.t. the specific reaction rates of the other steps (in this case k_S and k_D).

The rate of adsorption is

$$-r'_C = r_{ADS} = k_A \left(P_C C_v - \frac{C_{C \cdot S}}{K_A} \right) \tag{6-15}$$

Since we cannot measure either C_v or $C_{C \cdot S}$, we must replace these variables in the rate equation with measurable quantities in order for the equation to be meaningful.

(F9) For the elementary reaction step
$$C \cdot S \rightleftharpoons B \cdot S + P(g)$$
the rate law is
$$r_S = k_S C_{C \cdot S} - k_{-S} P_P C_{B \cdot S}$$

$$\boxed{r_S = k_S \left[C_{C \cdot S} - \frac{P_P C_{B \cdot S}}{K_S} \right],} \quad K_S = \frac{k_S}{k_{-S}}$$

Propylene is not adsorbed on the surface, consequently
$$C_{P \cdot S} = 0.$$

For steady-state operation we have:

$$r_{ADS} = r_S = r_D \tag{6-21}$$

and *for adsorptional control, k_A is small and k_S and k_D are large.*

Consequently, the ratios r_S/k_S and r_D/k_D are very small (approximately zero), while the ratio r_A/k_A is relatively large. The surface reaction rate expression is

$$r_S = k_S\left[C_{C \cdot S} - \frac{C_{B \cdot S} P_P}{K_S}\right]. \tag{6-22}$$

For adsorption control we set

$$\frac{r_S}{k_S} \simeq 0$$

and solve Equation (6-22) for $C_{C \cdot S}$:

$$C_{C \cdot S} = \frac{C_{B \cdot S} P_P}{K_S} \tag{6-23}$$

To be able to express $C_{C \cdot S}$ solely in terms of the partial pressures of the species present, we must evaluate $C_{B \cdot S}$. The rate of desorption is

$$r_D = k_D\left(C_{B \cdot S} - \frac{P_B C_v}{K_D}\right) \tag{6-17}$$

However, for adsorption control we can set

$$\frac{r_D}{k_D} \simeq 0$$

and then solve Equation (6-17) for $C_{B \cdot S}$:

$$C_{B \cdot S} = \frac{P_B C_v}{K_D} \tag{6-24}$$

After combining Equations (6-23) and (6-24),

$$C_{C \cdot S} = \frac{P_B P_P}{K_D K_S} C_v \tag{6-25}$$

replacing $C_{C \cdot S}$ in the rate equation by Equation (6-25), and then factoring C_v, we obtain

$$r_{ADS} = k_A\left(P_C - \frac{P_B P_P}{K_S K_D K_A}\right) \cdot C_v = k_A\left(P_C - \frac{P_B P_P}{K_e}\right) C_v. \tag{6-26}$$

It can be shown by setting $r_{ADS} = 0$ that the product $K_S K_D K_A$ is simply the overall equilibrium constant K_e for the reaction:

$$C \rightleftharpoons B + P \tag{6-9}$$

$$K_A K_S K_D = K_e \tag{6-27}$$

The equilibrium constant can be determined from thermodynamic data and is related to the change in the Gibbs free energy, $\Delta G°$, by the equation

$$RT \ln K_e = -\Delta G° \tag{6-28}$$

where R is the ideal gas constant and T is the absolute temperature. The concentration of vacant sites, C_v, can now be eliminated from Equation (6-26) by utilizing the equation to give the total concentration of sites C_t, which is assumed to be constant. Write an expression for the total concentration of sites in terms of the concentrations of species adsorbed on the surface and C_v.

(F10)

(F10)

Total sites = vacant sites + occupied sites

Since cumene, benzene and the inert inhibitor are all adsorbed on the surface, the concentration of occupied sites is $C_{C \cdot S} + C_{B \cdot S} + C_{I \cdot S}$, and the total concentration of sites is

$$C_t = C_v + C_{C \cdot S} + C_{B \cdot S} + C_{I \cdot S} \tag{F10-1}$$

Utilizing Equations (F10-1), (6-20), (6-24), and (6-25), solve for the concentration of vacant sites in terms of the partial pressures of the various species and C_t.

(F11)

The rate law for the catalytic decomposition of cumene, assuming that the adsorption of cumene is the rate-limiting step, is

$$-r'_C = r_{ADS} = \frac{C_t k_A \left(P_C - \dfrac{P_P P_B}{K_e}\right)}{1 + \dfrac{P_P P_B}{K_S K_P} + \dfrac{P_B}{K_D} + K_I P_I} \tag{6-29}$$

We now wish to sketch a plot of the initial rate as a function of the total pressure Π_0. Initially, no product is present; consequently $P_P = P_B = 0$. The initial rate is given by

$$-r'_{C,0} \equiv -r_0 = \frac{k_A y_{C0} \Pi_0 C_t}{1 + y_I K_I \Pi_0} \tag{6-30}$$

where y_{C0} and y_I are respectively, the initial mole fractions of cumene and inerts. In Figure 6-8 sketch $-r_0$ as a function of Π_0 for a system containing 80% cumene and 20% inhibitor.

(F12)

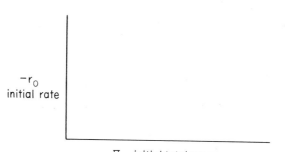

Figure 6–8

(F11)

$$C_t = C_{C \cdot S} + C_{B \cdot S} + C_{I \cdot S} + C_v \tag{F10-1}$$

Recalling,

$$C_{C \cdot S} = \frac{P_B P_P C_v}{K_D K_S} \tag{6-25}$$

$$C_{B \cdot S} = \frac{P_B C_v}{K_D} \tag{6-26}$$

$$C_{I \cdot S} = P_I K_I C_v \tag{6-27}$$

and substituting these into Equation (F10-1), we have:

$$C_t = \frac{P_B P_P C_v}{K_D K_S} + \frac{P_B C_v}{K_D} + P_I K_I C_v + C_v$$

Solving for C_v;

$$C_v = \frac{C_t}{\dfrac{P_B P_P}{K_D K_S} + \dfrac{P_B}{K_D} + P_I K_I + 1}$$

(F12)

At low pressures, we can neglect the second term in the denominator of Equation (6-30) w.r.t. 1,

$$y_I K_I \Pi_0 \ll 1$$

and observe that the initial rate increases linearly with pressure.

$$-r_0 \sim \Pi_0$$

At high pressures,

$$y_I K_I \Pi_0 \gg 1$$

and we see that the initial rate

$$-r_0 = \frac{k_A y_{C0} C_t}{y_I K_I}$$

is constant.

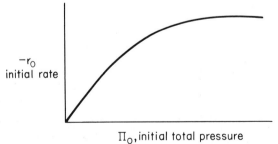

Figure 6-9

Next, in Figure 6-10, sketch $-r_0$ as a function of Π_0 when only cumene is initially present.

(F13)

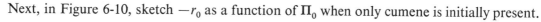

Figure 6–10

Before checking to see if Figure 6-11 is consistent with experimental observation, we shall derive the corresponding rate laws and initial rate plots when the surface reaction is rate limiting and then when the desorption of benzene is rate limiting.

6.2B Is the Surface Reaction Rate Limiting?

The rate of surface reaction is

$$r_S = k_S \left(C_{C \cdot S} - \frac{P_P C_{B \cdot S}}{K_S} \right) \tag{6-31}$$

Obtain the rate law for r_S in terms of the appropriate equilibrium constants, the appropriate partial pressures, and k_s and C_t when the surface reaction is assumed to be the rate limiting step.

(F14)

In Figure 6-12, sketch $-r_0$ vs. Π_0 for an initial mixture of (1) 50% cumene, 50% inhibitor and (2) pure cumene.

(F15)

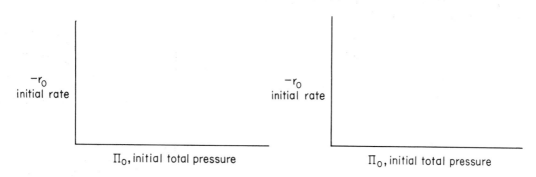

Figure 6–12

(F13)

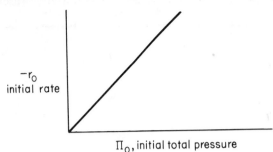

Figure 6–11

In Equation (6-30), when no inhibitors are present, $y_I = 0$ and $y_{C0} = 1$; then,
$$-r_0 = k_A \Pi_0 C_t$$
One observes that the initial rate increases linearly with total pressure.

(F14)

$$r_S = k_S \left(C_{C \cdot S} - \frac{C_{B \cdot S} P_P}{K_S} \right)$$

From the adsorption rate expression in Equation (6-15) and the condition $r_A/k_A \simeq 0$, we obtain a relationship for the surface concentration for adsorbed cumene:

$$C_{C \cdot S} = K_A P_C C_v$$

In a similar manner the surface concentration of adsorbed benzene can be evaluated from the desorption rate expression, Equation (6-17), along with the approximation:

$$\frac{r_D}{K_D} \simeq 0, \quad C_{B \cdot S} = \frac{P_B C_v}{K_D}$$

$$r_S = k_S \left(C_v P_C K_A - \frac{P_B P_P}{K_D K_S} C_v \right) = k_S K_A \left(P_C - \frac{P_B P_P}{K_e} \right) C_v$$

$$C_t = C_v + C_{B \cdot S} + C_{C \cdot S} + C_{I \cdot S}$$

$$C_v = \frac{C_t}{1 + \frac{P_B}{K_D} + K_A P_C + P_I K_I}$$

$$r_S = \frac{k_S C_t K_A \left(P_C - \frac{P_P P_B}{K_e} \right)}{1 + \frac{P_B}{K_D} + K_A P_C + P_I K_I} \tag{F14-1}$$

(F15)

For a mixture of 50% C and 50% I, the initial rate $P_P = P_B = 0$ can be determined from Equation (F14-1):

$$-r_0 = \frac{k_S C_t K_A (.5 \Pi_0)}{1 + (.5 K_A + .5 K_T) \Pi_0} \tag{F15-1}$$

For *pure* cumene, the initial rate equation is also obtained from Equation (F14-1):

$$-r_0 = \frac{k_S C_T K_A \Pi_0}{1 + K_A \Pi_0}$$

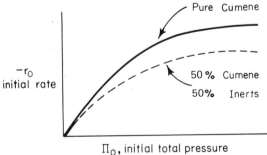

Figure 6–13

206

6.2C Is the Desorption of Benzene Rate Limiting?

The rate expression for the desorption of benzene is

$$r_D = k_D \left(C_{B \cdot S} - \frac{P_B C_v}{K_D} \right) \tag{6-17}$$

From the rate expression for the surface reaction we set

$$\frac{r_S}{k_S} \simeq 0 \quad \text{to obtain} \quad C_{B \cdot S} = \frac{C_{C \cdot S} K_S}{P_P} \tag{6-32}$$

From the rate expression for the adsorption of cumene we set

$$\frac{r_A}{k_A} \simeq 0 \quad \text{to obtain} \quad C_{C \cdot S} = K_A P_C C_v, \quad \text{then}$$

$$C_{B \cdot S} = \frac{K_A K_S P_C C_v}{P_P} \tag{6-33}$$

$$r_D = k_D K_A K_S \left(\frac{P_C}{P_P} - \frac{P_B}{K} \right) C_v \tag{6-34}$$

$$C_t = C_{C \cdot S} + C_{B \cdot S} + C_{I \cdot S} + C_v \tag{F14-1}$$

After substituting for the respective surface concentrations, we solve for C_v:

$$C_v = \frac{C_t}{1 + \frac{K_A K_S P_C}{P_P} + K_A P_C + K_I P_I} \tag{6-35}$$

Replacing C_v in Equation (6-34) by Equation (6-35) and multiplying the numerator and denominator by P_P, we obtain the rate expression for desorption control:

$$r_D = \frac{k_D C_t K_S K_A \left[P_C - \frac{(P_B P_P)}{K_e} \right]}{P_P + P_C K_A K_S + K_A P_P P_C + K_I P_I P_P} \tag{6-36}$$

To determine the dependence of the initial rate on the initial total pressure, we again set $P_P = P_B = 0$; and the rate law reduces to

$$-r_0 = k_D C_t$$

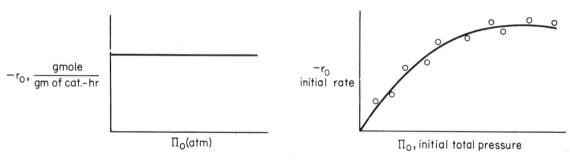

Figure 6–14   Figure 6–15

If desorption were controlling, we would see that the initial rate is independent of the initial composition and the initial pressure.

Summary of the Cumene Decomposition The experimental observations of $-r_0$ as a function of Π_0 are shown in Figure 6-15.

From the plot in Figure 6-15 we can clearly see that desorption is not the controlling step. However, to determine whether adsorption or surface reaction is controlling, we must use either pure cumene or at least a feed containing no inhibitors in our initial rate experiments. If we

carried out these experiments in the laboratory, we would find that the surface reaction is indeed the rate-limiting step.

6.3–Hydromethylation of Toluene Example

Hydrogen and toluene are reacted over a solid mineral catalyst containing clinoptilolite (a crystalline silica-alumina) to yield methane and benzene[†]:

$$C_6H_5CH_3 + H_2 \longrightarrow C_6H_6 + CH_4 \tag{6-37}$$

We wish to design tubular (packed bed) and CSTR (fluidized bed) reactors to produce 10 gmoles of benzene/min from a feed consisting of 20% toluene, 40% hydrogen, and 40% inerts at 600°C and 20 atm. To achieve this we must first determine the rate law from the differential reactor data presented in Table 6–2.

6.3A Searching for the Mechanism

In this table we find the rate of reaction of toluene as a function of the partial pressures of hydrogen (H), toluene (T), benzene (B), and methane (M). In the first two runs, methane was introduced into the feed along with hydrogen and toluene, while the other product, benzene, was only fed to the reactor along with the reactants in runs 3 and 4. In the remaining runs, neither of the products were present in the feed stream; consequently, since the conversion is less than 1% in the differential reactor, the partial pressures of methane and benzene in these runs are essentially zero, and the reaction rates are equivalent to initial rates of reaction.

TABLE 6–2

	Run	$-r'_T \times 10^{10}$ $\left(\dfrac{\text{gmoles toluene}}{\text{g of cat.-sec}}\right)$	Toluene P_T	Partial pressures (atm) Hydrogen (H$_2$)* P_H	Methane P_M	Benzene P_B
Set A	1	71.0	1	1	1	0
	2	71.0	1	1	4	0
Set B	3	41.6	1	1	0	1
	4	19.7	1	1	0	4
Set C	5	71.0	1	1	0	0
	6	142.0	1	2	0	0
	7	284.0	1	4	0	0
Set D	8	47.0	0.5	1	0	0
	9	71.0	1	1	0	0
	10	117.0	5	1	0	0
	11	127.0	10	1	0	0
	12	131.0	15	1	0	0
	13	133.0	20	1	0	0

*Note: $P_H \equiv P_{H_2}$

[†] J. Papp, D. Kallo and G. Schay, *Journal of Catalysis* 23, (1971), 168.

Assuming that this heterogeneous reaction is essentially irreversible, what qualitative conclusions can be drawn from the first set of runs (1 and 2) about the dependence of $-r_T$ on the partial pressure of methane and about the adsorption of methane on the surface of the catalyst?

(F16)

In Frame (F17) discuss qualitatively the dependence of $-r'_T$ on the partial pressure of benzene. (See Table 6–2, runs 3 and 4). Suggest a functional form of the rate expression that would explain the dependence of the rate of reaction on P_B.

(F17)

Referring to the final set of runs (8-13 in Table 6–2), discuss qualitatively the dependence of $-r'_T$ on P_T, and suggest a form of the rate expression, similar to Equation (F17-1), that would explain the dependence of the initial rate on the partial pressure of toluene.

(F18)

(F16) The rate of reaction is independent of the methane concentration. If methane were adsorbed on the surface, the partial pressure of methane would appear in the denominator of the rate expression and the rate would vary inversely with methane concentration, i.e.,

$$-r'_T \sim \frac{[\;]}{1 + K_M P_M + \cdots}$$

Consequently, we can conclude that methane is either very weakly adsorbed (i.e., $K_M P_M \ll 1$) or goes directly into the gas phase in a manner similar to propylene in the cumene decomposition.

(F17) In runs 3 and 4, we observed that for fixed concentrations (partial pressures) of hydrogen and toluene the rate decreases with increasing concentration of benzene. A rate expression in which the benzene partial pressure appears in the denominator could explain this dependency; i.e.,

$$-r'_T \sim \frac{1}{1 + K_B P_B + \cdots} \tag{F17-1}$$

The type of dependence of $-r'_T$ on P_B given by Equation (F17-1) suggests that benzene is adsorbed on the clinoptilolite surface.

(F18) At low concentrations of toluene (runs 8 and 9), the rate increases with increasing partial pressure of toluene, while at high toluene concentrations (runs 12 and 13), the rate is essentially independent of the toluene partial pressure.

A form of the rate expression that would describe this behavior is

$$-r'_T \sim \frac{P_T}{1 + K_T P_T + \cdots} \tag{F18-1}$$

A combination of Equations (F17-1) and (F18-1) suggests that the rate law may be in the form

$$-r'_T \sim \frac{P_T}{1 + K_T P_T + K_B P_B + \cdots} \tag{F18-2}$$

We now propose the following mechanism for the hydrodemethylation of toluene over clinoptilolite: i.e.,

$$T(g) + S \rightleftharpoons T \cdot S$$
$$H_2(g) + 2S \rightleftharpoons 2H \cdot S$$
$$2H \cdot S + T \cdot S \rightleftharpoons B \cdot S + 2S + M(g)$$
$$B \cdot S \rightleftharpoons B(g) + S$$

where T = toluene, B = benzene, H = hydrogen(H_2), and M = methane.

Assuming that *the rate of adsorption of toluene is controlling*, derive in Frame F19 the corresponding rate law and then determine if it is consistent with the data in Table 6–2.

(F19)

(F19) Assuming that the rate of adsorption of toluene is rate limiting, i.e.,

$$T(g) + S \rightleftharpoons T \cdot S$$

the rate equation is

$$-r_T = r_{ADS} = k_A \left(P_T C_v - \frac{C_{T \cdot S}}{K_A} \right) \tag{F19-1}$$

We must replace C_v and $C_{T \cdot S}$ in this equation by quantities that can be measured directly. For the following steps the respective rate expressions are:

$$H_2(g) + 2S \rightleftharpoons 2H \cdot S \qquad -r_H = k_H \left(P_H C_v^2 - \frac{C_{H \cdot S}^2}{K_H} \right) \tag{F19-2}$$

$$2H \cdot S + T \cdot S \rightleftharpoons B \cdot S + 2S + M(g) \qquad r_S = k_S \left(C_{T \cdot S} C_{H \cdot S}^2 - \frac{C_{B \cdot S} C_v^2 P_M}{K_S} \right) \tag{F19-3}$$

$$B \cdot S \rightleftharpoons B(g) + S \qquad r_D = k_D \left(C_{B \cdot S} - \frac{P_B C_v}{K_B} \right) \tag{F19-4}$$

Assuming that $-r_H/k_H \simeq 0$, we can solve Equation (F19-2) for $C_{H \cdot S}$:

$$C_{H \cdot S} = C_v \sqrt{K_H P_H} \tag{F19-5}$$

Equation (F19-4) can be solved in a similar manner to obtain the surface concentration of benzene after assuming $r_D/k_D \simeq 0$:

$$C_{B \cdot S} = \frac{P_B C_v}{K_B} \tag{F19-6}$$

The expression for the surface concentration of toluene is found from Equation (F19-3) by setting $r_S/k_S = 0$ and then rearranging:

$$C_{T \cdot S} = \frac{C_{B \cdot S} P_M}{K_S} \left(\frac{C_v}{C_{H \cdot S}} \right)^2 \tag{F19-7}$$

Substituting Equations (F19-5) and (F19-6) for $C_{B \cdot S}$ and $C_{H \cdot S}$,

$$C_{T \cdot S} = \frac{C_v P_B P_M}{K_B K_H K_S P_H} \tag{F19-8}$$

Since the inerts present do not act as inhibitors, $C_{I \cdot S} = 0$, and the total concentration of sites is

$$C_t = C_v + C_{B \cdot S} + C_{H \cdot S} + C_{T \cdot S} \tag{F19-9}$$

Substituting Equations (F19-5, F19-6, and F19-8) for the surface concentrations of the adsorbed species, we obtain the equation for the concentration of vacant sites:

$$C_v = \frac{C_t}{1 + \sqrt{K_H P_H} + \frac{P_B}{K_B} + \frac{P_B P_M}{P_H K_B K_H K_S}} \tag{F19-10}$$

Substituting for C_v and $C_{T \cdot S}$ in Equation (F19-1), we obtain the rate law for the proposed mechanism, assuming that the adsorption of toluene is the rate-controlling step.

$$-r'_T = \frac{k_A C_t \left(P_T - \frac{P_B P_M}{K_e P_H} \right)}{1 + \sqrt{K_H P_H} + \frac{P_B}{K_B} + \frac{P_B P_M K_A}{P_H K_e}} \tag{F19-11}$$

where

$$K_e = K_A K_H K_B K_S$$

If benzene and methane are absent in the feed, the initial rate equation

$$-r'_T = \frac{k_A C_t P_T}{1 + \sqrt{K_H P_H}} \qquad \text{(recall that } P_H \equiv P_{H_2}\text{)}$$

predicts that the rate will decrease with increasing hydrogen concentration. This prediction contradicts the information presented in Table 6-2 (set C) which shows the rate increases linearly with increasing hydrogen concentration. Consequently, for the suggested mechanism, we reject our hypothesis that the adsorption of toluene is rate limiting.

In continuing the search for a rate law that is consistent with experimental observations, there are several paths we might follow at this point. Circle the statement in Frame (F20) which best describes what you think should be the next step.

(F20)

1. Assume that the surface reaction is rate limiting and derive a rate law for the mechanism proposed on page 213.
2. Assume that the desorption of benzene is rate limiting and determine if the rate law for the mechanism on page 213 is consistent with the data in Table 6–2.
3. Select a different mechanism and test to see if adsorption control, surface-reaction control, or desorption control will result in the appropriate rate law.
4. Proceed along a path that is different from any of the above; explain.

Note that the rate law in Frame (F19) contained the *square root of the partial pressure* of one of the species, hydrogen. This type of dependence on the partial pressure results when one of the species, which is either a diatomic molecule (e.g., H_2, O_2) or a dimer, dissociates upon adsorption on the catalyst surface. We now propose a new mechanism; we assume that hydrogen is not adsorbed on the surface, but instead it reacts while it is in the gas phase with toluene, which is adsorbed on the surface. The sequence of steps for this mechanism is

$$T(g) + S \rightleftharpoons T \cdot S$$
$$H_2(g) + T \cdot S \rightleftharpoons B \cdot S + M(g)$$
$$B \cdot S \rightleftharpoons B(g) + S$$

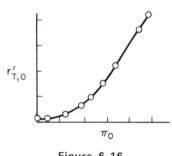

Figure 6–16

Assuming that the reaction between adsorbed toluene and gaseous hydrogen is rate limiting, derive the rate law and determine if it is consistent with the initial rate-total pressure plot for a fixed initial composition given in Figure 6-16, which was constructed *in part* from the data from Table 6–2 for an equal molar feed of $H_2(g)$ and $T(g)$.

(F21)

(F20) When we examine runs 5, 6, and 7 in Table 6-2, we see that the rate increases linearly with increasing hydrogen concentration and we conclude that the reaction is first order in H_2. In the light of this fact, it is doubtful that hydrogen is adsorbed on the surface. If it were adsorbed, the dependence of $-r'_T$ on P_H would be similar to the dependence of $-r'_T$ on P_T. Consequently, it appears that the mechanism is not correct, and we should choose a different one before proceeding further. Ans: 3

(F21)

$$T(g) + S \rightleftharpoons T \cdot S \qquad r_A = k_A \left(C_v P_T - \frac{C_{T \cdot S}}{K_T} \right) \tag{F21-1}$$

$$H_2(g) + T \cdot S \rightleftharpoons B \cdot S + M(g) \qquad r_S = k_S \left(P_{H_2} C_{T \cdot S} - \frac{C_{B \cdot S} P_M}{K_S} \right) \tag{F21-2}$$

$$B \cdot S \rightleftharpoons B(g) + S \qquad r_D = k_D \left(C_{B \cdot S} - \frac{P_B C_v}{K'_B} \right) \tag{F21-3}$$

$$\frac{r_A}{k_A} \simeq 0, \qquad C_{T \cdot S} = P_T C_v K_T \tag{F21-4}$$

$$\frac{r_D}{k_D} \simeq 0, \qquad C_{B \cdot S} = \frac{P_B C_v}{K'_B} \tag{F21-5}$$

$$C_t = C_v + C_{T \cdot S} + C_{B \cdot S} \tag{F21-6}$$

Substituting Equations (F21-4) and (F21-5) into Equation (F21-6) and rearranging, we obtain

$$C_v = \frac{C_t}{1 + P_T K_T + \frac{P_B}{K'_B}} \tag{F21-7}$$

Next, substitute for $C_{T \cdot S}$ and $C_{B \cdot S}$ and then substitute for C_v in Equation (F21-2) to obtain the rate law for the case of surface-reaction control:

$$-r'_T = r_S = \frac{C_t k_S K_T \left(P_H P_T - \frac{P_B P_M}{K_e} \right)}{1 + P_T K_T + \frac{P_B}{K'_B}} \tag{F21-8}$$

Neglecting the reverse reaction and letting $K_B = 1/K'_B$, and $k = k_S C_t$;

$$-r'_T = \frac{k K_T P_H P_T}{1 + K_B P_B + K_T P_T} \tag{F21-9}$$

Note that the equilibrium constant for adsorption of a given species is exactly the reciprocal of the equilibrium constant for the desorption of that species. For an equal molar feed consisting only of toluene and hydrogen, we find that the initial rate is the function of total pressure as described by

$$P_{T,0} = P_{H,0} = .5 \Pi_0 \tag{F21-10}$$

$$-r_{T_0} = \frac{.25 k K_T \Pi_0^2}{1 + .5 K_T \Pi_0} \tag{F21-11}$$

A sketch of $-r_{T_0}$ vs. Π_0, obtained from Equation (F21-11), indicates that the mechanism and rate-limiting step are in agreement with the initial rate-total pressure plot given in Figure 6-16. Equation (F21-9) is also consistent with all of the data in Table 6-2.

6.3B Evaluation of The Rate Law Parameters

In the original work on this reaction by Papp, et al.,[†] over 25 models were tested against the experimental data, and it was concluded that the mechanism and rate-limiting step discussed in Frame (F21) (i.e., the surface reaction between adsorbed toluene and H_2 gas) is the correct one. Assuming that the reaction is essentially irreversible, the rate law for the reaction on clinoptilolite is

$$-r'_T = kK_T \frac{P_H P_T}{1 + K_B P_B + K_T P_T} \tag{6-38}$$

How would you plot the data given in Table 6–2 to evaluate the constants k, K_T, and K_B? Select your choice(s) from Frame (F22) and explain in detail how you would determine k, K_T, and K_B from the plot(s) you have chosen.

(F22)

1. $-r'_T$ vs. $P_T P_H$ for constant P_B
2. $-r'_T$ vs. P_T for constant P_H and P_B
3. $\dfrac{\Pi_0}{-r'_{T_0}}$ vs. $\dfrac{1}{\Pi_0}$ [see Equation (F21-11)]
4. $\dfrac{P_T}{-r'_T}$ vs. P_H for constant P_B
5. $\dfrac{P_T}{-r'_T}$ vs. P_B for constant P_H
6. $\dfrac{P_T P_H}{-r'_T}$ vs. P_T for constant P_B
7. $\dfrac{P_T P_H}{-r'_T}$ vs. P_B for constant P_T

Note: One could also use the multiple regression techniques described in Problem 5-12 to determine the rate law parameters (e.g., Let $Y = (P_H P_T / -r'_T)$).

[†] J. Papp, D. Kallo and G. Schay, *Journal of Catalysis* 23, (1971), 168.

(F22)

We see from a rearrangement of Equation (F21-9) that a plot of $(\Pi_0/-r'_{T_0})$ vs. $1/\Pi_0$ would enable us to determine K_T and k, but not K_B. If both sides of Equation (6-38) are divided by $P_H P_T$ and the equation is then inverted, i.e.,

$$\frac{P_H P_T}{-r'_T} = \frac{1}{kK_T} + \frac{K_B}{kK_T}P_B + \frac{P_T}{k} \qquad (F22-1)$$

we see that a plot of $P_H P_T / -r'_T$ vs. P_T at constant P_B will be linear with slope $1/k$.

Run	$-r'_T \times 10^{10}$	P_T	P_H	$\dfrac{P_T P_H}{-r'_T} \times 10^{-8}$	P_B
3	41.6	1	1	2.40	1
4	18.5	1	1	5.40	4
5	71	1	1	1.41	0
7	284	1	4	1.41	0
8	47	0.5	1	1.06	0
10	117	5	1	4.27	0
11	127	10	1	7.87	0
12	131	15	1	11.45	0
13	133	20	1	15.03	0

We can evaluate k as a function of P_T (for constant P_B) from the slope of a plot of the left-hand side of Equation (F22-1) (See Figure 6-17.).

$$\text{Slope} = \frac{15.0 - .8}{20} \times 10^8$$

$$\text{Slope} = .71 \times 10^8$$

$$k = \frac{1}{\text{slope}}$$

$$k = 1.41 \times 10^{-8} \frac{\text{gmole}}{\text{g of cat. sec atm}}$$

Rearranging Equation (F22-1) and taking $P_B = 0$,

$$\frac{P_H P_T}{-r'_T} = \frac{P_T}{k} + \frac{1}{kK_T} \qquad (F22-2)$$

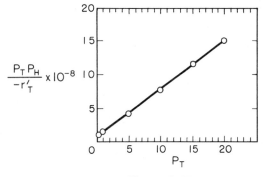
Figure 6-17

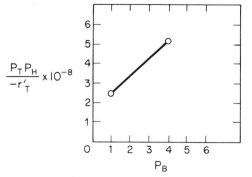
Figure 6-18

From Figure 6-17, $(P_H P_T / -r'_T) = 10^9$ when $P_T = 13.1$. Substituting these values in Equation (F22-2) we find that

$$10^9 - \frac{13}{1.41 \times 10^{-8}} = .7 \times 10^8 = \frac{1}{kK_T}$$

$$K_T = 1.01 \text{ (atm}^{-1}\text{)}$$

The constant K_B can be evaluated from the slope of the plot of $P_H P_T / -r'_T$ vs. P_B for constant P_T:

$K_B = \text{Slope} \cdot kK_T$ $\qquad K_B = (1.02 \times 10^8)(1.41 \times 10^{-8})(1.01)$ $\qquad K_B = 1.45 \text{ (atm}^{-1}\text{)}$

After substituting the numerical values for k, K_B, and K_T at 600°C into Equation (6-38), the rate law is

$$-r'_T = \frac{1.41 \times 10^{-8} P_H P_T}{1 + 1.45 P_B + 1.01 P_T}, \qquad \frac{\text{gmoles toluene}}{\text{g of cat./sec}} \tag{6-39}$$

Express $-r'_T$ as a function of the conversion X, numerically evaluating all symbols for the specified feed conditions to the CSTR (fluidized) and to the tubular (fixed bed) reactors, i.e., 40% H_2, 20% T, 40% Inerts at 10 atm, 600°C). Neglect pressure drop in both reactors and assume the reaction is carried out isothermally. ($P_A = C_A RT = \;?$)

(F23)

By combining terms involving X in the denominator of Equation (F23-6), the rate expression becomes

$$-r'_T = \frac{5.64 \times 10^{-8}(1-X)(2-X)}{3.02 + .88X} \tag{6-40}$$

for the specified reactor entrance conditions.

6.3C Design of Gas-Solid Catalytic Reactors

Before proceeding further we shall derive the design equation for a catalytic reactor, which will give the catalyst weight needed to achieve a specified conversion X. This derivation will be carried out in a manner analogous to the development of the tubular reaction design equation. To accomplish this, we simply replace the volume coordinate in Equation (A20-1) by the catalyst weight coordinate W. It might be helpful at this point to review Frames (A19), (D12), and (E16). In a mole balance on a system of catalyst mass ΔW, we let

$$F_{A0} \longrightarrow F_A(W) \qquad \frac{\text{moles } A}{\text{time}}$$

$$F_A \longrightarrow F_A(W + \Delta W) \qquad \frac{\text{moles } A}{\text{time}}$$

The generalized mole balance on species A then results in the equation

$$F_A(W) - F_A(W + \Delta W) = -r'_A \Delta W \tag{6-41}$$

The dimensions of the right-hand side of Equation (6-41) are

$$(-r'_A)(\Delta W) = \left(\frac{\text{moles } A}{\text{time} \cdot \text{mass cat.}}\right)(\text{mass cat.}) = \frac{\text{moles } A}{\text{time}}$$

After dividing by ΔW and taking as the limit $\Delta W \to 0$,

$$-\frac{dF_A}{dW} = -r'_A \tag{6-42}$$

(F23) For the hydrodemethylation of toluene over clinoptilolite at 600°C,

$$C_6H_5CH_3 + H_2 \longrightarrow C_6H_6 + CH_4$$

the rate law is

$$-r'_T = \frac{1.41 \times 10^{-8} P_H P_T}{1 + 1.45 P_B + 1.01 P_T}$$

$$P_T = C_T RT = C_{T,0} RT_0 \left(\frac{1-X}{1+\epsilon X}\right) = P_{T,0} \left(\frac{1-X}{1+\epsilon X}\right) \tag{F23-1}$$

$$\epsilon = y_{T0} \delta = .2(0) = 0$$

$$P_{T,0} = y_{T,0} \Pi_0 = (.2)(10) = 2 \text{ atm}$$

$$P_T = 2(1 - X) \tag{F23-2}$$

$$P_H = P_{T,0}(\theta_H - X),$$

$$\theta_H = \frac{.4}{.2} = 2$$

$$P_H = 2(2 - X) \tag{F23-3}$$

$$P_B = C_B RT = C_{T,0} RT_0 (\theta_B + X) = P_{T,0}(\theta_B + X) \tag{F23-4}$$

$$\theta_B = 0$$

$$P_B = 2X \tag{F23-5}$$

Substituting for P_B, P_H, and P_T in Equation (6-39),

$$-r'_T = \frac{5.64 \times 10^{-8}(1-X)(2-X)}{1 + 2.90(X) + 2.02(1-X)} \tag{F23-6}$$

we substitute for F_A in terms of the conversion and the molar feed rate entering the reactor, F_{A0}:

$$F_{A0}\frac{dX}{dW} = -r'_A \tag{6-43}$$

The integral form of the packed-catalyst bed design equation is

$$W = F_{A0}\int_0^X \frac{dX}{-r'_A} \tag{6-44}$$

Deduce the design equation for a perfectly mixed fluidized-bed catalytic reactor that can be modeled as a CSTR. (What formula will give the catalyst weight necessary to achieve the desired conversion X?) Also, given the bulk density of the fluidized bed, ρ_B, determine the bed volume.

(F24)

The bulk density of the fluidized bed will be a function of the velocity of the fluid through the bed.[†] If the bulk density in the *fluidized bed* is 0.4 g/cu cm, determine the catalyst weight and reactor volume necessary to produce 10 gmoles/min of benzene from a volumetric feed rate of 400 *l*/min.

(F25)

Feed conditions are as before: 10 atm, 600°C, 20% toluene 40% H_2, and $R = .082\frac{\text{liter atm}}{\text{gmole}°K}$

$$-r'_T = \frac{5.64 \times 10^{-8}(1-X)(2-X)}{3.02 + .88X}, \quad \frac{\text{gmoles T}}{\text{g of cat. sec}}$$

[†]For further information see D. Kunii and O. Levenspiel, *Fluidization Engineering*, (New York: John Wiley and Sons, 1969).

(F24)

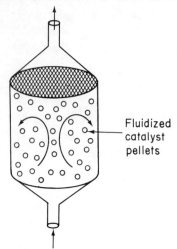

Figure 6–19. Fluidized Bed Reactor

For a well-mixed fluidized bed reactor (see Figure 6-19) in which there are no spatial variations in reaction rate, the CSTR-type design equation is

$$W = \frac{F_{A0}X}{-r'_A} \tag{F24-1}$$

The reactor volume is merely the catalyst weight divided by the bulk density of the catalyst.

$$V = \frac{W}{\rho_B} = \frac{F_{A0}X}{-r'_A \rho_B} = \frac{F_{A0}X}{-r_A} \tag{F24-2}$$

(F25)

$$W = \frac{F_{A0}X}{-r'_A}$$

Let $A \equiv T$:

$$F_{A0} = C_{A0}v_0 = \left(\frac{P_{A0}}{RT_0}\right)(v_0) = \left[\frac{2}{(.082)(873)}\right]400$$

$$F_{A0} = 11.175 \text{ gmoles toluene/min}$$

$$F_B = F_{A0}X = 10 \frac{\text{gmoles}}{\text{min.}}$$

The desired conversion is

$$X = \frac{10.0}{11.175} = .895$$

$$-r'_T = \frac{5.64 \times 10^{-8}(1-X)(2-X)}{3.02 + .88X} \times \frac{1000 \text{ g}}{1 \text{ kg}} \times \frac{60 \text{ sec}}{1 \text{ min}}$$

$$= \frac{3.384 \times 10^{-3}(1-X)(2-X)}{3.02 + .88X} \quad \frac{\text{gmole } T}{\text{kg of cat. min}}$$

$$X = .895$$

$$-r'_T = 1.03 \times 10^{-4} \frac{\text{gmole } T}{\text{kg of cat. min}}$$

$$W = \frac{F_{A0}X}{-r'_T} = \frac{10 \text{ gmole } T/\text{min}}{1.030 \times 10^{-4} \frac{\text{gmole } T}{\text{kg of cat. min}}}$$

$$= 97{,}087 \text{ kg of catalyst}$$

$$V = \frac{97{,}087 \text{ kg}}{.4 \frac{\text{kg}}{l}} = 2.43 \times 10^5 \, l$$

These quantities for the catalyst weight and reactor volume are quite high, especially for the low feed rates given. The catalyst weight could probably be reduced significantly by increasing the temperature in the bed.

In Frame (F26) calculate and compare the catalyst weight and volume for a tubular packed bed reactor with that for a fluidized bed reactor. The feed conditions to the tubular reactor are identical with those for the fluidized bed. The bulk density of the packed bed reactor is 2.3 g/cu cm.[†]

(F26)

[†]For further discussion on the catalytic sequence of adsorption, surface reaction, and desorption see
1. Chap. 7 of *Reaction Kinetics for Chemical Engineers*, by S. M. Walas, (New York: McGraw-Hill, 1959).
2. Chap. 9 of *Chemical Engineering Kinetics*, 2nd ed., by J. M. Smith, (New York: McGraw-Hill, 1970).
3. Chap. 19 of *Chemical Process Principles: Part 3–Kinetics and Mechanism* by O. A. Hougen and K. M. Watson, (New York: John Wiley and Sons, 1943).

(F26)

$$-r'_T = \frac{3.384 \times 10^{-3}(1-X)(2-X)}{3.02 + .88X}, \quad \frac{\text{gmole } T}{\text{kg of cat. min}}$$

X	$-r'_T$	$1/-r'_T$
.0	2.24×10^{-3}	446
.2	1.52×10^{-3}	656
.4	9.63×10^{-4}	1038
.6	5.34×10^{-4}	1872
.8	2.18×10^{-4}	4585
.9	$.98 \times 10^{-4}$	10240

$$W = F_{A0} \int_0^{0.895} \frac{dX}{-r'_T} = (11.175)(\text{Area})$$

Using the graphical form of Simpson's Rule we find

$$\text{Area} = 1957 \frac{\text{kg of cat. min}}{\text{gmole } T}$$

$$W = (11.175)(1957)$$
$$= 21{,}869 \text{ kg of catalyst}$$
$$V = 9508 \text{ liters}$$

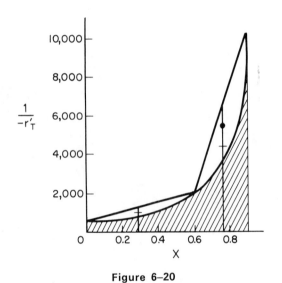

Figure 6–20

6.4–Houdry Hydrogenation Reactor

In addition to the tubular packed bed and fluidized bed reactors, other shapes and sizes of heterogeneous reactors are in common use. One example is the Houdry dehydrogenation reactor shown in Figure 6-21[†].

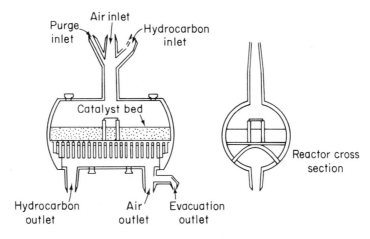

Figure 6–21. Houdry Dehydrogenation Reactor.

This type of configuration enables the reaction to be carried out with a very low pressure drop through the bed.[‡] One of the reactions which can be carried out in reactors of this type which are 10-20 ft in diameter and 20-40 ft long and may contain as much as 300,000 lb. of catalyst, involves the production of propylene from propane.

$$C_3H_8 \longrightarrow H_2 + C_3H_6$$

The reactor is operated at temperatures of about 1100°F and the superficial gas velocities at the entrance to the reactor range may be several hundred feet/second. Even so, the reactor configuration permits low bed velocities.

As the reaction proceeds, carbon (coke) is deposited on the catalyst causing it to become deactivated. The coke is removed through a regeneration process in which the propane feed line in the reactor is closed off while air is fed to the reactor to burn off the coke deposited on the catalyst, i.e.,

$$C + O_2 \longrightarrow CO_2$$

The excess air and carbon dioxide leave through an outlet at the bottom of the reactor different from the product outlet. After the catalyst has been regenerated the air flow through the bed is shut off, the reactor evacuated, and the propane feed resumed. The photograph of the plant in Figure 6-22 shows five reactors side by side. While two of the reactors are undergoing regeneration, the others are either producing propylene (i.e., they are onstream) or in transition between being onstream and regeneration.

The cost of the entire operation, including regeneration, product separation, cost of feed

---

[†]J. J. McKetta, ed., *Advances in Petroleum Chemistry and Refining*, Vol. 4, (New York: Wiley Interscience, 1961), Chap. 10.

[‡]A discussion of pressure drop through packed bed reactors can be found in *Reaction Kinetics for Chemical Engineers* by S. Walas, (New York: McGraw-Hill Book Co., 1959), Chap. 7.

Figure 6-22

stock (cents/gal),[†] is about 2 cents/lb of propylene produced. The propylene usually sells for about 3 cents/lb, yielding a payout for the plant of approximately 3.5 years.

The formation of carbon deposits on the catalyst particle is a type of catalyst deactivation referred to as *fouling*. The other common types of deactivation are *ageing* and *poisoning*. A thorough discussion of catalyst deactivation can be found in Levenspiel[‡] and in a recent review by Butt.[§]

6.5–Classification of Catalysts

To conclude our discussion of the sequence of steps in catalytic reactions (i.e., adsorption, surface reaction, and desorption), we shall briefly discuss several classes of reactions and the catalysts used in each class.[¶]

6.5A Alkylation and Dealkylation Reactions

Alkylation is the addition of an alkyl group to an organic compound. This type of reaction is commonly carried out in the presence of the Friedel-Crafts catalysts, $AlCl_3$ along with a trace of HCl. One such reaction is

$$C_4H_8 + i\text{-}C_4H_{10} \xrightleftharpoons{AlCl_3} i\text{-}C_8H_{18}$$

A similar alkylation is the formation of ethyl benzene from benzene and ethylene.

$$C_6H_6 + C_2H_4 \longrightarrow C_6H_5C_2H_5$$

[†]February 1971 prices.
[‡]O. Levenspiel, *Chemical Reaction Engineering*, 2nd ed., (New York: John Wiley and Sons, 1972), Chap. 15.
[§]J. Butt, *Advances in Chemistry*, *109* (1972).
[¶]J. H. Sinfelt, *I.E.C.* 62 No. 2, (1970), 23. J. H. Sinfelt, *I.E.C.* 62 No. 10, (1970), 66. W. B. Innes, *Catalysis*, vol. 2, ed. P. H. Emmett, New York: Reinhold Publishing Co., (1955), p. 1.

The cracking of petroleum products is probably the most common dealkylation reaction. Silica-alumina, silica-magnesia, and a clay (montmorillonite) are common dealkylation catalysts.

6.5B Isomerization Reactions

In petroleum production, the conversion of normal hydrocarbon chains to branched chains is important, since the latter has a higher octane number. Acid-promoted $AlCl_3$ is a catalyst used in such isomerization reactions. Although this and other acid catalysts are used in isomerization reactions, it has been found that the conversion of normal paraffins to isoparaffins is easiest when both acid sites and hydrogenation sites are present.

What similarities and what differences exist in the catalysts used for isomerization, alkylation, and dealkylation reactions?

(F27)

6.5C Hydrogenation and Dehydrogenation Reactions

The bonding strength between hydrogen and metal surfaces increases with an increase in vacant d-orbitals. Maximum catalytic activity will not be realized if the bonding is too strong and the products are not readily released. Consequently, this maximum in catalytic activity occurs when there is approximately one vacant d-orbital per atom. The most active metals for reactions involving hydrogen are generally Co, Ni, Rh, Pd, Ir, and Pt. On the other hand, V, Cr, Cb, Mo, Ta, and W, each of which has a large number of vacant d-orbitals, are relatively inactive as a result of the strong adsorption for the reactants or the products or both. However, the oxides of Mo, (MoO_2), and Cr, (Cr_2O_3), are quite active for most reactions involving hydrogen. Dehydrogenation reactions are favored at high temperatures (at least 600°C), and hydrogenation reactions are favored at lower temperatures. Industrial butadiene, which has been used to produce synthetic rubber, can be obtained by the dehydrogenation of the butenes:

$$CH_3CH=CHCH_3 \xrightarrow{cat} CH_2=CHCH=CH_2 + H_2$$

(possible catalysts: calcium nickel phosphate, Cr_2O_3, etc.)

These same catalysts could also be used in the dehydrogenation of ethyl benzene to form styrene:

$$\phi CH_2CH_3 \xrightarrow{cat} H_2 + \phi CH=CH_2$$

An example of cyclization, which may be considered to be a special type of dehydrogenation, is the formation of cyclohexane from *n*-hexane and is given by Equation (6-2). List three hydrogenation catalysts and state at what temperatures they would become dehydrogenation catalysts.

(F27) In both alkylation and dealkylation reactions, $AlCl_3$ is used as a catalyst. Dealkylation, or cracking, usually requires high temperatures. Various isomerization reactions can be carried out over acid-promoted $AlCl_3$.

(F28)

6.5D Oxidation Reactions

The transition group elements (Group VIII) and Subgroup I are used extensively in oxidation reactions. Ag, Cu, Pt, Fe, Ni, and their oxides are generally good oxidation catalysts. In addition, V_2O_5 and MnO_2 are frequently used for oxidation reactions. A few of the principal types of catalytic oxidation reactions are:

1. Oxygen addition:

$$2C_2H_4 + O_2 \xrightarrow{Ag} 2C_2H_4O$$

$$2SO_2 + O_2 \xrightarrow{V_2O_5} 2SO_3$$

$$2CO + O_2 \xrightarrow{Cu} 2CO_2$$

2. Oxygenolysis of carbon-hydrogen bonds:

$$2C_2H_5OH + O_2 \xrightarrow{Cu} 2CH_3CHO + 2H_2O$$

$$2CH_3OH + O_2 \xrightarrow{Ag} 2HCHO + 2H_2O$$

3. Oxygenation of nitrogen-hydrogen bonds:

$$5O_2 + 4NH_3 \xrightarrow{Pt} 4NO + 6H_2O$$

4. Complete combustion:

$$2C_2H_6 + 7O_2 \xrightarrow{Ni} 4CO_2 + 6H_2O$$

What catalyst(s) can be used for both oxidation reactions and hydrogenation reactions?

(F29)

6.5E Hydration and Dehydration Reactions

Hydration and dehydration catalysts have a strong affinity for water. One such catalyst is Al_2O_3, which is used in the dehydration of alcohols to form olefins. For reactions of this type, the following sequence (in which the surface reaction is usually rate controlling) has been suggested:

$$A = \text{alcohol},\ S = \text{site},\ 0 = \text{olefin},\ W = \text{water}$$

(F28)
$$Co, Ni, Pt, Cr_2O_3.$$
Dehydrogenation usually occurs at temperatures greater than 600°C for straight chain hydrocarbons.

(F29) Pt and Ni.

Adsorption: $\quad A + S \rightleftharpoons A \cdot S$
Surface reaction: $\quad A \cdot S + S \rightleftharpoons 0 \cdot S + W \cdot S$
Desorption of:
1. olefins: $\quad 0 \cdot S \rightleftharpoons 0 + S$
2. water: $\quad W \cdot S \rightleftharpoons W + S$

In addition to alumina, silica-alumina gels, clays, phosphoric acid, and phosphoric acid salts on inert carriers have also been used for hydration-dehydration reactions. An example of an industrial catalytic hydration reaction is the synthesis of ethanol from ethylene:

$$CH_2{=}CH_2 + H_2O \longrightarrow CH_3CH_2OH$$

6.5F Halogenation and Dehalogenation Reactions

Usually reactions of this type take place readily without utilizing catalysts. However, when selectivity of the desired product is low or it is necessary to run the reaction at a lower temperature, the use of a catalyst is desirable. Supported copper and silver halides can be used for the halogenation of hydrocarbons. Hydrochlorination reactions can be carried out with mercury, copper, or zinc halides.

What catalysts and intermediate products would you use to form styrene from an equimolar mixture of ethylene and benzene?

(F30)

For each type of reaction or specific reaction listed in Frame (F31), fill in which catalyst (or catalysts) you would use.

(F31)

Reaction: Catalyst

1. Halogenation-dehalogenation   ___ ___ ___ ___
2. Hydration-dehydration   ___ ___ ___ ___
3. Alkylation-dealkylation   ___ ___ ___ ___
4. Hydrogenation-dehydrogenation ___ ___ ___ ___
5. Oxidation   ___ ___ ___ ___
6. Isomerization   ___ ___ ___ ___

$AlCl_3$	Al_2O_3	MnO_2	Co	Cu	Pt
$CuCl_2$	Cr_2O_3	MoO_2	W	Ni	U_2O_5
AgCl	Cr	Ag	Rh	V_2O_5	Pd

(F30) First we could carry out an alkylation reaction to form ethyl benzene, which is then dehydrogenated to form styrene. We will need both an alkylation catalyst and a dehydrogenation catalyst:

$$C_2H_4 + C_6H_6 \xrightarrow[\text{trace HCl}]{\text{AlCl}_3} C_6H_5C_2H_5 \xrightarrow{\text{Ni}} C_6H_5CH=CH_2 + H_2$$

(F31)

1. Halogenation-dehalogenation	$CuCl_2$	AgCl		
2. Hydration-dehydration	Al_2O_3			
3. Alkylation-dealkylation	$AlCl_3$			
4. Hydrogenation-dehydrogenation	Co	Pt	Cr_2O_3	Ni
5. Oxidation	Cu	Ag	Ni	V_2O_5
6. Isomerization	$AlCl_3$			

SUMMARY

1. Types of adsorption:
   a. chemisorption.
   b. physical adsorption.

2. The Langmuir isotherm relating the concentration of species A on the surface to the partial pressure of A in the gas phase is

$$C_{A \cdot S} = \frac{K_A C_t P_A}{1 + K_A P_A} \tag{S6-1}$$

3. The different types of catalysts used in accelerating the rates of halogenation, hydration, hydrogenation, oxidation, and isomerization reactions were discussed.

4. The sequence of steps for solid catalyzed isomerization is

$$A \longrightarrow B \tag{S6-2}$$

   a. Mass transfer of A from the bulk fluid to the external surface of the pellet.
   b. Diffusion of A into the interior of the pellet.
   c. Adsorption of A onto the catalytic surface.
   d. Surface reaction of A to form B.
   e. Desorption of B from the surface.
   f. Diffusion of B from the pellet interior to the external surface.
   g. Mass transfer of B away from the solid surface to the bulk fluid.

5. Assuming that mass transfer is not rate limiting, the rate of adsorption is

$$r_{ADS} = k_A \left(C_v P_A - \frac{C_{A \cdot S}}{K_A} \right) \tag{S6-3}$$

   The rate of surface reaction is

$$r_S = k_S \left(C_{A \cdot S} - \frac{C_{B \cdot S}}{K_S} \right) \tag{S6-4}$$

   The rate of desorption is

$$r_D = k_D \left(C_{B \cdot S} - \frac{P_B C_v}{K_D} \right) \tag{S6-5}$$

   At steady state,

$$r_{ADS} = r_S = r_D \tag{S6-6}$$

6. If there are no inhibitors present, the total concentration of sites is

$$C_t = C_v + C_{A \cdot S} + C_{B \cdot S} \tag{S6-7}$$

7. If we assume that the surface reaction is rate limiting, we set

$$\frac{r_D}{k_D} \simeq 0, \qquad \frac{r_{ADS}}{k_A} \simeq 0$$

   and solve for $C_{A \cdot S}$ and $C_{B \cdot S}$ in terms of P_A and P_B. After substitution of these quantities in Equation (S6-4), the concentration of vacant sites is eliminated with the aid of Equation (S6-7):

$$-r'_A = r_S = \frac{C_t k_S K_A \left(P_A - \frac{P_B}{K_e} \right)}{\left(1 + K_A P_A + \frac{P_B}{K_D} \right)} \tag{S6-8}$$

$$-r'_A = \frac{C_t k_S K_A \left(P_A - \frac{P_B}{K_e} \right)}{(1 + K_A P_A + K_B P_B)} \tag{S6-9}$$

Recall that the equilibrium constant for desorption of species B is the reciprocal of the equilibrium constant for the adsorption of species B, i.e., $\quad K_B = \dfrac{1}{K_D} \tag{S6-10}$

QUESTIONS AND PROBLEMS

P6-1. In your plant, the reversible isomerization of compound A, A$\rightleftharpoons$B, is carried out over a supported metal catalyst in an isothermal fixed bed flow reactor.

A and B are liquid at the process conditions, and volume change on reaction is negligible. The equilibrium constant for the reaction is 8.5 at 350°F and 6.0 at 400°F. The catalyzed reaction is pseudo-first order in A, with an apparent Arrhenius activation energy of 26,900 Btu/lb-mole.

Side reactions have a negligible effect on yield, but they slowly deactivate the catalyst. To increase the reaction rate and to partly offset deactivation, the temperature is raised on a schedule from 350 to 400°F. At the end of an operating cycle, the catalyst is dumped and replaced.

The nominal fresh feed of A is 300 gal/hr, measured at 90°F. Unconverted A is separated from the reactor product and recycled to the feed. Total feed rate thus depends on catalyst activity.

The ACC Corporation offers an alternative, in which catalyst is to be regenerated in place by treatment with a solvent of proprietary composition. The cost is attractively low compared to catalyst replacement. ACC guarantees to achieve at least (1) 90% of fresh catalyst activity or (2) 2.25 times the activity of spent catalyst, whichever is less.

Following the use of the ACC regeneration procedure a test run was made. Unfortunately, the feed rate was limited by trouble in another part of the plant.

You are asked to review the following data and determine whether the guarantee was met. Support your conclusions by appropriate calculations.

Catalyst	Temp (°F)	Fresh feed rate (gal/hr)	Concentration of A in reactor product (%)
Fresh	350	310	25.0
Spent	400	300	32.2
Regenerated	350	220	19.0

(Calif. PECEE, August 1969)

P6-2. For the heterogeneous gas-solid reaction

$$A(g) + S \rightleftharpoons A \cdot S$$
$$A \cdot S + S \rightleftharpoons B \cdot S + S$$
$$B \cdot S \rightleftharpoons B(g) + S$$

plot the initial rate, $-r_0$, as a function of the initial total pressure Π_0 for a given initial mole fraction if:

a. The adsorption A is the controlling step.
b. The surface reaction is the controlling step.
c. The desorption of B is the controlling step.

Initially there is no product in the system.

Repeat parts a, b, and c, assuming that there is an inhibitor present in the system.

P6-3. The following heterogeneous reaction is taking place in a system in which there is an inert adsorbing gas (inhibitor) present:

$$A(g) + S \rightleftharpoons A \cdot S$$
$$A \cdot S + B(g) \rightleftharpoons C \cdot S$$
$$C \cdot S \rightleftharpoons C(g) + S$$

Assuming that the adsorbtion of A is controlling, sketch the initial reaction rate as a function of the total pressure.
[Quiz U of M]

P6-4. The initial-rate total-pressure data for the heterogeneous catalytic reaction

$$A + B \underset{}{\overset{cat}{\rightleftharpoons}} R$$

is given in the following table:

Π_0 (atm)	$-r'_0$ (gmole/g of cat./hr)
.25	.00004
.4	.00015
1.0	.00084
4.0	.00890
10.0	.03333
40.0	.17800
60.0	.27800
80.0	.37700

The following reaction sequence is believed to be the mechanism by which reaction proceeds:

$$A(g) + S \rightleftharpoons A \cdot S$$
$$A \cdot S + B(g) \rightleftharpoons R \cdot S$$
$$R \cdot S \rightleftharpoons R(g) + S$$

Determine the tubular reactor catalyst weight and volume and also the fluidized bed reactor volume (CSTR) and catalyst weight necessary to achieve 50% conversion. The volumetric flow rate to each reactor is 60 l/min at 300°K and 10 atm. The feed consists of 25% A, 50% B, and 25% inerts; however, the inerts do not adsorb onto the surface.

Additional information: The bulk density of the catalyst in the tubular packed bed reactor is 2.3 g/cu cm, and the bulk density in the fluidized bed is .7 g/cu cm. The change in the Gibbs free energy for the reaction is −750 cal./gmole of A. Assume that the equilibrium constant for the adsorption of A, K_A, is approximately equal to the equilibrium constant for the adsorption of R, K_R.

P6-5. The dehydration of *n*-butyl alcohol (butanol) over an alumina-silica catalyst was investigated by J. F. Maurer.[†] The data in Figure P6-5 were obtained at 750°F. in a modified differential reactor. The feed consisted of pure butanol.

a) Suggest a mechanism and rate-controlling step that is consistent with the experimental data.

b) If a catalyst poison is inadvertently passed through the bed and poisons two-thirds of the active catalyst sites, by what percent must the entering flow rate of butanol be reduced in order to maintain the same outlet conversion before poisoning occurred in:
  1) packed bed reactor
  2) fluidized bed reactor?

[†] J. F. Maurer, The Effect of Pressure on the Dehydration of Butanol 1 over an Alumina-Silica Catalyst. (Ph.D. thesis, University of Michigan, 1950).

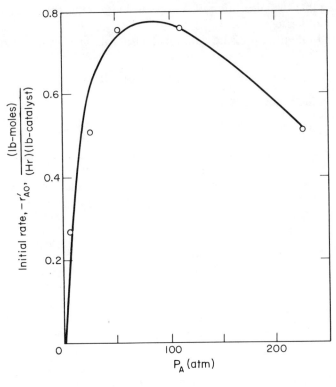

Figure P6–5

P6-6. The dehydrogenation of ethylbenzene is one of the most important methods for the manufacture of styrene. This reaction has been studied using a Shell 105 catalyst (95 wt. % Fe_2O_3, 5% Cr_2O_3, 2% KOH) in a differential reactor [*I.E.C. Process Design Development* 4, (1965), 281]. It was observed that the rate of reaction was unaffected by the concentration of hydrogen but decreased when styrene was added to the feed stream. As the partial pressure of ethylbenzene is increased to moderate values, the rate of reaction becomes independent of the partial pressure of ethylbenzene. Suggest a mechanism that is consistent with the experimental observations and derive the corresponding rate law. Using the information from the following table,

	T = 630°C		
$-r_A \times 10^3$ (gmole/min/g of cat.)	1.415	.214	.166
P_{EtB}, atm	1.0	.01	1.0
P_{Sty}, atm	.0	.0	1.0

evaluate all constants in your model and then determine the catalyst weight necessary to produce 2000 kg of styrene per day in (a) a fluidized bed reactor and (b) a fixed bed reactor. Steam is used as a diluent in the feed stream, which consists of 75 wt. % H_2O and 25 wt. % ethylbenzene. The conversion of ethylbenzene is to be 33%.

P6-7. The irreversible reaction

$$A_2 \longrightarrow 2B$$

takes place in a fluidized bed according to the following mechanism in which the surface reaction is rate limiting:

$$A_2(g) + 2S \rightleftharpoons 2A \cdot S \qquad K_A = 10 \text{ atm}^{-1}$$
$$A \cdot S \longrightarrow B \cdot S \qquad k_S = .1 \text{ hr}^{-1}$$
$$B \cdot S \rightleftharpoons B(g) + S \qquad K_D = 1 \text{ atm}$$
$$C_t = .01 \text{ gmole/g of cat.}$$

If the feed consists of 40% A_2, 20% inerts (nonadsorbing), and 40% B, determine the total pressure at which the rate will be at a maximum when 50% conversion is achieved in the reactor. If the entering molar flow rate of A is 10 gmoles of A/hr, determine the corresponding minimum catalyst weight.

P6-8. To remove oxides of nitrogen (assumed to be NO) from automobile exhaust, a scheme has been proposed that uses unburned carbon monoxide (CO) in the exhaust to reduce the NO over a solid catalyst, according to the reaction

$$CO + NO \longrightarrow \text{products } (N_2, CO_2)$$

Experimental data for a particular solid catalyst indicate that the reaction rate can be well represented over a large range of temperatures by

$$r = \frac{kP_N P_C}{(1 + K_1 P_N + K_2 P_C)^2}$$

where
P_N = gas phase partial pressure of NO,
P_C = gas phase partial pressure of CO, and
k, K_1, K_2 = coefficients depending only on temperature

a. Based on your experience with other such systems, you are asked to propose an adsorption/surface reaction/desorption mechanism that will explain the experimentally observed kinetics.
b. A certain engineer thinks that it would be desirable to operate with a very large stoichiometric excess of CO in order to minimize catalytic reactor volume. Do you agree or disagree? Explain briefly.

P6-9. The use of a differential reactor to study the formation of methane from hydrogen and carbon monoxide over a nickel catalyst was discussed in Chapter 5. Suggest a mechanism and rate-limiting step that is consistent with the experimental observation. It is desired to produce 2 tons/day of CH_4. Calculate the catalyst weights necessary to achieve 80% conversion in
a. fixed bed and
b. fluidized bed.

The feed consists of 75% H_2 and 25% CO at a temperature of 500°F and a pressure of 10 atms. Assume both molecular and atomic hydrogen are absorbed on the surface. See page 163.

P6-10. The catalytic oxidation of ethanol has been carried out over a tantalum oxide catalyst. [*J. of Catalysis* 25, (1972) 194]. The overall reaction

$$CH_3CH_2OH + \tfrac{1}{2}O_2 \longrightarrow CH_3CHO + H_2O$$

is believed to proceed by the following mechanism in which ethanol is adsorbed on one type of site, S, and oxygen is adsorbed on a different type of site, S'. Ethanol undergoes dissociative adsorption on two type S sites:

$$C_2H_5OH + 2S \rightleftharpoons C_2H_5O \cdot S + H \cdot S$$

oxygen also undergoes dissociative adsorption on two type S' sites (presumably metal ions):

$$O_2 + 2S' \rightleftharpoons 2O \cdot S'$$

a. Assuming that the rate limiting step is

$$C_2H_5O \cdot S + O \cdot S' \longrightarrow C_2H_4O + OH \cdot S' + S$$

and that the adsorbed hydroxyl may be desorbed into water by

$$OH \cdot S' + H \cdot S \rightleftharpoons H_2O + S + S'$$

show that the initial rate law is

$$-r_{A0} = \frac{kP_A^{1/2}P_{O_2}^{1/2}}{(1 + K_{O_2}P_{O_2}^{1/2})(1 + 2K_A^{1/2}P_A^{1/2})} \quad \text{(P6-10-1)}$$

Note: $C_{A \cdot S} = C_{H \cdot S}$

where $A \equiv C_2H_5OH$, $A \cdot S \equiv C_2H_5O \cdot S$, $w \equiv H_2O$.

b. Assuming that the reaction is irreversible, show that

$$-r_A = \frac{-r_{A0}}{\left[1 + \dfrac{\mu K_w P_w}{(K_A P_A)(1 + K_{O_2} P_{O_2}^{1/2})}\right]^{1/2}}$$

where $-r_{A0}$ is given by Equation (P6-10-1) and μ is the ratio of the total number of sites of type S to the total number of sites of type S'.

c. For an equal molar mixture of oxygen and ethanol, calculate the catalyst weight necessary to achieve 40% conversion of oxygen in a fluidized bed reactor operated at 300°C and 1 atm.

   Additional Information:

   $k = 1.61 \times 10^{-5}$ gmole/hr/gmcat/torr

   $K_A^{1/2} = .048(\text{torr})^{-1/2}$

   $K_{O_2} = .109(\text{torr})^{-1/2}$

   Take $\mu = 1$ (1 torr = 1 mm Hg)

SUPPLEMENTARY READING

1. A discussion of heterogeneous catalytic mechanisms and rate controlling steps may be found in Chapter 9 of

   SMITH, J. M., *Chemical Engineering Kinetics*, 2nd ed. New York: McGraw-Hill, Inc., 1970.

   or in Chapter 19 of

   HOUGEN, O. A., and K. M. WATSON, *Chemical Process Principles: Part 3, Kinetics and Catalysis.* New York: John Wiley and Sons, 1947.

   or in Chapter 4 of

   BOUDART, M., *Kinetics of Chemical Processes.* Englewood Cliffs: Prentice-Hall, Inc., 1968.

2. A discussion on the types and rates of adsorption along with techniques used in measuring catalytic surface areas is presented in Chapter 8 of

   SMITH, J. M. *Chemical Engineering Kinetics*, 2nd ed. New York: McGraw-Hill, Inc., 1970.

   The properties of catalytic surfaces are discussed in

   PARRAVANO, G. *Ind. Eng. Chem*, 58, No. 10, (1966) 45.

   HAYWARD, D. O. and B. M. W. TRAPNELL, *Chemisorption*, 2nd ed. London: Butterworth, 1964.

   ROSS, S. and J. P. OLIVIER, *On Physical Adsorption.* New York: Wiley Interscience, 1964.

   A number of the topics discussed in this chapter can also be found in

   THOMAS, J. M. and W. J. THOMAS, *Introduction to the Principles of Heterogeneous Catalysis.* New York: Academic Press, Inc., 1967.

3. A discussion of the types of catalysis, methods of catalyst selection, method of preparation, and classes of catalysts can be found in Chapter 9 of

WALAS, S. M., *Reaction Kinetics for Chemical Engineers*, New York: McGraw-Hill, Inc, 1959.

and in an article by

DEMAIO, D. and A. NAGLIERI, *Chemical Engineering*, 75, No. 16, (1968) 127.

Some broad principles for classifying catalysts have been suggested by

COUGHLIN, R. W., *I.E.C.* 59, No. 9, (1967), 45.

and by

INNES, W. B., *Catalysis II*, p. 1, ed. P. H. EMMETT. New York: Reinhold Publishing Co., 1955.

4. A recent advance in heterogeneous catalysis can be found in

SINFELT, J. H., *I.E.C.* 62, No. 2, (1970), 23.

SINFELT, J. H., *I.E.C.* 62, No. 10, (1970), 66.

and in the following journals:

Advances in Catalysis
Journal of Catalysis
Catalysis Reviews

5. An introductory discussion of homogeneous catalysis is presented in Chapter 9 of

FROST, A. A. and RALPH G. PEARSON, *Kinetics and Mechanisms*, 2nd ed. New York: John Wiley and Sons, 1961.

and Chapter 6 of

STEVENS, B., *Chemical Kinetics*, 2nd ed. London: Chapman and Hall, 1970.

A more advanced presentation is given in Chapters 15 and 16 of

BENSON, S. W., *Foundations of Chemical Kinetics*, New York: McGraw-Hill, Inc., 1960.

7
Nonelementary Homogeneous Reactions

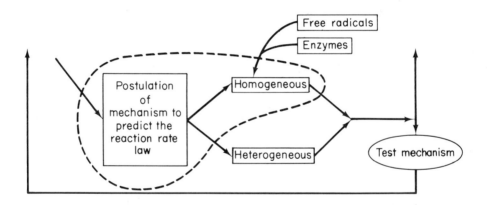

Until now, we have been discussing homogeneous reaction rate laws in which the concentration is raised to some power n, which is an integer. That is, the rate law (i.e. kinetic rate expression) is

$$-r_A = kC_A^n \tag{7-1}$$

where $n = 0, 1, 2, 3$

We said that if $n = 1$, the reaction is first order w.r.t. A, and that if $n = 2$, the reaction is second order w.r.t. A, etc. However, a large number of homogeneous reactions involve the formation and subsequent reaction of an intermediate species. When this is the case it is not uncommon to find a reaction order that is not an integer. For example, the rate law for the decomposition of acetaldehyde,

$$CH_3CHO \longrightarrow CH_4 + CO \tag{3-26}$$

243

at approximately 500°C, is

$$-r_{CH_3CHO} = kC_{CH_3CHO}^{3/2} \qquad (7\text{-}2)$$

Another common form of the rate law resulting from reactions involving active intermediates is one in which the rate is directly proportional to the reactant concentration and inversely proportional to the sum of a constant and the reactant concentration. An example of this type of kinetic expression is observed for the formation of hydrogen iodide,

$$H_2 + I_2 \longrightarrow 2HI$$

The rate law for this reaction is

$$r_{HI} = \frac{k_1 k_3 C_{I_2} C_{H_2}}{k_2 + k_3 C_{H_2}} \qquad (7\text{-}3)$$

For rate expressions similar or equivalent to those given by Equation (7-3), reaction orders cannot be defined. That is, for rate laws where the denominator is a polynomial in the species concentrations, reaction orders are described only for the limiting values of reactant and/or product concentrations. Reactions of this type are *nonelementary* in that there is no correspondence between reaction order and stoichiometry.

7.1–Fundamentals

To illustrate the principles that delineate the manner in which rate laws similar to those in Equations (7-2) and (7-3) are formed, we shall first consider the gas phase decomposition of azomethane, AZO, to give ethane, Et, and nitrogen, N_2:

$$(CH_3)_2N_2 \longrightarrow C_2H_6 + N_2 \qquad (7\text{-}4)$$

Experimental observations show that the rate law for N_2 is first order w.r.t. AZO at pressures greater than 1 atm (relatively high concentrations);

$$r_{N_2} \propto C_{AZO}$$

and to be second order at pressures below 50 mm Hg (low concentrations);[†]

$$r_{N_2} \propto C_{AZO}^2$$

7.1A Active Intermediates

This apparent change in reaction order can be explained by the theory developed by Lindemann.[‡] An activated molecule, $[(CH_3)_2 N_2]^*$, results from the collision or interaction between two azomethane molecules, i.e.,

$$(CH_3)_2N_2 + (CH_3)_2N_2 \xrightarrow{k_1} (CH_3)_2N_2 + [(CH_3)_2N_2]^* \qquad (7\text{-}5)$$

This activation can occur when translational kinetic energy is transferred into energy stored in internal degrees of freedom, particularly vibrational degrees of freedom.[§] An unstable molecule, i.e., active intermediate, is *not* formed solely as a consequence of the molecule moving at a high velocity, (high translational kinetic energy). For a reaction to occur, the energy must be absorbed into the chemical bonds where high amplitude oscillations will lead to bond ruptures, mole-

[†]H. C. Ramsperger, *J. Am. Chem. Soc.* 49, (1927), 912.
[‡]F. A. Lindemann, *Trans. Faraday Soc.* 17, (1922), 598.
[§]W. J. Moore, *Physical Chemistry*, 2nd ed., (Englewood Cliffs, N.J.: Prentice-Hall, Inc. 1955).

cular rearrangement, and decomposition. In the absence of photochemical effects or similar phenomena, this transfer of translational energy to vibrational energy to produce an active intermediate can occur only as a consequence of molecular collision or interaction. Other types of active intermediates which can be formed are free radicals (one or more unpaired electrons, e.g., H·), ionic intermediates (e.g., carbonium ion), and enzyme-substrate complexes, to mention a few.

In Lindemann's theory of active intermediates, decomposition does not occur instantaneously after internal activation of the molecule; rather, there is a time lag during which the species remains activated. There are two reaction paths which the active intermediate (activated complex) may follow after being formed by the reaction

$$(CH_3)_2N_2 + (CH_3)_2N_2 \xrightarrow{k_1} (CH_3)_2N_2 + [(CH_3)_2N_2]^* \tag{7-5}$$

In one path the activated molecule may become deactivated through collision with another molecule,

$$[(CH_3)_2N_2]^* + (CH_3)_2N_2 \xrightarrow{k_2} (CH_3)_2N_2 + (CH_3)_2N_2 \tag{7-6}$$

which is just the reverse reaction of that given by Equation (7-5). In the alternative path the active intermediate decomposes spontaneously to form ethane and nitrogen.

$$[(CH_3)_2N_2]^* \xrightarrow{k_3} C_2H_6 + N_2 \tag{7-7}$$

In our analysis of nonelementary reactions we assume that the individual reactions in the mechanism are elementary. That is, the overall reaction, Equation (7-4), for which the rate expression is nonelementary, consists of the sequence of elementary reactions, Equations (7-5), (7-6), and (7-7). Using this assumption, write rate laws for
1. The rate of formation of nitrogen
2. The rate of disappearance of azomethane.

(G1)

The concentration of the active intermediate, AZO*, is very difficult to measure, since it is highly reactive and very short lived. Consequently, evaluation of the reaction rate expressions, (G1-1) and (G1-3), in their present forms becomes quite difficult, if not impossible. To overcome this difficulty we need to express the concentration of the active intermediate, C_{AZO^*}, in terms of the concentration of azomethane, C_{AZO}. As mentioned in Chapter 3, the total or net rate of formation of a particular species involved in many simultaneous reactions is the sum of the rates of formation of each individual reaction for that species. For example, in the elementary reaction sequence

$$\begin{aligned} A &\xrightarrow{k_1} B \quad &\text{Reaction 1} \\ C &\xrightarrow{k_2} B \quad &\text{Reaction 2} \\ D &\xrightarrow{k_3} B \quad &\text{Reaction 3} \end{aligned} \tag{7-8}$$

(G1)

Since the reaction given by Equation (7-7) is elementary, we can write

$$r_{N_2} = k_3 C_{AZO^*} \tag{G1-1}$$

Azomethane is depleted by the reaction in Equation (7-5) and is formed from the reverse reaction, Equation (7-6). Therefore, the net rate of reaction for azomethane,

$$AZO + AZO \underset{k_2}{\overset{k_1}{\rightleftharpoons}} AZO^* + AZO \tag{G1-2}$$

is given by the kinetic expression

$$-r_{AZO} = k_1 C_{AZO}^2 - k_2 C_{AZO} C_{AZO^*} \tag{G1-3}$$

where
$$AZO \equiv [(CH_3)_2 N_2]$$

the net rate of formation of B, r_B, is

$$r_B = \begin{pmatrix} \text{Rate of} \\ \text{formation} \\ \text{of B from} \\ \text{reaction 1} \end{pmatrix} + \begin{pmatrix} \text{Rate of} \\ \text{formation} \\ \text{of B from} \\ \text{reaction 2} \end{pmatrix} + \begin{pmatrix} \text{Rate of} \\ \text{formation} \\ \text{of B from} \\ \text{reaction 3} \end{pmatrix} \quad (7\text{-}9)$$

$$= r_{B_1} + r_{B_2} + r_{B_3}$$

Since each reaction is elementary, the rate law follows directly from Equation (7-8)

$$r_B = k_1 C_A + k_2 C_C + k_3 C_D$$

We can generalize the rate of formation of species j occurring from n different reactions as

$$r_j = \sum_{i=1}^{n} r_{ji}. \quad (7\text{-}10)$$

Write the net rate of formation of the active intermediate AZO* for the azomethane decomposition reaction sequence given in Equations (7-5), (7-6), and (7-7).

(G2)

7.1B Pseudo-Steady-State Hypothesis (PSSH)

In most instances it is not possible to eliminate the concentration of the active intermediate in the differential forms of the mole balance equations to obtain closed form solutions. However, approximate solutions may be obtained. The active intermediate molecule has a very short life time because of its high reactivity (i.e., large specific reaction rates). We shall also consider it to be present only in low concentrations. These two conditions lead to the psuedo-steady-state approximation, in which the rate of formation of the active intermediate is assumed to be equal to its rate of disappearance.[†] As a result, the net rate of formation of the active intermediate, r^*, is zero:

$$r^* = 0 \quad (7\text{-}11)$$

In Frame (G1) we found that the rate of formation of the product, nitrogen, was

$$r_{N_2} = k_3 C_{AZO^*}. \quad (G1\text{-}1)$$

and in Frame (G2) we foun d that the rate of formation of AZO* was

$$r_{AZO^*} = k_1 C_{AZO}^2 - k_2 C_{AZO} C_{AZO^*} - k_3 C_{AZO^*}. \quad (G2\text{-}7)$$

Using the pseudo-steady-state hypothesis (*P.S.S.H.*), combine Equations (G2-7) and (G1-1) to obtain a rate expression for N_2 solely in terms of the concentration of azomethane. Then show how the resulting rate law approximates first-order kinetics at high concentrations of azomethane and second-order kinetics at low concentrations.

[†]For further elaboration on this section, see Rutherford Aris, *American Scientist* 58, (1970), 419.

(G2) Since the active intermediate AZO* is present in all three reactions in the decomposition mechanism, the net rate of formation of AZO* is the sum of the rates of each of the reaction Equations (7-5), (7-6,) and (7-7), i.e.,

$$\begin{Bmatrix}\text{Net rate} \\ \text{of} \\ \text{formation} \\ \text{of AZO*}\end{Bmatrix} = \begin{Bmatrix}\text{Rate of} \\ \text{formation} \\ \text{of AZO* in} \\ \text{Equation (7-5)}\end{Bmatrix} + \begin{Bmatrix}\text{Rate of} \\ \text{formation} \\ \text{of AZO* in} \\ \text{Equation (7-6)}\end{Bmatrix} + \begin{Bmatrix}\text{Rate of} \\ \text{formation} \\ \text{of AZO* in} \\ \text{Equation (7-7)}\end{Bmatrix} \quad \text{(G2-1)}$$

$$r_{AZO^*} = r_{AZO^* \, (7\text{-}5)} + r_{AZO^* \, (7\text{-}6)} + r_{AZO^* \, (7\text{-}7)} \quad \text{(G2-2)}$$

Since each reaction is elementary for reaction (7-5),

$$2\text{AZO} \xrightarrow{k_1} \text{AZO} + \text{AZO*} \quad (7\text{-}5)$$

the rate expression is

$$r_{AZO^* \, (7\text{-}5)} = k_1 C_{AZO}^2 \quad \text{(G2-3)}$$

For reaction (7-6),

$$\text{AZO} + \text{AZO*} \xrightarrow{k_2} 2\text{AZO} \quad (7\text{-}6)$$

$$-r_{AZO^* \, (7\text{-}6)} = k_2 C_{AZO} C_{AZO^*} \quad \text{(G2-4)}$$

or, multiplying by -1,

$$r_{AZO^* \, (7\text{-}6)} = -k_2 C_{AZO} C_{AZO^*} \quad \text{(G2-5)}$$

For reaction (7-7),

$$\text{AZO*} \longrightarrow \text{N}_2 + \text{ethane} \quad (7\text{-}7)$$

$$r_{AZO^* \, (7\text{-}7)} = -k_3 C_{AZO^*} \quad \text{(G2-6)}$$

Substituting Equations (G2-3), (G2-5), and (G2-6) into Equation (G2-2), we obtain

$$r_{AZO^*} = k_1 C_{AZO}^2 - k_2 C_{AZO} C_{AZO^*} - k_3 C_{AZO^*} \quad \text{(G2-7)}$$

(G3)

We have seen that the rate law developed for the mechanism

$$2A \underset{k_2}{\overset{k_1}{\rightleftharpoons}} A + A^*$$
$$A^* \xrightarrow{k_3} B + C \tag{7-12}$$

can explain the apparent first-order dependence of the reaction rate on concentration at high reactant concentrations, as well as the apparent second-order dependence on concentration at low reactant concentrations.

In the following reaction scheme, two A molecules react to form an energized dimer, which is an active intermediate. This dimer either decomposes back to two A molecules or reacts with a molecule of species C to give the desired product D.

$$2A \underset{k_2}{\overset{k_1}{\rightleftharpoons}} B^*$$
$$B^* + C \xrightarrow{k_3} D \tag{7-13}$$

Assuming that each reaction step in the mechanism is elementary and that the pseudo-steady-state hypothesis holds, derive a rate law for the rate of formation of D in terms of the concentrations of A and C.

(G4)

7.2–Searching for a Mechanism

In many instances the rate data are correlated before a mechanism is suggested. Suppose, for example, that the following expression has been obtained from correlation of experimental data

$$r_{N_2} = \frac{.2C_{AZO}^2}{4 + 1.2C_{AZO}} \tag{7-14}$$

It is the normal procedure to reduce the additive constant in the denominator to 1. We therefore divide the numerator and denominator of Equation (7-14) by 4 to obtain

$$r_{N_2} = \frac{.05C_{AZO}^2}{1 + .3C_{AZO}} \tag{7-15}$$

(G3)

$$r_{N_2} = k_3 C^*_{AZO} \tag{G1-1}$$

Using the pseudo-steady-state approximation for the active intermediate in the azomethane decomposition,

$$r_{AZO^*} = 0, \tag{G3-1}$$

$$r_{AZO^*} = k_1 C^2_{AZO} - k_2 C_{AZO} C_{AZO^*} - k_3 C_{AZO^*} = 0 \tag{G3-2}$$

we can solve Equation (G3-2) for C_{AZO^*} in terms of C_{AZO}:

$$C^*_{AZO} = \frac{k_1 C^2_{AZO}}{k_3 + k_2 C_{AZO}} \tag{G3-3}$$

Substituting Equation (G3-3) into Equation (G1-1),

$$r_{N_2} = \frac{k_1 k_3 C^2_{AZO}}{k_3 + k_2 C_{AZO}} \tag{G3-4}$$

At low concentrations

$$k_2 C_{AZO} \ll k_3$$

for which case we obtain the second-order rate law

$$r_{N_2} = k_1 C^2_{AZO} \tag{G3-5}$$

At high concentrations

$$k_2 C_{AZO} \gg k_3$$

in which case, the rate expression follows first order kinetics.

$$r_{N_2} = \frac{k_1 k_3}{k_2} C_{AZO} \tag{G3-6}$$

(G4)

The reaction mechanism is believed to be given by the following individual reaction steps.

$$2A \xrightarrow{k_1} B^* \tag{G4-1}$$

$$B^* \xrightarrow{k_2} 2A \tag{G4-2}$$

$$B^* + C \xrightarrow{k_3} D \tag{G4-3}$$

Since each reaction step is elementary, the rate expression for reaction (G4-3) is

$$r_D = k_3 C_{B^*} C_C \tag{G4-4}$$

The active intermediate B is highly unstable and its concentration is quite small and difficult to measure. Therefore, to evaluate r_D, C_{B^*} must be eliminated from the rate expression. To accomplish this we write a rate expression for the rate of formation of B*, and then, using the psuedo-steady-state hypothesis, we set $r_{B^*} = 0$:

$$r_{B^*} = 0 = k_1 C^2_A - k_2 C_{B^*} - k_3 C_{B^*} C_C \tag{G4-5}$$

Solving for C_{B^*},

$$C_{B^*} = \frac{k_1 C^2_A}{k_2 + k_3 C_C} \tag{G4-6}$$

Substituting Equation (G4-6) into (G4-4) we obtain

$$r_D = \frac{k_1 k_3 C_C C^2_A}{k_2 + k_3 C_C} \tag{G4-7}$$

7.2A General Considerations

The rules of thumb listed in Table 7-1 may be of some help in the development of a mechanism that is consistent with the experimental rate law.

TABLE 7-1

1. Species having the concentration(s) appearing in the denominator of the rate law probably collide with the active intermediate, e.g.,
$$A + A^* \longrightarrow [\ ]$$
2. If a constant appears in the denominator, one of the reaction steps is probably the spontaneous decomposition of the active intermediate, e.g.,
$$A^* \longrightarrow [\ ]$$
3. Species having the concentration(s) appearing in the numerator of the rate law probably produce the active intermediate in one of the reaction steps, e.g.,
$$[\text{Reactant}] \longrightarrow A^* + [\ ]$$

Upon application of Table 7–1 to the azomethane example just discussed, we see from the rate Equation (7-15) that

1. AZO* decomposes spontaneously [Equation. (7-7)], resulting in a constant in the denominator of the rate expression.

2. The active intermediate, AZO*, collides with azomethane AZO [Equation (7-6)], resulting in the appearance of the concentration of AZO in the denominator;

3. The appearance of AZO in the numerator suggests that the active intermediate AZO* is formed from AZO. Referring to Equation (7-5), we see that this is indeed the case.

For the reaction given by the stoichiometric equation

$$2A + C \longrightarrow D$$

the rate law is

$$r_D = \frac{.001 C_A^2 C_C}{1 + 6C_C} \tag{7-16}$$

Referring to Table 7–1, what statements can you make about the mechanism, assuming that it involves an active intermediate? Suggest a possible mechanism.

(G5)

To determine which of the mechanisms in Frame (G5) is the most plausible, one should derive rate expressions for each of the mechanisms to see which, if any, produces a rate expression that is the most consistent with the experimental data.

Now, let us proceed to some slightly more complex examples involving chain reactions.

(G5)

$$r_D = \frac{.001 C_A^2 C_C}{1 + 6C_C} \tag{7-16}$$

The denominator suggests that

1. The active intermediate (call it B*) decomposes spontaneously:

$$B^* \xrightarrow{k_2} 2A \tag{G4-2}$$

2. The active intermediate reacts with C.

$$B^* + C \xrightarrow{k_3} D \tag{G4-3}$$

The numerator suggests that

1. The active intermediate is formed from A. Since the concentration of A is raised to second power and since we assume that each step in the mechanism is elementary, the intermediate could be formed by the reaction

$$2A \longrightarrow [\text{intermediate}] + \tag{G5-1}$$

$$2A \longrightarrow B^* \tag{G4-1}$$

2. From the guidelines for deducing a mechanism that are given in Table 7–1, we might suppose that, since the concentration of species C appears in the numerator, the active intermediate is formed from C.

$$C \longrightarrow [\text{intermediate}] + \tag{G5-2}$$

However, since C reacts with the intermediate we might question the likelihood of this possibility.

3. The active intermediate might be formed from the reaction

$$2A + C \longrightarrow [\text{intermediate}] + \tag{G5-3}$$

Here, one might also question the likelihood of the occurrence of a termolecular reaction in a single step.

To determine which of the mechanistic steps in Equations (G4-1), (G5-2), or (G5-3) is most reasonable, the pseudo-steady-state hypothesis should be applied to each mechanism to obtain its rate expression. A mechanism that is consistent with the rate expression in Equation (7-16) is

$$\begin{aligned} 2A &\rightleftharpoons B^* \\ B^* + C &\longrightarrow D \end{aligned} \tag{G5-4}$$

The chain reaction consists of the following sequence:

1. *Initiation*—formation of an active intermediate
2. *Propagation or chain transfer*—interaction of an active intermediate with the reactant or product to produce another active intermediate.
3. *Termination*—deactivation of the active intermediate.

7.2B Hydrogen Bromide Illustrative Example

One of the most studied chain reactions is that between hydrogen and bromine

$$H_2 + Br_2 \longrightarrow 2HBr$$

In studying this reaction we first deduce the rate law from reaction rate-concentration data and then formulate a mechanism and rate equation that is consistent with experimental observations. The relative rates of the reaction of bromine, hydrogen, and hydrogen bromide are

$$-r_{H_2} = -r_{Br_2} = \frac{r_{HBr}}{2}$$

7.2B.1 Synthesis of the Rate Law from Experimental Data. The log-log plots in Frames (G6), (G7), and (G8) show the rate of formation of HBr as a function of the concentration of H_2, Br_2, and HBr respectively. The data in Figure 7-1 were obtained by varying the hydrogen concentration while holding the concentrations of all other species concentrations constant. Determine the reaction order w.r.t. H_2.

(G6)

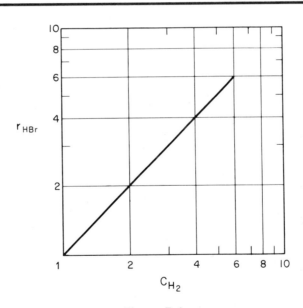

Figure 7-1

The plot in Figure 7-3 relating the rate of formation of hydrogen bromide to the bromide concentration, was obtained a similar manner.

(G6)

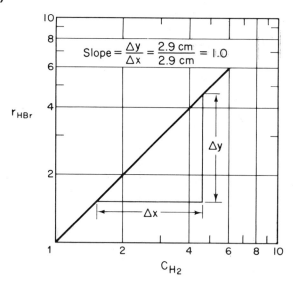

Figure 7-2

$$r_{HBr} \propto C_{H_2}^{\alpha}$$
$$\ln r_{HBr} = \alpha \ln C_{H_2} + [\quad]$$

From the slope of the plot in Figure 7-2, $\alpha = 1.0$. The rate of formation of HBr is first order w.r.t. the concentration of hydrogen, i.e.,

$$r_{HBr} \propto C_{H_2}$$

1. What is the apparent reaction order w.r.t. Br_2 at very low concentrations of Br_2?
2. What is the apparent reaction order w.r.t. Br_2 at very high concentrations of Br_2?

(G7)

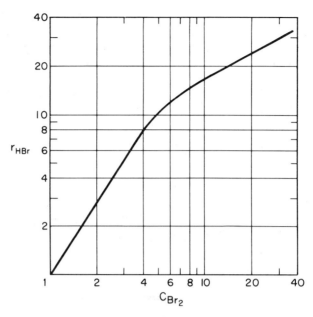

Figure 7-3

From the information presented in Figure 7-5, what observations and statements can you make about the dependence of r_{HBr} on the concentration of HBr?

(G8)

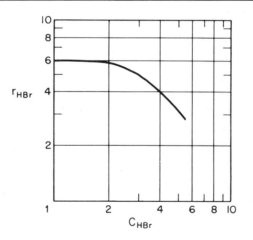

Figure 7-5

In summarizing the interpretation of the data in Figures 7-1, 7-3, and 7-5, we can say that the rate of HBr formation is

1. 1st order w.r.t. to the H_2 concentration.

2. Nearly independent of the HBr concentration at low HBr concentrations and decreases with increasing HBr concentration at high HBr concentrations.

3. $\frac{3}{2}$ order in Br_2 at low Br_2 concentrations and $\frac{1}{2}$ order in Br_2 at high Br_2 concentrations.

257

(G7)

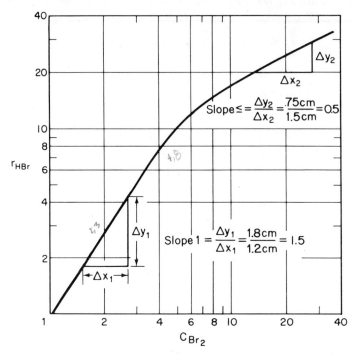

Figure 7-4

From Figure 7-4 we find that at low concentrations of bromine, the reaction is three-halves order w.r.t. bromine, i.e.,

$$r_{HBr} \propto C_{Br_2}^{3/2}$$

and at high concentrations of bromine, the reaction is half-order w.r.t. bromine, i.e.,

$$r_{HBr} \propto C_{Br_2}^{1/2}$$

(G8)

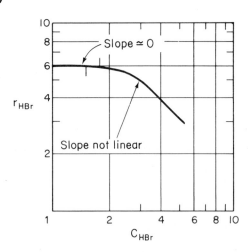

Figure 7-6

At low concentrations of HBr the rate of formation of HBr is essentially independent of the HBr concentration; at high concentrations the rate of formation decreases with increasing HBr concentration in a nonlinear manner on a log-log plot (see Figure 7-6).

Suggest a partial form of the rate law which will explain only the concentration dependence of Br_2. [Refer to Frame (G3) if necessary.]

(G9)

Suggest a partial form of the rate expression for the concentration dependence of HBr that could explain previously stated observations w.r.t. HBr.

(G10)

(G9) It is customary, in discussions of reactions of this sort, to let the species formula enclosed by parentheses represent its concentration, e.g., $(Br_2) \equiv C_{Br_2}$. In Frame (G3), we saw that a rate law of the form

$$-r_A = \frac{k_1(A)^2}{1 + k_2(A)} \tag{G9-1}$$

$$\text{At low } C_A: \quad -r_A = k_1(A)^2 \tag{G9-2}$$

$$\text{At high } C_A: \quad -r_A = \frac{k_1(A)^2}{k_2(A)} = k'(A) \tag{G9-3}$$

could account for the change in the apparent reaction order as one goes from low to high concentration.

For the HBr reaction

$$\text{At low } C_{Br_2}: \quad r_{HBr} \propto C_{Br_2}^{3/2} \tag{G9-4}$$

and

$$\text{At high } C_{Br_2}: \quad r_{HBr} \propto C_{Br_2}^{1/2} \tag{G9-5}$$

By analogy,

$$r_{HBr} \propto \frac{k'_1(Br_2)^{3/2}}{[\ ] + k'_2(Br_2)} \tag{G9-6}$$

where [ ] represents a reaction constant or the concentration of one of the species that will be determined later.

$$\text{At high } C_{Br_2}: \quad k'_2(Br) \gg [\ ]$$

$$r_{HBr} \propto \frac{k'_1}{k'_2}(Br_2)^{1/2} \tag{G9-7}$$

and

$$\text{At low } C_{Br_2}: \quad [\ ] \gg k'_2 Br$$

$$r_{HBr} \propto \frac{k'_1(Br_2)^{3/2}}{[\ ]} \tag{G9-8}$$

Consequently, we see that the concentration dependence of bromine given in Equation (G9-6) is consistent with the experimental data.

(G10)

$$\text{At low } C_{HBr}: \quad r_{HBr} \neq f(HBr) \tag{G10-1}$$

$$\text{At high } C_{HBr}: \quad r_{HBr} \text{ decreases as (HBr) increases} \tag{G10-2}$$

One rate expression in which the concentration dependence of HBr is consistent with the experimental observation is

$$r_{HBr} \propto \frac{1}{(HBr) + [\ ]} \tag{G10-3}$$

At low C_{HBr}: $(HBr) \ll [\ ] \quad r_{HBr} \propto \frac{1}{[\ ]} \neq f(HBr)$

At high C_{HBr}: As (HBr) increases, the quantity $1/((HBr) + [\ ])$ decreases as does the rate of formation of HBr.

Consequently the concentration dependence of HBr denoted by Equation (G10-3) is consistent with the experimental observation.

Combine the results of Frames (G6), (G9), and (G10) to produce a rate law for the formation of HBr.

(G11)

7.2B.2 Finding the Reaction Mechanism. Now that a rate law has been synthesized from the experimental data, we shall try to propose a mechanism that is consistent with this equation. The method of attack will be as given in Table 7–2.

TABLE 7–2

1. Assume an activated intermediate(s).
2. Postulate a mechanism, utilizing the rate law obtained from experimental data, if possible.
3. Model each reaction in the mechanism sequence as an elementary reaction.
4. After writing rate expressions for the rate of formation of the desired product, write the rate expressions for each of the active intermediates.
5. Utilize the pseudo-steady-state hypothesis.
6. Eliminate the concentration of the intermediate species in the rate expressions by solving the simultaneous equations developed in steps 4 and 5 above.
7. If the derived rate law does not agree with experimental observation, assume a new mechanism and/or intermediates and go to step 3. A strong background in organic and inorganic chemistry is helpful in predicting the activated intermediates for the reaction under consideration.

We first postulate that the active intermediates are bromine and hydrogen free radicals (Br· and H·). Referring to the denominator of the rate equation

$$r_{HBr} = \frac{k'_1(H_2)(Br_2)^{3/2}}{(HBr) + k'_2(Br_2)}, \tag{7-17}$$

and to Table 7–1, we see that the active intermediates Br· and H· collide with HBr and Br_2. The four reactions (two intermediates times two species) delineating these collisions or interactions are

$$H\cdot + HBr \longrightarrow H_2 + Br\cdot \tag{7-18}$$

$$H\cdot + Br_2 \longrightarrow HBr + Br\cdot \tag{7-19}$$

$$Br\cdot + HBr \longrightarrow H\cdot + Br_2 \tag{7-20}$$

$$Br\cdot + Br_2 \longrightarrow Br_2 + Br\cdot$$

The last reaction has no effect on the overall reaction since one Br_2 and one Br· are produced for every Br_2 and Br· that are consumed. However, it is logical to assume that Br· and H_2 probably react instead of Br· and Br_2:

$$Br\cdot + H_2 \longrightarrow H\cdot + HBr \tag{7-21}$$

Note that our proposed scheme of reactions, Equations (7-18) through (7-21), produces one of the active intermediates in each step. Now we only need a reaction yielding the initial formation of Br· or H· Experience and, possibly, a good guess tell us that the fractional reaction order of Br_2 in the numerator suggests that the primary reactive intermediate is Br· formed from Br_2 and terminated by the reaction of two bromine free radicals

$$Br_2 \longrightarrow 2Br\cdot \tag{7-22}$$

$$2Br\cdot \longrightarrow Br_2 \tag{7-23}$$

(G11) The concentration dependence of the species H_2, Br_2, and HBr in the rate expression are, respectively,

$$r_{HBr} \propto (H_2) \tag{G6-1}$$

$$r_{HBr} \propto \frac{k'_1(Br_2)^{3/2}}{[\ ] + k'_2(Br_2)} \tag{G9-6}$$

$$r_{HBr} \propto \frac{1}{(HBr) + [\ ]} \tag{G10-3}$$

Comparing Equations (G9-6) and (G10-3) we see that [ ] in Equation (G9-6) could be HBr and that [ ] in Equation (G10-3) could be $k'_2 Br_2$. Therefore, Equations (G9-6) and (G10-3) are combined to give

$$r_{HBr} \propto \frac{k'_1((Br_2)^{3/2}}{(HBr) + k'_2(Br_2)} \tag{G11-1}$$

If we now incorporate dependence of the reaction rates on H_2 into Equation (G11-1), we arrive at the rate expression

$$r_{HBr} = \frac{k'_1(H_2)(Br_2)^{3/2}}{(HBr) + k'_2(Br_2)} \tag{G11-2}$$

or, identically,

$$r_{HBr} = \frac{k'_1 C_{H_2} C_{Br_2}^{3/2}}{C_{HBr} + k'_2 C_{Br_2}} \tag{G11-3}$$

which is consistent with *all* of the experimental observations.

As a first approximation we shall assume that the mechanism consists of reactions (7-18), (7-19), (7-21), (7-22), and (7-23). If this mechanism does not produce a rate equation that is consistent with the experimental observations in Equation (7-17), we could propose a different mechanism that might include Equation (7-20) along with $H_2 \rightarrow 2H\cdot$ and similar reactions. The mechanism we first propose is

$$\text{Initiation:} \quad Br_2 \xrightarrow{k_1} 2Br\cdot \quad (7\text{-}22)$$

$$\text{Propagation:} \begin{cases} Br\cdot + H_2 \xrightarrow{k_2} HBr + H\cdot & (7\text{-}21) \\ H\cdot + Br_2 \xrightarrow{k_3} HBr + Br\cdot & (7\text{-}19) \\ H\cdot + HBr \xrightarrow{k_4} H_2 + Br\cdot & (7\text{-}18) \end{cases}$$

$$\text{Termination:} \quad 2Br\cdot \xrightarrow{k_5} Br_2 \quad (7\text{-}23)$$

The specific reaction rates k_1 and k_5 are defined w.r.t. Br_2.

Remembering that each step in the mechanism sequence is elementary, write rate expressions for HBr, Br·, and H· in terms of the appropriate species concentrations.

(G12)

The intermediate species (Br·) and (H·) are free radicals, which are present in low concentrations and highly reactive. Consequently, the use of the pseudo-steady-state hypothesis is justified. The application of this hypothesis to Equations (G12-3) and (G12-4) gives

$$r_{HBr} = k_2(H_2)(Br\cdot) + k_3(H\cdot)(Br_2) - k_4(H\cdot)(HBr) \quad (G12\text{-}2)$$

$$r_{H\cdot} = 0 = k_2(H_2)(Br\cdot) - k_3(H\cdot)(Br_2) - k_4(H\cdot)(HBr) \quad (7\text{-}24)$$

$$r_{Br\cdot} = 0 = 2k_1(Br_2) - k_2(H_2)(Br\cdot) + k_3(H\cdot)(Br_2) + k_4(H\cdot)(HBr) - 2k_5(Br\cdot) \quad (7\text{-}25)$$

Adding (7-25) and (7-24) gives

$$2k_5(Br\cdot)^2 = 2k_1(Br_2)$$

$$(Br\cdot) = \left(\frac{k_1}{k_5}\right)^{1/2}(Br_2)^{1/2} \quad (7\text{-}26)$$

Subtracting (7-24) from (G12-2) gives

$$r_{HBr} = 2k_3(H\cdot)(Br_2) \quad (7\text{-}27)$$

Now write r_{HBr} solely in terms of the concentration of H_2, Br_2, and HBr.

(G13)

We can rewrite Equation (G13-4) in the form

$$r_{HBr} = \frac{k_1' C_{H_2} C_{Br_2}^{3/2}}{C_{HBr} + k_2' C_{Br_2}} \quad (7\text{-}28)$$

By comparing Equations (7-17) and (7-28) we see that the mechanism delineated by the series

(G12)

Each reaction is to be considered elementary. First we write down the net rate of formation of the product resulting from all the reaction steps:

$$r_{HBr} = \begin{Bmatrix} \text{Rate} \\ \text{HBr is formed} \\ \text{in reaction (7-21)} \end{Bmatrix} + \begin{Bmatrix} \text{Rate} \\ \text{HBr is formed} \\ \text{in reaction (7-19)} \end{Bmatrix} + \begin{Bmatrix} \text{Rate} \\ \text{HBr is formed} \\ \text{in reaction (7-18)} \end{Bmatrix}$$

$$r_{HBr} = r_{HBr,\,(7\text{-}21)} + r_{HBr,\,(7\text{-}19)} + r_{HBr,\,(7\text{-}18)} \tag{G12-1}$$

$$r_{HBr} = k_2(Br\cdot)(H_2) + k_3(H\cdot)(Br_2) - k_4(HBr)(H\cdot) \tag{G12-2}$$

We next write the net rate of formation of the intermediate species.

For $(H\cdot)$: $\quad r_{H\cdot} = k_2(H_2)(Br\cdot) - k_3(H\cdot)(Br_2) - k_4(H\cdot)(HBr) \tag{G12-3}$

For $(Br\cdot)$: $\quad r_{Br\cdot} = 2k_1(Br_2) - k_2(Br\cdot)(H_2) + k_3(Br_2)(H\cdot) + k_4(H\cdot)(HBr) - 2k_5(Br\cdot)^2 \tag{G12-4}$

(G13)

$$r_{HBr} = 2k_3(H\cdot)(Br_2) \tag{7-27}$$

To eliminate $(H\cdot)$ from this equation, rewrite Equation (7-24),

$$k_2(H_2)(Br\cdot) = k_3(H\cdot)(Br_2) + k_4(H\cdot)(HBr) \tag{G13-1}$$

and solve for $(H\cdot)$.

$$(H\cdot) = \frac{k_2(H_2)(Br\cdot)}{k_3(Br_2) + k_4(HBr)} \tag{G13-2}$$

To eliminate $(Br\cdot)$ recall

$$(Br\cdot) = \left(\frac{k_1}{k_5}\right)^{1/2}(Br_2)^{1/2} \tag{7-26}$$

and $\quad (H\cdot) = \dfrac{k_2\left(\dfrac{k_1}{k_5}\right)^{1/2}(H_2)(Br_2)^{1/2}}{k_3(Br_2) + k_4(HBr)} \tag{G13-3}$

Substituting Equation (G13-3) into Equation (7-27), we see that

$$r_{HBr} = \frac{2k_3 k_2 \left(\dfrac{k_1}{k_5}\right)^{1/2}(H_2)(Br_2)^{3/2}}{k_3(Br_2) + k_4(HBr)} \tag{G13-4}$$

of Equations (7-22), (7-21), (7-19), (7-18), and (7-23) produces a rate expression that is consistent with the experimental observations.

Another example of a gas phase chain reaction involving free radicals concerns hydrodealkylation, in which alkyl aromatics are converted to unsubstituted aromatics. A process for the hydrodealkylation of toluene from petroleum stocks to produce benzene and methane has recently been developed, utilizing a solid Detol catalyst.† A hydrodealkylation of this type had previously been observed to occur in high temperature homogeneous gas phase reactions involving free radicals. The free radical mechanism is believed to proceed by the sequence

$$\text{Initiation:} \quad H_2 \xrightarrow{k_1} 2H \cdot \quad (7\text{-}29)$$

$$\text{Propagation:} \quad \begin{cases} H \cdot + C_6H_5CH_3 \xrightarrow{k_2} C_6H_5 \cdot + CH_4 & (7\text{-}30) \\ C_6H_5 \cdot + H_2 \xrightarrow{k_3} C_6H_6 + H \cdot & (7\text{-}31) \end{cases}$$

$$\text{Termination:} \quad 2H \cdot \xrightarrow{k_4} H_2 \quad (7\text{-}32)$$

The specific reaction rates k_1 and k_4 are defined w.r.t. H_2. Derive the reaction rate expression for the rate of formation of benzene based on this mechanism.

(G14)

Before proceeding further it might be beneficial for the reader to work through some of the following problems: Pb. 7-1 through 7-7 and 7-12, 7-14.

7.3–Enzymatic Reaction Fundamentals

7.3A Definitions and Mechanisms

Another class of reactions in which the pseudo-steady-state hypothesis is used is the enzymatically catalyzed reaction, which is characteristic of most biological reactions. An *enzyme*, E, is a protein or protein-like substance with catalytic properties. A *substrate*, S, is the substance that is chemically transformed at an accelerated rate because of the action of an enzyme on it. Most enzymes are usually named in terms of the reactions they catalyze. It is a customary practice to add the suffix-*ase* to a major part of the name of the substrate on which the enzyme acts. For example, the enzyme that catalyzes the decomposition of urea is urease, the enzyme that attacks tyrosine is tyrosinase, and uric acid is attacked by uricase. There are three major types of enzyme reactions:

  I. Soluble enzyme–insoluble substrate.
  II. Insoluble enzyme–soluble substrate.
  III. Soluble enzyme–soluble substrate.

An example of a Type I reaction is the use of enzymes such as proteases or amylases in laundry detergents; this enzyme reaction has caused some controversy in relation to water pollution. Once in solution, the soluble enzyme may digest, (i.e., break down) an insoluble substrate such as a blood stain.

†A. H. Weiss and L. Freedman, *I.E.C. Process Design Development* 2, (1963), 165.

(G14)

For the elementary Equation (7-31) we write the rate of formation of benzene as:

$$r_{C_6H_6} = k_3(H_2)(C_6H_5\cdot) \tag{G14-1}$$

We need to eliminate the concentration of the free radical ($C_6H_5\cdot$) by expressing it in terms of the concentrations of toluene and hydrogen. Under the psuedo-steady-state hypothesis we set the rates of formation of ($H\cdot$) and ($C_6H_5\cdot$) equal to zero,
i.e.,

$$r_{C_6H_5\cdot} = 0 = k_2(H\cdot)(C_6H_5CH_3) - k_3(C_6H_5\cdot)(H_2) \tag{G14-2}$$

$$r_{H\cdot} = 0 = 2k_1(H_2) - k_2(H\cdot)(C_6H_5CH_3) + k_3(C_6H_5\cdot)(H_2) - 2k_4(H\cdot)^2 \tag{G14-3}$$

Adding Equations (G14-2) and (G14-3),

$$2k_1(H_2) - 2k_4(H\cdot)^2 = 0$$

Solving for ($H\cdot$),

$$(H\cdot) = \left(\frac{k_1}{k_4}\right)^{1/2}(H_2)^{1/2} \tag{G14-4}$$

Solving Equation (G14-2) for $C_6H_5\cdot$;

$$(C_6H_5\cdot) = \frac{k_2}{k_3}\frac{(H\cdot)(C_6H_5CH_3)}{(H_2)} \tag{G14-5}$$

Replacing ($H\cdot$) by Equation (G14-4),

$$(C_6H_5\cdot) = \left(\frac{k_1}{k_4}\right)^{1/2}\frac{k_2}{k_3}\frac{(C_6H_5CH_3)}{H_2^{1/2}} \tag{G14-6}$$

We can now eliminate ($C_6H_5\cdot$) from the reaction rate expression (G14-1) to obtain

$$r_{C_6H_6} = k_2\left(\frac{k_1}{k_4}\right)^{1/2}(H_2)^{1/2}(C_6H_5CH_3) \tag{G14-7}$$

The reaction rate expression for the mechanism given by Equations (7-29) through (7-32) is

$$r_{C_6H_6} = k(C_{H_2})^{1/2}\cdot(C_T) \tag{G14-8}$$

where

$$k = 10^{10.5}\exp\left[\frac{-50,000}{RT}\right] \tag{G14-9}$$

C_T = concentration of toluene

A major research effort is currently being directed at Type II reactions. By attaching active enzyme groups to solid surfaces, continuous processing units similar to the packed catalytic bed reactor discussed in Chapter 6 can be developed.

Clearly, the greatest activity in the study of enzymes has been in relation to biological reactions, since virtually every synthetic and degradation reaction in all living cells has been shown to be controlled and catalyzed by specific enzymes.[†] Many of these reactions are homogeneous in the liquid phase; that is, Type III reactions (soluble enzyme–soluble substrate). In the following brief presentation we shall limit our discussion to Type III reactions, although the resulting equations have been found to be applicable to the Type I and Type II reactions in certain instances.

In developing some of the elementary principles of the kinetics of enzyme reactions, we shall discuss an enzymatic reaction that has been suggested as *part* of a system that would reduce the size of an artificial kidney.[‡] The desired result is the production of an artificial kidney that could be worn by the patient and would incorporate a replaceable unit for the elimination of the nitrogenous waste products such as uric acid and creatinine. In the microencapsulation scheme proposed by Levine and LaCourse, the enzyme urease would be used in the removal of urea from the blood stream. Here, the catalytic action of urease would cause urea to decompose into ammonia and carbon dioxide. The mechanism of the reaction is believed to proceed by the following sequence of elementary reactions:

1. The enzyme urease reacts with the substrate urea to form an enzyme-substrate complex, E·S:

$$NH_2CONH_2 + \text{urease} \xrightarrow{k_1} [NH_2CONH_2 \cdot \text{urease}]^* \tag{7-33}$$

2. This complex can decompose back to urea and urease:

$$[NH_2CONH_2 \cdot \text{urease}]^* \xrightarrow{k_2} \text{urease} + NH_2CONH_2 \tag{7-34}$$

3. Or it can react with water to give ammonia, carbon dioxide, and urease:

$$[NH_2CONH_2 \cdot \text{urease}]^* + H_2O \xrightarrow{k_3} 2NH_3 + CO_2 + \text{urease} \tag{7-35}$$

Letting E, S, W, E·S, and P represent the enzyme, substrate, water, the enzyme-substrate complex, and the reaction products, respectively, we can write reactions (7-33), (7-34), and (7-35) symbolically in the form

$$E + S \xrightarrow{k_1} E \cdot S \tag{7-36}$$

$$E \cdot S \xrightarrow{k_2} E + S \tag{7-37}$$

$$E \cdot S + W \xrightarrow{k_3} P + E \tag{7-38}$$

Here, $P = 2NH_3 + CO_2$.

[†]R. G. Denkewalter and R. Hirschmann, *Am. Scientist* 57, no. 4, (1969), 389.
[‡]S. N. Levine and W. C. LaCourse, *J. Biomed. Mater. Res.* 1, (1967), 275.

Assuming that each step is elementary, write the rate laws for the rate of reaction of the substrate S, $-r_S$, and the rate of formation of the enzyme-substrate complex, $r_{E \cdot S}$.

(G15)

We note from the reaction sequence in Frame (G15) that the enzyme is not consumed by the reaction. The total concentration of the enzyme in the system, E_t, is constant and equal to the sum of the concentrations of the free or unbonded enzyme E and the enzyme-substrate complex E·S.

$$E_t = (E) + (E \cdot S) \tag{7-39}$$

Apply the pseudo-steady-state hypothesis to the complex E·S, and derive a rate law for the substrate $-r_S$ in terms of the substrate concentration, (S), the total enzyme concentration, E_t, and the appropriate specific reaction rates.

(G16)

7.3B Michaelis-Menten Equation

Since the reaction of urea and urease is carried out in aqueous solution, water is, of course, in excess, and the concentration of water is therefore considered constant. Let

$$k'_3 = k_3(W), \qquad K_m = \frac{k'_3 + k_2}{k_1}$$

Dividing the numerator and denominator by k_1, we obtain a form of the *Michaelis-Menten Equation*.

$$-r_S = \frac{k'_3(S)E_t}{(S) + K_m} \tag{7-40}$$

K_m is the Michaelis constant. If in addition we let V_{max} represent the maximum rate of reaction for a given total enzyme concentration,

$$V_{max} = k'_3(E_t) \tag{7-41}$$

then the Michaelis-Menten Equation takes the familiar form,

$$\boxed{-r_s = \frac{V_{max}(S)}{K_m + (S)}} \tag{7-42}$$

For a given enzyme concentration, sketch the rate of disappearance of the substrate as a function of the substrate concentration.

(G15)

The rate of disappearance of the substrate $-r_s$ is

$$-r_s = k_1(E)(S) - k_2(E \cdot S) \qquad \text{(G15-1)}$$

The net rate of formation of the complex is

$$r_{E \cdot S} = k_1(E)(S) - k_2(E \cdot S) - k_3(W)(E \cdot S) \qquad \text{(G15-2)}$$

(G16)

Rearranging Equation (7-39), the enzyme concentration is

$$(E) = E_t - (E \cdot S) \qquad \text{(G16-1)}$$

Substituting Equation (G16-1) into Equation (G15-2) and using the pseudo-steady-state hypothesis,

$$r_{E \cdot S} = 0 = k_1[E_t - (E \cdot S)](S) - k_2(E \cdot S) - k_3(E \cdot S)(W) \qquad \text{(G16-2)}$$

Solving for $(E \cdot S)$,

$$(E \cdot S) = \frac{k_1(E_t)(S)}{k_1(S) + k_2 + k_3(W)} \qquad \text{(G16-3)}$$

$$-r_s = k_1[E_t - (E \cdot S)](S) - k_2(E \cdot S) \qquad \text{(G16-4)}$$

Subtracting Equation (G16-2) from (G16-4) we get

$$-r_s = k_3(W)(E \cdot S) \qquad \text{(G16-5)}$$

Substituting for $(E \cdot S)$,

$$-r_s = \frac{k_1 k_3(W)(E_t)(S)}{k_1(S) + k_2 + k_3(W)} \qquad \text{(G16-6)}$$

Note: Throughout $E_t \equiv (E_t) =$ total concentration of enzyme with typical units (gmole/l).

(G17)

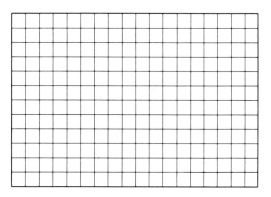

Figure 7-7

The parameters V_{max} and K_m are used to characterize the enzymatic reactions that are described by Michaelis-Menten kinetics. In biology texts dealing with the kinetics of enzymatic reactions, the symbol v is used instead of $-r_s$ to denote the rate of reaction. For the reaction we have been discussing,

$$\text{urea} + \text{urease} \underset{k_2}{\overset{k_1}{\rightleftharpoons}} [\text{urea} \cdot \text{urease}]^* \xrightarrow[+H_2O]{k_3} 2NH_3 + CO_2 + \text{urease}$$

the rate of reaction is given as a function of urea concentration in Table 7–3.[†]

TABLE 7–3

C_{urea} (gmoles/l)	.2	.02	.01	.005	.002
$-r_{urea}$ (gmoles/l/sec)	1.08	.55	.38	.2	.09

Explain how you would plot the data in Table 7–3 to obtain a straight line from which you could determine the Michaelis-Menten parameters K_m and V_{max}.

(G18)

[†]*J. Biomed. Mater. Res.*, *1*, p 275. Also, D. J. Fink (Unpublished notes, University of Michigan, 1971.)

(G17)

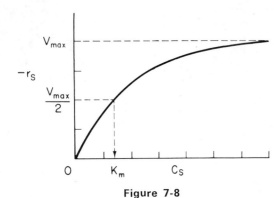

Figure 7-8

$$-r_s = \frac{V_{max}(S)}{(S) + K_m} \quad (7\text{-}42)$$

At low substrate concentrations:
$$-r_s \simeq \frac{V_{max}(S)}{K_m} \quad (G17\text{-}1)$$

At high substrate concentrations: $(S) \gg K_m$
and
$$-r_s = V_{max} \quad (G17\text{-}2)$$

Consider when the substrate concentration is such that the reaction rate is equal to one-half the maximum rate,
$$-r_s = \frac{V_{max}}{2} \quad (G17\text{-}3)$$

then
$$\frac{V_{max}}{2} = \frac{V_{max}(S_{1/2})}{K_m + (S_{1/2})} \quad (G17\text{-}4)$$

Solving Equation (G17-4) for the Michaelis constant,
$$K_m = (S_{1/2}) \quad (G17\text{-}5)$$

The Michaelis constant is equal to the substrate concentration at which the rate of reaction is equal to one-half the maximum rate.

(G18)

Inverting Equation (7-42),

$$\frac{1}{-r_s} = \frac{(S) + K_m}{V_{max}(S)} = \frac{1}{V_{max}} + \frac{K_m}{V_{max}}\left[\frac{1}{(S)}\right] \quad (G18\text{-}1)$$

or
$$\frac{1}{-r_{urea}} = \frac{1}{V_{max}} + \frac{K_m}{V_{max}}\left[\frac{1}{C_{urea}}\right] \quad (G18\text{-}2)$$

A plot of the reciprocal reaction rate vs. the reciprocal urea concentration should be a straight line with to intercept $1/V_{max}$ and slope K_m/V_{max}. This type of plot is called a *Lineweaver-Burk Plot*.

TABLE (G18)

C_{urea} (gmoles/l)	$-r_{urea}$ (gmoles/l/sec)	$\frac{1}{C_{urea}}$	$\frac{1}{-r_{urea}}$
.20	1.08	5.0	.93
.02	.55	50.0	1.82
.01	.38	100.0	2.63
.005	.20	200.0	5.00
.002	.09	500.0	11.10

The data in Table G18 is presented in Figure 7-9 in the form of a *Lineweaver-Burk* plot. The intercept is .75; therefore the maximum reaction velocity

$$\frac{1}{V_{max}} = .75\ l\text{-sec/gmole}$$

is

$$V_{max} = 1.33\ \text{gmoles}/l/\text{sec}$$

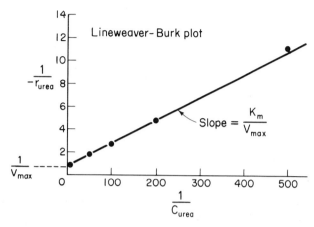

Figure 7-9

From the slope, which is .02 sec, we can calculate the Michaelis constant, K_m:

$$\frac{K_m}{V_{max}} = \text{slope} = .02\ \text{sec}$$

$$K_m = .0266\ \text{gmole}/l$$

Substituting K_m and V_{max} into Equation (7-42),

$$-r_s = \frac{1.33\ C_{urea}}{.0266 + C_{urea}} \tag{7-43}$$

Levine and LaCourse suggest that the total concentration of urease, (E_t), corresponding to the above value of V_{max} is approximately 5 g/l.

7.3C Batch Reactor Calculations

Calculate the time needed to convert 80% of the urea to ammonia and carbon dioxide in a 500 ml batch reactor. The initial concentration of urea is 0.1 gmole/l, and the urease concentration is .001 g/l. The reaction is to be carried out isothermally at the same temperature at which the data in Table 7–3 was obtained.

(G19)

The parameters K_m and V_{max} can readily be determined from batch reactor data by using the integral method of analysis. By dividing both sides of Equation (G19-5) by tK_m/V_{max} and rearranging, i.e.,

$$\frac{1}{t}\ln\left(\frac{1}{1-X}\right) = \frac{V_{max}}{K_m} - \frac{C_{urea0}X}{K_m t} \tag{7-44}$$

273

(G19)

A mole balance on urea in the batch reactor gives

$$-\frac{dN_{urea}}{dt} = -r_{urea}V \tag{G19-1}$$

Since this is a liquid phase reaction, the mole balance can be put in the form

$$-\frac{dC_{urea}}{dt} = -r_{urea} \tag{G19-2}$$

The rate law for urea decomposition is

$$-r_{urea} = \frac{V_{max}C_{urea}}{K_m + C_{urea}} \tag{G19-3}$$

Substituting Equation (G19-3) into Equation (G19-2), and than rearranging and integrating, we get

$$t = \int_{C_{urea}}^{C_{urea0}} \frac{dC_{urea}}{-r_{urea}} = \int_{C_{urea}}^{C_{urea0}} \frac{K_m + C_{urea}}{V_{max}C_{urea}} dC_{urea}$$
$$= \frac{K_m}{V_{max}} \ln \frac{C_{urea0}}{C_{urea}} + \frac{C_{urea0} - C_{urea}}{V_{max}} \tag{G19-4}$$

We next write Equation (G19-4) in terms of conversion:

$$C_{urea} = C_{urea0}(1 - X)$$
$$t = \frac{K_m}{V_{max}} \ln \frac{1}{1-X} + \frac{C_{urea0}X}{V_{max}} \tag{G19-5}$$

where $K_m = .0266$ gmole/l, $X = .8$, $C_{urea0} = .1$ gmole/l.

When the total concentration of the enzyme urease, E_t, was 5 g/l, V_{max} was 1.33. However, for the conditions in the batch reactor, the enzyme concentration is only .001 g/l. Since $V_{max} = E_t \cdot k_3$, V_{max} for the second enzyme concentration is

$$V_{max_2} = \frac{E_{t_2}}{E_{t_1}} V_{max_1} = \frac{.001}{5} \cdot 1.33 = 2.66 \times 10^{-4} \frac{\text{gmole}}{\text{sec}}$$
$$X = .8$$
$$t = \left(\frac{.0266}{.000266}\right) \ln \left(\frac{1}{.2}\right) + \frac{(.8)(.1)}{.000266}$$
$$= 160.9 + 300.7$$
$$= 461.6 \text{ sec}$$

we see that K_m and V_{max} can be determined from the slope and intercept of a plot of $1/t \ln[1/(1-X)]$ vs. X/t. Rather than writing the integrated form of the Michaelis-Menten equation in terms of conversion, we could express it in terms of the substrate concentration S;

$$\frac{1}{t}\ln\frac{S_0}{S} = \frac{V_{max}}{K_m} - \frac{S_0 - S}{K_m t} \quad (7\text{-}45)$$

where S_0 is the initial concentration of the substrate. *In cases similar to Equation (7-45) where there is no possibility of confusion, we shall not bother to enclose the substrate or other species in parentheses in order to represent concentration, i.e., $C_S \equiv (S) \equiv S$*. The corresponding plot in terms of substrate concentration is

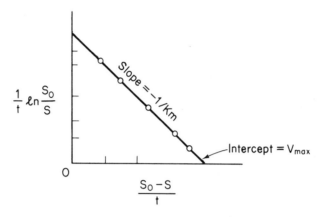

Figure 7-10

The maximum velocity, V_{max}, in addition to being a function of the total enzyme concentration, may also be a function of other variables such as ionic strength, pH, temperature, and inhibitor concentrations. For a number of enzymes V_{max} exhibits an optimum in pH. (At a low pH the V_{max} increases with increasing pH, while at a high pH it decreases with increasing pH, with the maximum value lying in the intermediate pH range.) One mathematical model that has been used to express this dependency on hydrogen ion concentration is[†] (See Problem 7-18)

$$V_{max} = \frac{V_0}{\left(1 + \dfrac{H}{K_a} + \dfrac{K_b}{H}\right)} \quad (7\text{-}46)$$

where H is hydrogen ion concentration and K_a and K_b are the ionization constants of the enzyme-substrate complex in the acidic and basic solutions, respectively. The denominator of Equation (7-46) is referred to as the *Michaelis pH function*. The Michaelis constant, K_m, has been observed to exhibit a similar dependence on pH.

7.4–Inhibition of Enzyme Reactions

In addition to pH, another factor that greatly influences the rates of enzyme catalyzed reactions is the presence of an inhibitor. The most dramatic consequences of enzyme inhibition are found in living organisms where the inhibition of any particular enzyme involved in a primary metabolic sequence will render the entire sequence inoperative, resulting in either serious damage or death of the organism. For example, the inhibition of a single enzyme, cytochrome oxidase, by cyanide will cause the aerobic oxidation process to stop; death occurs in a very few

[†]R. G. Carbonell and M. D. Kostin, *AIChE J.* 18, p. 1, (1972).

minutes. There are also beneficial inhibitors such as the ones used in the treatment of leukemia and other neoplastic diseases.

The three most common types of reversible inhibition occurring in enzymatic reactions are *competitive*, *uncompetitive*, and *noncompetitive*. The enzyme molecule is analogous to the heterogeneous catalytic surface in that it contains active sites. When competitive inhibition occurs, the substrate and inhibitor are usually similar molecules that compete for the same site on the enzyme. Uncompetitive inhibition occurs when the inhibitor deactivates the enzyme-substrate complex, usually by attaching itself to both the substrate and enzyme molecules of the complex. Noncompetitive inhibition occurs with enzymes containing at least two different types of sites. The inhibitor attaches only to one type of site and the substrate only to the other.

7.4A Competitive Inhibition

Competitive inhibition is of particular importance in pharmacokinetics (drug therapy). If a patient were administered two or more drugs simultaneously which react within the body with a common enzyme, cofactor, or active species, this could lead to competitive inhibition of the formation of the respective metabolites and produce serious consequences.

In this type of inhibition another substance, I, competes with the substrate for the enzyme molecules to form an inhibitor-enzyme complex, (E·I).

Typical steps in an enzymatic reaction of this type of inhibition are

$$\begin{aligned} E + S &\rightleftharpoons E \cdot S \\ E + I &\rightleftharpoons E \cdot I \\ E \cdot S &\longrightarrow E + P \end{aligned} \qquad (7\text{-}47)$$

Show that the rate law for the mechanism in Equation (7-47), involving competitive inhibition is

$$r_p = \frac{V_{max} S}{S + K_m \left(1 + \dfrac{I}{K_i}\right)} \qquad (7\text{-}48)$$

where I is the inhibitor concentration and r_p is the rate of formation of product P. (*Hint:* Apply the pseudo-steady-state hypothesis to (E·S) and (I·E).)

(G20)

Letting $K'_m = K_m (1 + I/K_i)$, we see that the effect of a competitive inhibition is to increase the Michaelis constant, K'_m.

7.4B Uncompetitive Inhibition

Here, the inhibitor does not compete with the substrate for the enzyme; instead it ties up the enzyme-substrate complex by forming an inhibitor-enzyme-substrate complex, thereby restricting the breakdown of the (E·S) complex to produce the desired product. Typical steps in

(G20)

$$E + S \underset{k_2}{\overset{k_1}{\rightleftharpoons}} E \cdot S$$

$$E + I \underset{k_4}{\overset{k_3}{\rightleftharpoons}} E \cdot I$$

$$E \cdot S \overset{k_5}{\longrightarrow} E + P$$

The rate of formation of product P corresponding to the last step in this reaction sequence is

$$r_p = k_5(E \cdot S) \tag{G20-1}$$

The uncomplexed or free enzyme concentration is

$$(E) = E_t - (E \cdot S) - (E \cdot I) \tag{G20-2}$$

Using the pseudo-steady-state approximation for the enzyme-substrate complex:

$$r_{E \cdot S} \simeq 0 = k_1[E_t - (E \cdot S) - (E \cdot I)](S) - k_2(E \cdot S) - k_5(E \cdot S) \tag{G20-3}$$

For the enzyme-inhibitor complex,

$$r_{E \cdot I} \simeq 0 = k_3[E_t - (E \cdot S) - (E \cdot I)](I) - k_4(E \cdot I) \tag{G20-4}$$

Dividing Equation (G20-3) by $k_1(S)$ and Equation (G20-4) by $k_3(I)$ and rearranging each we obtain

$$E_t - (E \cdot S) - (E \cdot I) = \frac{k_2 + k_5}{k_1(S)}(E \cdot S) \tag{G20-5}$$

$$E_t - (E \cdot S) - (E \cdot I) = \frac{k_4}{k_3(I)}(E \cdot I) \tag{G20-6}$$

Substracting Equation (G20-5) from (G20-6) and solving for $E \cdot I$, we find

$$(E \cdot I) = \frac{k_2 + k_5}{k_1}\left(\frac{I}{S}\right)\frac{k_3}{k_4}(E \cdot S) \tag{G20-7}$$

After substituting Equation (G20-7) into (G20-5) to solve for $E \cdot S$, i.e.,

$$(E \cdot S) = \frac{E_t(S)}{(S) + \dfrac{k_2 + k_5}{k_1}\left(1 + \dfrac{k_3}{k_4}(I)\right)} \tag{G20-8}$$

we substitute for $E \cdot S$ in Equation (G20-1) to obtain the rate law for the inhibition of the competitive type:

$$r_p = \frac{V_{max} \cdot S}{S + K_m\left(1 + \dfrac{I}{K_i}\right)} = \frac{V_{max} \cdot S}{S + K'_m} \tag{7-48}$$

where

$$V_{max} = E_t k_5, \qquad K_i = \frac{k_4}{k_3},$$

$$K_m = \frac{k_2 + k_5}{k_1}, \qquad K'_m = K_m\left(1 + \frac{I}{K_i}\right)$$

an uncompetitive type enzymatic reaction are

$$E + S \underset{k_2}{\overset{k_1}{\rightleftharpoons}} E \cdot S$$

$$I + E \cdot S \underset{k_4}{\overset{k_3}{\rightleftharpoons}} IES$$

$$E \cdot S \overset{k_5}{\longrightarrow} P + E$$

The rate of formation of the product is

$$r_p = k_5(E \cdot S) \tag{7-49}$$

Application of the pseudo-steady-state hypothesis to E·S and IES yields the respective equations

$$r_{E \cdot S} \simeq 0 = k_1(E)(S) - k_2(E \cdot S) - k_3(I)(E \cdot S) + k_4(IES) - k_5(E \cdot S) \tag{7-50}$$

$$r_{IES} \simeq 0 = k_3(I)(E \cdot S) - k_4(IES) \tag{7-51}$$

To obtain E·S, we solve Equations (7-50) and (7-51) together with the mole balance for the total enzyme initially present,

$$E_t = (E) + (E \cdot S) + (IES) \tag{7-52}$$

and substitute the result into Equation (7-49) to obtain the rate law for the above mechanism involving the uncompetitive inhibition of an enzymatic reaction:

$$r_p = \frac{V_{max}S}{K_m + S\left(1 + \dfrac{I}{K_i}\right)} \tag{7-53}$$

A Lineweaver-Burk plot of Equation (7-53) for different inhibitor concentrations will result in a family of parallel lines all with a slope of K_m/V_{max}.

7.4C Noncompetitive Inhibition

In noncompetitive inhibition, the substrate and inhibitor molecules react with different types of sites on the enzyme molecule, and consequently the deactivating complex, IES, can be formed by two reversible reaction paths:

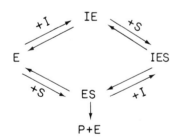

Figure 7-11

1. After a substrate molecule attaches to the enzyme molecule at the substrate site, the inhibitor molecule attaches to the enzyme at the inhibitor site.

2. After an inhibitor molecule attaches to the enzyme molecule at the inhibitor site, the substrate molecule attaches to the enzyme at the substrate site.

These paths, along with the formation of the product, P, are shown in Figure 7-11.

This reaction sequence can also be written as

Reaction 1	$E + S \rightleftharpoons E \cdot S$		(7-54)
Reaction 2	$E + I \rightleftharpoons I \cdot E$		(7-55)
Reaction 3	$I + E \cdot S \rightleftharpoons IES$		(7-56)
Reaction 4	$I \cdot E + S \rightleftharpoons IES$		(7-57)
Reaction 5	$E \cdot S \rightleftharpoons P + E$		(7-58)

An alternative method of finding the rate law for the sequence presented in Equations (7-54)–(7-58) is to assume that the last reaction, (7-58), is rate limiting and that each of the other reactions, (7-54) through (7-57), is essentially in equilibrium. This method is analogous to that used to derive the rate law in heterogeneous catalysis, where one of the steps (e.g., adsorption) is rate controlling. Assuming the rate limiting step (Reaction 5) is irreversible, the rate of formation of the product is

$$r_p = k_p (\text{E} \cdot \text{S}) \tag{7-59}$$

Using this equilibrium technique, the rate law for the first reaction, Equation (7-54),

$$r_1 = k_1 \left[(\text{E})(\text{S}) - \frac{(\text{E} \cdot \text{S})}{K_s} \right] \tag{7-60}$$

is rearranged and the ratio of the rate of reaction to the specific reaction rate is set equal to zero.

$$\frac{r_1}{k_1} \simeq 0 = (\text{E})(\text{S}) - \frac{(\text{E} \cdot \text{S})}{K_s}$$

The resulting equation is solved for the concentration of the enzyme-substrate complex.

$$(\text{E} \cdot \text{S}) = K_s (\text{S})(\text{E}) \tag{7-61}$$

Solve for the concentrations of the complexes (I·E) and (IES) in terms of the free enzyme concentration, (E), and substrate concentration (S) in a manner similar to that for (E·S); substitute these expressions in the total enzyme balance; and then obtain the rate law for noncompetitive inhibition. (Let K_s, K_I, K'_I, and K'_s represent the equilibrium constants for reactions 1 through 4, respectively.)

(G21)

Since the substrate and inhibitor attach at different sites, a reasonable assumption is that the equilibrium between the enzyme and the substrate is the same whether or not an inhibitor is attached to the enzyme, i.e., $K_s = K'_s$; then

$$(\text{EIS}) = K_s (\text{E} \cdot \text{I})(\text{S})$$

(G21)

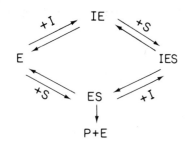

For reaction 2 shown in Equation (7-55),
$$(E \cdot I) = K_1(I)(E) \tag{G21-1}$$
We can use either Equation (7-56)
$$(EIS) = K_1'(E \cdot S)(I) = K_s K_1'(S)(E)(I) \tag{G21-2}$$
or (7-57)
$$(EIS) = K_s'(I \cdot E)(S) = K_s' K_1(I)(E)(S) \tag{G21-3}$$
to express IES in terms of $I \cdot E$, and S.

The mole balance on the total amount of enzyme, bound and free (E), is
$$E_t = E + E \cdot S + I \cdot E + IES \tag{G21-4}$$
Substituting Equations (7-61), (G21-1) and (G21-2) into (G21-4) and rearranging,
$$(E) = \frac{E_t}{1 + K_s(S) + K_1(I) + (I)(S)(K_1' K_s)} \tag{G21-5}$$
The rate of formation of the product for a noncompetitive inhibition of an enzymatic reaction is
$$r_p = \frac{k_p K_s E_t(S)}{1 + K_s(S) + K_1(I) + K_1' K_s(I)(S)} \tag{G21-6}$$

and
$$(E \cdot S) = K_s(E)(S)$$

By making a similar assumption for the equilibrium between the enzyme and the inhibitor, $K_I = K'_I$, we can rearrange Equation (G21-6) so that

$$r_p = \frac{V_{max} S}{(S + K_m)\left(1 + \dfrac{1}{K_i}\right)} \tag{7-62}$$

where

$$K_m = \frac{1}{K_s}, \qquad K_i = \frac{1}{K_I}, \qquad V_{max} = k_p E_t$$

Equation (7-62) is in the form of the rate law that is usually given for an enzymatic reaction exhibiting noncompetitive inhibition. Heavy metal ions such as Pb^{++}, Ag^+, Hg^{++}, and others, as well as inhibitors which react with the enzyme to form chemical derivatives, are typical examples of noncompetitive inhibitors. The various types of inhibition are compared with a reaction in which no inhibitors are present (dark line) on the Lineweaver-Burk plot shown below. We observe the following relationships:

(1) In *competitive inhibition* the slope increases with increasing inhibitor concentration while the intercept remains fixed;

(2) In *uncompetitive inhibition* the y-intercept increases with increasing inhibitor concentration while the slope remains fixed;

(3) *In noncompetitive inhibition* both the intercept and slope will increase with increasing inhibitor concentration.

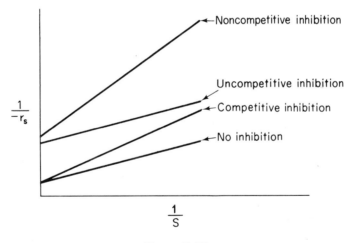

Figure 7-12

In addition to the Lineweaver-Burk plot, the *Eadie plot* is also used to present data of enzymatic reactions. The Eadie plot magnifies departures from linearity which might not be observed in the Lineweaver-Burk plot. In Figure 7-13a, b, and c, the Eadie plot is used to present the data for three enzymatic reactions; one with the competitive type of inhibition, one with the uncompetitive type of inhibition, and the third with the noncompetitive type of inhibition. Each figure contains two lines corresponding to two different inhibitor concentrations; line *a* represents the higher inhibitor concentration and line *b* the lower inhibitor concentration. Determine the figure that corresponds to each of the three types of inhibition: competitive, uncompetitive, and noncompetitive.

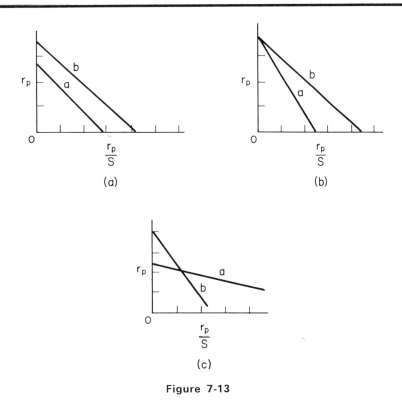

Figure 7-13

7.5–Multiple Enzyme and Substrate Systems

In the last section we discussed how the addition of a second substrate, I, to enzyme catalyzed reactions could deactivate the enzyme and greatly inhibit the reaction. In the present section, we look not only at systems in which the addition of a second substrate is necessary to activate the enzyme, but also other multiple-enzyme and multiple-substrate systems in which cyclic regeneration of the activated enzyme occurs.

7.5A Enzyme Regeneration

As a first example we shall consider the oxidation of glucose (S_r) with the aid of the enzyme glucose oxidase [represented as either G.O. or (E_o)] to give δ-gluconolactone (P)

$$\text{glucose} + \text{G.O.} \rightleftharpoons (\text{glucose} \cdot \text{G.O.}) \rightleftharpoons (\delta\text{-lactone} \cdot \text{G.O.H}_2) \rightleftharpoons \delta\text{-lactone} + \text{G.O.H}_2$$

In this reaction, the reduced form of glucose oxidase (G.O.H$_2$) which will be represented by E_r, cannot catalyze further reactions until it is oxidized back to E_o. This oxidation is usually carried out by adding molecular oxygen to the system so that glucose oxidase, E_o, is regenerated. Hydrogen peroxide is also produced in this oxidation regeneration step.

$$\text{G.O.H}_2 + \text{O}_2 \longrightarrow \text{G.O.} + \text{H}_2\text{O}_2$$

(G22) *For competitive inhibition:*

$$r_p = \frac{V_{max}S}{S + K'_m} \tag{G22-1}$$

Multiplying by $(S + K'_m)$ and rearranging, we obtain

$$r_p S + K'_m r_p = V_{max} S \tag{G22-2}$$

$$r_p = V_{max} - K'_m \frac{r_p}{S} \tag{G22-3}$$

We see that a plot of r_p vs. r_p/S will be a straight line with a slope K'_m, where

$$K'_m = \left(1 + \frac{I}{K_i}\right) K_m$$

A change in the inhibitor concentration will change the slope of the Eadie plot but not the intercept. The greater the inhibitor concentration, the greater (i.e., steeper) the slope.

Ans.: Figure 7-13b.

For uncompetitive inhibition:

$$r_p = \frac{V_{max}S}{K_m + S\left(1 + \dfrac{I}{K_i}\right)} \tag{7-53}$$

After rearranging (7-53) so that

$$r_p = \frac{V_{max}}{\left(1 + \dfrac{I}{K_i}\right)} - \frac{K_m}{\left(1 + \dfrac{I}{K_i}\right)} \frac{r_p}{S} \tag{G22-4}$$

we see that increasing the inhibitor concentration will decrease the slope and the intercept of an Eadie plot.

Ans.: Figure 7-13c.

For noncompetitive inhibition:

$$r_p = \frac{V_{max}S}{(S + K_m)\left(1 + \dfrac{I}{K_i}\right)} \tag{G22-5}$$

Again, multiplying by $(S + K_m)$ and rearranging, we obtain

$$r_p = \frac{V_{max}}{\left(1 + \dfrac{I}{K_i}\right)} - K_m \frac{r_p}{S} \tag{G22-6}$$

In the case of noncompetitive inhibition, increasing the inhibitor concentration decreases the intercept but does not affect the slope of an Eadie plot.

Ans.: Figure 7-13a.

Overall, the reaction is written

$$\text{glucose} + O_2 \xrightarrow{\text{glucose oxidase}} H_2O_2 + \delta\text{-gluconolactone}$$

In biochemistry texts, reactions of this type involving regeneration are usually written in the form

$$\text{glucose }(S_r) \diagdown \quad \diagup \text{G.O. }(E_o) \diagdown \quad \diagup H_2O_2(P_2)$$
$$\delta\text{-lactone }(P_1) \diagup \quad \diagdown \text{G.O.H}_2(E_r) \diagup \quad \diagdown O_2(S_2)$$

The reaction is believed to proceed by the following sequence of elementary reactions:

$$E_o + S_r \underset{k_{-1}}{\overset{k_1}{\rightleftharpoons}} E_o \cdot S_r$$

$$E_o \cdot S_r \underset{k_{-2}}{\overset{k_2}{\rightleftharpoons}} E_r \cdot S_o$$

$$E_r \cdot S_o \xrightarrow{k_3} P_1 + E_r$$

$$E_r + O_2 \xrightarrow{k_4} E_o + H_2O_2$$

We shall assume that reaction involving the dissociation between reduced glucose oxidase and δ-lactone is rate limiting. The rate of formation of δ-lactone (P_1) is given by the equation

$$r_P = k_3(E_r \cdot S_o) \tag{7-63}$$

After applying the pseudo-steady state hypothesis to the rates of formation of $(E_o \cdot S_r)$, $(E_r \cdot S_o)$ and (E_r),

$$r_{E_o \cdot S_r} = 0 = k_1(E_o)(S_r) - k_{-1}(E_o \cdot S_r) - k_2(E_o \cdot S_r) + k_{-2}(E_r \cdot S_o)$$

$$r_{E_r \cdot S_o} = 0 = k_2(E_o \cdot S_r) - k_{-2}(E_r \cdot S_o) - k_3(E_r \cdot S_o)$$

$$r_{E_r} = 0 = k_3(E_r \cdot S_o) - k_4(E_r)(O_2)$$

we can solve for the following concentrations of the active intermediates in terms of the concentrations of glucose, oxygen, and unbound oxidized enzyme.

$$E_o \cdot S_r = \frac{k_1(E_o)(S_r)(k_{-2} + k_3)}{k_{-1}k_{-2} + k_{-1}k_3 + k_2k_3} \tag{7-64}$$

$$E_r \cdot S_o = \frac{k_1 k_2 (E_o)(S_r)}{k_{-1}k_{-2} + k_{-1}k_3 + k_2k_3} \tag{7-65}$$

$$E_r = \frac{k_3(E_r \cdot S_o)}{k_4(O_2)} = \frac{k_1 k_2 k_3 (E_o)(S_r)}{k_4(O_2)(k_{-1}k_{-2} + k_{-1}k_3 + k_2k_3)} \tag{7-66}$$

After substituting Equation (7-65) into Equation (7-63) the rate law is written as

$$r_p = \frac{k_1 k_2 k_3 (E_o)(S_r)}{k_{-1}k_{-2} + k_{-1}k_3 + k_2k_3} \tag{7-67}$$

The total enzyme initially present is given by the sum

$$(E_t) = (E_o) + (E_o \cdot S_r) + (E_r \cdot S_o) + (E_r) \tag{7-68}$$

After using Equations (7-64), (7-65), and (7-66) to substitute for the active intermediate in

Equation (7-68), one can then solve for the unbound oxidized enzyme concentration, E_o, and substitute it into Equation (7-67) to obtain the form of the rate law

$$r_p = \frac{V_{max}(S_r)(O_2)}{K_1(S_r) + K_2(O_2) + (S_r)(O_2)} \tag{7-69}$$

Construct a Lineweaver-Burk plot for this system in which each line corresponds to a different oxygen concentration. Also discuss the effect of total gas pressure of fixed oxygen concentration on the rate of this liquid phase reaction, r_p. (Hint: Recall Henry's Law.)

(G23) ───

───

The above reaction illustrates how an enzyme can be regenerated through the addition of another substrate, in this case O_2.

7.5B Enzyme Cofactors

In many enzymatic reactions, and in particular biological reactions, a second substrate (i.e., species) must be introduced in order to activate the enzyme. This substrate, which is referred to as a *cofactor* or *coenzyme* even though it is not an enzyme as such, attaches to the enzyme and is most often either reduced or oxidized during the course of the reaction. The enzyme-cofactor complex is referred to as a *holoenzyme*. An example of the type of system in which a cofactor is used is the formation of ethanol from acetaldehyde in the presence of the enzyme alcohol dehydrogenase (ADH) and the cofactor nicotinamide adenine dinucleotide (NAD). After the enzyme is activated by combination with the cofactor in its reduced state, NADH,

$$ADH + NADH \underset{k_{-1}}{\overset{k_1}{\rightleftharpoons}} ADH \cdot NADH,$$

$$(\text{Inactive}) \rightleftharpoons (\text{Active})$$

the holoenzyme (ADH·NADH) reacts with acetaldehyde in acid solution to produce ethanol and the oxidized form of the enzyme-cofactor coupling (ADH·NAD$^+$)

$$ADH \cdot NADH + CH_3CHO + H^+ \underset{k_{-2}}{\overset{k_2}{\rightleftharpoons}} ADH \cdot NAD^+ + CH_3CH_2OH$$

The inactive form of the enzyme-cofactor complex for a specific reaction and reaction direction is called an *apoenzyme*. This reaction is followed by dissociation of the apoenzyme (ADH·NAD$^+$) which is usually relatively slow.

$$ADH \cdot NAD^+ \underset{k_{-3}}{\overset{k_3}{\rightleftharpoons}} ADH + NAD^+$$

The values of the specific rates are:[†] $k_1 = 3.7 \times 10^6$ l/gmole/sec, $k_{-1} = 1.6$ sec^{-1}, $k_2(H^+) = 2.4 \times 10^5$ l/gmole/sec, $k_{-2} = 3.5 \times 10^3$ l/gmole/sec, $k_3 = 37$ sec^{-1}, $k_{-3} = 3 \times 10^5$ l/gmole/sec.

[†]H. Theorell, et. al, *Acta Chem. Scand.*, 9 1148 (1955).

(G23)

We first invert Equation (7-69) to find

$$\frac{1}{r_p} = \left[\frac{1}{V_{max}} + \frac{K_1}{V_{max}(O_2)}\right] + \left[\frac{K_2}{V_{max}(S_r)}\right] \qquad \text{(G23-1)}$$

From Equation (G23-1) we see that changes in the oxygen concentration will only affect the intercept of a Lineweaver-Burk plot. As the oxygen concentration increases the intercept decreases.

Assuming equilibrium between oxygen in the gas phase and oxygen in the liquid phase, we obtain the following relationship (*Henry's Law*).

$$C_{O_2} = H'P_{O_2} = H'y_{O_2}\Pi$$

where

P_{O_2} = partial pressure of O_2 in the gas phase
C_{O_2} = liquid phase concentration of O_2
y_{O_2} = gas phase mole fraction of O_2 (fixed)
Π = total pressure
H' = Henry's Law constant

$$r_p = \frac{V_{max}y_{O_2}H'\Pi(S_r)}{K_1(S_r) + K_2y_{O_2}H'\Pi(S_r)(y_{O_2}H'\Pi)} \qquad \text{(G23-2)}$$

at high pressures

$$r_p \simeq V_{max} \times S_r/(K_2 + S_r)$$

at low pressures

$$r_p \simeq \left(y_{O_2}H'\frac{V_{max}}{K_1}\right)\Pi$$

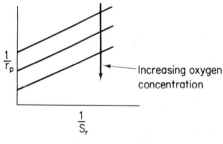

Figure 7-14

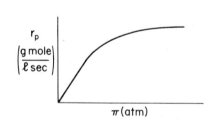

Figure 7-15

Typical initial concentrations for a small laboratory batch reactor experiment might be [acetaldehyde]$_0$ = 10^{-1} gmoles/l, [ADH]$_0$ = 10^{-7} gmole/l, and [NADH]$_0$ = 10^{-4} gmole/l. The overall reaction is often written in the form

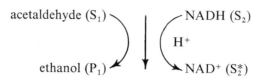

Alcohol dehydrogenase

The ADH enzyme molecule produced by the dissociation of (ADH·NAD$^+$) can participate in subsequent reactions involving the formation of ethanol, while the nicotinamide adenine dinucleotide from the dissociation cannot participate until it is reduced back to NADH. Since the initial concentration of NADH is usually several orders of magnitude greater than the initial concentration of enzyme, the consumption of NADH will not limit the overall rate of formation of ethanol nearly so much as the slow dissociation of the (ADH·NAD$^+$) complex. This apoenzyme essentially ties up the enzyme preventing it from becoming free (unbound) to combine with NADH to form the holoenzyme which reacts with acetaldehyde to produce ethanol. We note that the reaction rate might be increased considerably if we had a way of going directly from (ADH·NAD$^+$) to (ADH·NADH), i.e.

$$\text{ADH·NAD}^+ \xrightarrow{?} \text{ADH·NADH}$$

rather than having the enzyme ADH go through the steps of dissociating from

$$\text{ADH·NAD}^+ \longrightarrow \text{ADH} + \text{NAD}^+$$

and then combining with NADH

$$\text{ADH} + \text{NADH} \longrightarrow \text{ADH·NADH}$$

Before considering the possibility of going directly from the apoenzyme to the holoenzyme, assume that the rate of dissociation of the complex (ADH·NAD$^+$)

$$\text{ADH·NAD}^+ \xrightarrow{k_3} \text{ADH} + \text{NAD}^+$$

is irreversible, and show that the initial rate law for ethanol in the enzyme cofactor reaction sequence discussed on page 289 is of the form

$$r_p = \frac{V_{\max}(S_1)(S_2)}{K_{12} + K_1(S_1) + K_2(S_2) + (S_1)(S_2)} \quad (7\text{-}70)$$

(G24)

Equation (7-70) is of a form which is often used in the interpretation of initial rate data for enzymatic reactions involving two substrates. The parameters K_{12}, K_1, K_2, and $V_{\max}$ in Equation

(G24)

$$ADH + NADH \rightleftharpoons ADH \cdot NADH$$
$$ADH \cdot NADH + CH_3CHO + H^+ \rightleftharpoons ADH \cdot NAD^+ + CH_3CH_2OH$$
$$ADH \cdot NAD^+ \longrightarrow ADH + NAD^+$$

Let $E = ADH$, $S_1 = CH_3CHO$, $S_2 = NADH$, $S_2^* = NAD^+$, $P_1 = CH_3CH_2OH$,
Then

$$E + S_2 \underset{k_{-1}}{\overset{k_1}{\rightleftharpoons}} E \cdot S_2$$

$$E \cdot S_2 + S_1 + H^+ \underset{k_{-2}}{\overset{k_2}{\rightleftharpoons}} E \cdot S_2^* + P_1$$

$$E \cdot S_2^* \overset{k_3}{\longrightarrow} E + S_2^*$$

By adding the rate law for the rate of formation of ethanol (P_1)
$$r_p = k_2(H^+)(S_1)(E \cdot S_2^*) - k_{-2}(E \cdot S_2^*)P_1$$
to the equation for $-r_{E \cdot S_2^*} = 0$, (below) we see that the rate law for ethanol can be written as
$$r_p = k_3(E \cdot S_2^*) \tag{G24-1}$$

Application of the PSSH to the holoenzyme ($E \cdot S_2$) and the apoenzyme ($E \cdot S_2^*$)
$$r_{E \cdot S_2} = 0 = k_1(E)(S_2) - k_{-1}(E \cdot S_2) - k_2(H^+)(S_1)(E \cdot S_2) + k_{-2}(E \cdot S_2^*)(P_1)$$
$$r_{E \cdot S_2^*} = 0 = k_2(H^+)(S_1)(E \cdot S_2) - k_{-2}(E \cdot S_2^*)P_1 - k_3(E \cdot S_2^*)$$

allows one to solve for the concentrations of the cofactor-enzyme complexes in terms of S_1, S_2, S_2^* and E.

$$(E \cdot S_2) = \frac{k_1(E)(S_2)(k_{-2}(P_1) + k_3)}{k_{-1}k_{-2}(P_1) + k_{-1}k_3 + k_2(H^+)k_3(S_1)} \tag{G24-2}$$

$$(E \cdot S_2^*) = \frac{k_1 k_2 H^+(S_1)(S_2)(E)}{k_{-1}k_{-2}(P_1) + k_{-1}k_3 + k_2(H^+)k_3(S_1)} \tag{G24-3}$$

The total concentration of bound and unbound enzyme is
$$(E_t) = (E) + (E \cdot S_2) + (E \cdot S_2^*) \tag{G24-4}$$

Substituting Equations (G24-2) and (G24-3) into Equation (G24-4), the unbound enzyme concentration is

$$(E) = \frac{(E_t)(k_{-1}k_{-2}(P_1) + k_{-1}k_3 + k_2(H^+)k_3(S_1))}{k_{-1}k_{-2}(P_1) + k_{-1}k_3 + k_2(H^+)k_3(S_1) + k_1(S_2)(k_{-2}(P_1) + k_3) + k_1 k_2(H^+)(S_1)(S_2)} \tag{G24-5}$$

Setting $(P_1) = 0$, we obtain the initial rate law by combining Equations (G24-1), (G24-3) and (G24-5).

$$r_p = \frac{V_{max}(S_1)(S_2)}{K_{12} + K_1(S_1) + K_2(S_2) + (S_1)(S_2)} \tag{G24-6}$$

where

$$V_{max} = k_3(E_t) = (37 \text{ sec}^{-1})(10^{-7} \text{ gmole}/l) = 3.7 \times 10^{-6} \text{ gmole}/l/\text{sec}.$$

$$K_{12} = \frac{k_{-1}k_3}{k_1 k_2 H^+} = \frac{(1.6 \text{ sec}^{-1})(37 \text{ sec}^{-1})}{(3.7 \times 10^6 \; l/\text{gmole/sec})(2.4 \times 10^{+5} \; l/\text{gmole/sec})}$$
$$= 6.66 \times 10^{-11} \; (\text{gmole}/l)^2$$

$$K_1 = \frac{k_3}{k_1} = \frac{37 \text{ sec}^{-1}}{3.7 \times 10^6 \; l/\text{gmole/sec}} = 10^{-5} \text{ gmole}/l$$

$$K_2 = \frac{k_3}{k_2 H^+} = \frac{37 \text{ sec}^{-1}}{2.4 \times 10^5 \; l/\text{gmole/sec}} = 1.54 \times 10^{-4} \text{ gmole}/l$$

$$\boxed{r_p = \frac{3.7 \times 10^{-6}(S_1)(S_2)}{6.66 \times 10^{-11} + 10^{-5}(S_1) + 1.54 \times 10^{-4}(S_2) + (S_1)(S_2)}}$$

(7-70), which was first developed by Dalziel,† may be evaluated through a series of Lineweaver-Burk plots.

After substituting the numerical values for K_1, K_2, K_{12} and recalling that the initial concentrations specified ($S_{1,0} = .1$ gmole/l, $S_{2,0} = 10^{-4}$ gmole/l) we see that we can neglect K_{12} and $K_2(S_2)$ w.r.t. the other terms in the denominator, in which case Equation (G25-7) becomes

$$r_p = \frac{3.7 \times 10^{-6}(S_2)}{10^{-5} + (S_2)}$$

The initial rate is $r_p = 3.7 \times 10^{-6}$ gmole/l/sec. In the next section we shall compare the above rate with one in which a third substrate is added to the system.

7.5C Multiple Substrate Systems

In the previous section we stated that the rate of formation of ethanol might be increased if the (ADH·NAD$^+$) complex could be converted by some means directly to the enzyme-cofactor complex (ADH·NADH) without having the enzyme ADH go through a series of reactions. This can be achieved by the addition of a third substrate, S_3, (e.g. propanediol) which during reaction (to form DL-lactaldehyde) will also regenerate the cofactor (NADH). The overall reaction sequence for this case is

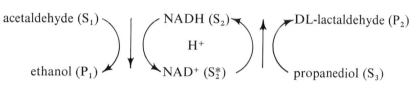

As a first approximation, this sequence of reactions could be represented by the following elementary steps:

$$E + S_2 \underset{k_{-1}}{\overset{k_1}{\rightleftarrows}} E \cdot S_2$$

$$E \cdot S_2 + S_1 \overset{(k_2 H^+)}{\longrightarrow} E \cdot S_2^* + P_1$$

$$E \cdot S_2^* + S_3 \underset{k_{-4}}{\overset{k_4}{\rightleftarrows}} E \cdot S_2 + P_2$$

Derive the rate law for this sequence assuming that the rate of reaction between the (ADH·NADH) complex and acetaldehyde is rate limiting.

(G25)

This same reaction has been carried in the *reverse direction* by Gupta and Robinson.‡ They

---

†K. Dalziel, *Acta. Chem. Scand. 11*, 1706 (1957).
‡N.K. Gupta, and W.G. Robinson, *Biochim. Biophys. Acta 118*, 431 (1966).

(G25) Assuming that the reaction between the cofactor-enzyme complex and acetaldehyde is rate limiting, the rate law is

$$r_{p_1} = k_2(E \cdot S_2)(S_1) \tag{G25-1}$$

(Note: $k_2 \equiv k_2(H^+)$)

After applying the PSSH to $(E \cdot S_2)$ and $(E \cdot S_2^*)$ to obtain the equations

$$r_{E \cdot S_2} = 0 = k_1(E)(S_2) - k_{-1}(E \cdot S_2) - k_2(E \cdot S_2)(S_1) + k_4(S_3)(E \cdot S_2^*) - k_{-4}(E \cdot S_2)(P_2) \tag{G25-2}$$

$$r_{E \cdot S_2^*} = 0 = k_2(E \cdot S_2)(S_1) - k_4(E \cdot S_2^*)(S_3) + k_{-4}(E \cdot S_2)(P_2) \tag{G25-3}$$

we add equations (G25-2) and (G25-3) and rearrange to obtain

$$(E \cdot S_2) = \frac{k_1}{k_{-1}}(E)(S_2) \tag{G25-4}$$

$$(E \cdot S_2^*) = \frac{(k_2(S_1) + k_{-4}(P_2))}{k_4(S_3)}(E \cdot S_2) = \left[\frac{k_1}{k_{-1}}\right]\frac{(k_2(S_1) + k_{-4}(P_2))(E)(S_2)}{k_4(S_3)} \tag{G25-5}$$

The total enzyme concentration is again constant at (E_t)

$$(E_t) = (E) + (E \cdot S_2) + (E \cdot S_2^*) \tag{G25-6}$$

$$(E_t) = (E) + \frac{k_1}{k_{-1}}(E)(S_2) + \frac{k_1 k_2}{k_{-1} k_4}\frac{(S_1)}{(S_3)}(E)(S_2) + \frac{k_1 k_{-4}(P_2)(E)(S_2)}{k_{-1} k_4 (S_3)} \tag{G25-7}$$

After solving Equation (G25-7), for E, we combine Equations (G25-1) and (G25-4) to obtain the rate law

$$r_{p_1} = \left[\frac{k_1 k_2}{k_{-1}}\right]\frac{(S_1)(S_2)(S_3)(E_t)}{(S_3) + \frac{k_1(S_2)(S_3)}{k_{-1}} + \frac{k_1 k_2 (S_1)(S_2)}{k_{-1} k_4} + \frac{k_1 k_{-4}}{k_{-1} k_4}(P_2)(S_2)} \tag{G25-8}$$

Setting $P_2 = 0$ and rearranging, we obtain the initial rate law

$$r_{p_1} = \frac{V_{max}(S_1)(S_2)(S_3)}{K_3(S_3) + K_{23}(S_2)(S_3) + (S_1)(S_2)} \tag{G25-9}$$

where:

$$V_{max} = k_4(E_t)$$

$$K_3 = \frac{k_{-1} k_4}{k_1 k_2}$$

$$K_{23} = \frac{k_4}{k_2}$$

measured the initial rate of conversion of DL-lactaldehyde to propanediol in the presence of NAD$^+$ and ADH. The rate of dissociation of the enzyme cofactor complex (ADH·NADH) is believed to be rate limiting. This is confirmed by the fact that when ethanol was added to the system, the reaction rate increased 100 fold by having the ethanol convert the (ADH·NADH) directly back to (ADH·NAD$^+$).

Assume that the specific reaction rates for the conversion of propanediol to DL-lactaldehyde are the same order of magnitude as the *corresponding* specific reaction rates for the conversion of ethanol to acetaldehyde. Calculate the initial rate of formation of ethanol in the presence of propanediol and compare this rate with the initial rate when propanediol is absent from the system.

(G26)

In analyzing multiple reactions in this manner, one should always question the validity of the application of the PSSH to the various active intermediates.

Since the nicotinamide adenine dinucleotide is continually regenerated and the total concentration of the cofactor (in its *oxidized, reduced, bound,* and *unbound* forms) remains constant throughout the course of the reaction, it might be desirable to replace S_2 in the rate law in terms of the total cofactor concentration, S_t. Neglecting any unbound S_2^*, the total (initial) cofactor concentration is

$$S_t = S_2 + E \cdot S_2 + E \cdot S_2^* \tag{7-71}$$

The total concentration of enzyme is

$$E_t = E + E \cdot S_2 + E \cdot S_2^* \tag{7-72}$$

Subtracting Equation (7-71) from Equation (7-72) one obtains

$$(E_t) = (E) - (S_2) + (S_t) \tag{7-73}$$

Equation (G25-7) can be rewritten in the form

$$(E_t) = (E) + \beta(S_2)(E) \tag{7-74}$$

where

$$\beta = \frac{k_1}{k_{-1}} + \frac{k_1 k_2 (S_1)}{k_{-1} k_4 (S_3)} + \frac{k_1 k_{-3} P_2}{k_{-1} k_4 (S_3)}$$

After adding Equations (7-73) and (7-74) and rearranging, we obtain

$$\beta(E)^2 - [\beta(E_t) - 1 - \beta(S_t)](E) - (E_t) = 0 \tag{7-75}$$

which is solved for the unbound enzyme concentration in terms of S_1, S_3, P_2, E_t, and S_t.

$$E = \frac{(\beta(E_t) - 1 - \beta(S_t)) \pm \sqrt{(\beta(E_t) - 1 - \beta(S_t))^2 + 4\beta(E_t)}}{2\beta} \tag{7-76}$$

We can substitute Equation (7-76) into (7-71) and rearrange to determine the concentration of the unbound cofactor in its reduced form, i.e.,

$$(S_2) = \frac{-1 \pm \sqrt{(\beta(E_t) - 1 - \beta(S_t))^2 + 4\beta(E_t)}}{2\beta} + (S_t - E_t)/2 \tag{7-77}$$

The rate law was given by

$$r_{p_1} = k_2(E \cdot S_2)(S_1) = \frac{k_1 k_2}{k_{-1}}(S_1)(S_2)(E) \tag{7-78}$$

(G26) Assuming the corresponding specific reaction rates to be equal,

$$k_4 = O(k_{-2})$$
$$k_4 \simeq 3.5 \times 10^3 \ l/\text{gmole/sec}$$
$$k_{-4} = O(k_2 H^+)$$
$$k_{-4} \simeq 2.4 \times 10^5 \ l/\text{gmole/sec}$$

We can then obtain numerical values for V_{max}, K_3 and K_{23} in Equation (G25-9)

$$V_{max} = (k_4)(E_t) = (3.5 \times 10^3)(10^{-7}) = 3.5 \times 10^{-4} \ \text{sec}^{-1}$$

$$K_3 = \frac{k_{-1}k_4}{k_1 k_2} = \frac{(1.6 \ \text{sec}^{-1})(3.5 \times 10^3 \ l/\text{gmole/sec})}{(3.7 \times 10^6 \ l/\text{gmole/sec})(2.4 \times 10^5 \ l/\text{gmole/sec})} = 6.3 \times 10^{-9} \ \frac{\text{gmole}}{l}$$

$$K_{23} = \frac{k_4}{k_2} = \frac{3.5 \times 10^3 \ l/\text{gmole/sec}}{2.4 \times 10^5 \ l/\text{gmole/sec}} = 1.45 \times 10^{-2}$$

$$r_p = \frac{3.5 \times 10^{-4}(S_1)(S_2)(S_3)}{6.3 \times 10^{-9}(S_3) + 1.45 \times 10^{-2}(S_2)(S_3) + (S_1)(S_2)}$$

initially

$$(S_1) = .1 \ \text{gmole}/l$$
$$(S_2) = 10^{-4} \ \text{gmole}/l$$

$$r_p = \frac{3.5 \times 10^{-9} (S_3)}{6.3 \times 10^{-9}(S_3) + 1.45 \times 10^{-6}(S_3) + 10^{-5}} \qquad \text{(G26-2)}$$

If

$$S_3 < 1$$

we can neglect the first two terms in the denominator with respect to the last term.

$$r_p = 3.5 \times 10^{-4} \ \text{sec}^{-1} \ (S_3) \qquad \text{(G26-3)}$$

When no propanediol was present we found

$$r_{p_1} = 3.7 \times 10^{-6} \ \text{gmole}/l/\text{sec}$$

If the propanediol and acetaldehyde are present in the same initial concentration

$$S_3 = .1 \ \text{gmole}/l$$

then

$$r_p = 3.5 \times 10^{-5} \ \text{gmole}/l/\text{sec}$$

and we see that the addition of the third substrate increases the rate of reaction by a factor of 10!!

One could substitute Equations (7-76) and (7-77) for S_2 and E into Equation (7-78) to arrive at a reasonably complicated rate law involving S_1, S_t, S_3, P_2, and E_t. However, a computer solution would be used in most reaction sequences that are this involved algebraically, in which case further substitution would not be necessary and one could use Equations (7-76), (7-77), and (7-78) directly.

7.5D Multiple Enzymes Systems

We shall again consider the production of ethanol from acetaldehyde which uses the cofactor NADH. However the regeneration of NAD^+ to NADH is brought about in a reaction catalyzed by acetaldehyde dehydrogenease (E_2) which produces acetic acid from acetaldehyde

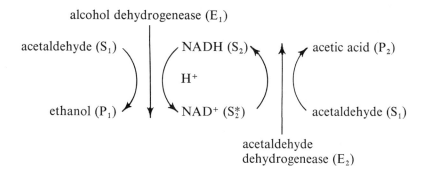

This sequence can be written in abstract notation as

$$S_2 + E_1 \rightleftharpoons E_1 \cdot S_2$$
$$E_1 \cdot S_2 + S_1 + H^+ \rightleftharpoons E_1 \cdot S_2^* + P_1$$
$$E_1 \cdot S_2^* \rightleftharpoons E_1 + S_2^*$$
$$S_2^* + E_2 \rightleftharpoons E_2 \cdot S_2^*$$
$$E_2 \cdot S_2^* + S_1 \rightleftharpoons E_2 \cdot S_2 + P_2$$
$$E_2 \cdot S_2 \rightleftharpoons E_2 + S_2$$

One could apply the PSSH to $E_1 \cdot S_2$, $E_1 \cdot S_2^*$, $E_2 \cdot S_2^*$ and $E_2 \cdot S_2$ and represent the total enzyme concentrations as

$$E_{1t} = E_1 + E_1 \cdot S_2 + E_1 \cdot S_2^*$$

$$E_{2t} = E_2 + E_2 \cdot S_2 + E_2 \cdot S_2^*$$

in deriving the rate law for this system. We shall not carry through the necessary algebraic manipulation to obtain the rate law here as all the principles for determining the rate law have been presented and there would be little more to be accomplished by doing this.

The above sections on enzymatic reactions were meant to serve as a brief yet somewhat encompassing discussion of enzyme kinetics. Further discussion can be found in the references in the supplementary reading.

SUMMARY

1. The azomethane decomposition is

$$2\text{AZO} \underset{k_2}{\overset{k_1}{\rightleftharpoons}} \text{AZO} + \text{AZO}^* \tag{S7-1}$$

$$\text{AZO}^* \xrightarrow{k_3} \text{N}_2 + \text{ethane}$$

The rate expression for the mechanism of this decomposition,

$$r_{\text{N}_2} = \frac{k(\text{AZO})^2}{1 + k'(\text{AZO})} \tag{S7-2}$$

exhibits first-order dependence w.r.t. AZO at high AZO concentrations and second-order dependence w.r.t. AZO at low AZO concentrations.

2. In the pseudo-steady-state hypothesis, we set the net rate of formation of the active intermediates equal to zero. If the actual intermediate A* is involved in m different reactions we set

$$r_{\text{A}^*,\text{net}} \equiv \sum_{i=1}^{m} r_{\text{A}^*i} = 0 \tag{S7-3}$$

This approximation is justified when the active intermediate is highly reactive and present in low concentrations.

3. The enzymatic reaction for the decomposition of urea, S, catalyzed by urease, E,

$$\text{E} + \text{S} \underset{k_2}{\overset{k_1}{\rightleftharpoons}} \text{E} \cdot \text{S} \tag{S7-4}$$

$$\text{W} + \text{E} \cdot \text{S} \xrightarrow{k_3} \text{P} + \text{E}$$

follows Michaelis-Menten kinetics; the rate expression is

$$r_\text{S} = \frac{V_{\max}\text{S}}{\text{S} + K_m} \tag{S7-5}$$

where $V_{\max}$ is the maximum rate of reaction for a given enzyme concentration and K_m is the Michaelis constant.

4. The total amount of a given enzyme in the system is the sum of the free enzyme, E, and the bound enzyme E·S.

$$E_t = \text{E} \cdot \text{S} + \text{E} \tag{S7-6}$$

To arrive at Equation (S7-5) we treat each reaction as elementary, apply the pseudo-steady-state hypothesis to the complex, and utilize Equation (S7-6).

5. For competitive inhibition:

$$\text{E} + \text{S} \rightleftharpoons \text{E} \cdot \text{S}$$
$$\text{E} + \text{I} \rightleftharpoons \text{E} \cdot \text{I}$$
$$\text{E} \cdot \text{S} \longrightarrow \text{E} + \text{P}$$

The rate law is

$$r_p = \frac{V_{\max}\text{S}}{\text{S} + K_m\left(1 + \dfrac{\text{I}}{K_i}\right)}$$

6. For uncompetitive inhibition:

$$\text{E} + \text{S} \rightleftharpoons \text{E} \cdot \text{S}$$
$$\text{I} + \text{E} \cdot \text{S} \rightleftharpoons \text{IES}$$
$$\text{E} \cdot \text{S} \longrightarrow \text{P} + \text{E}$$

The rate law is

$$r_p = \frac{V_{max}S}{K_m + S\left(1 + \dfrac{I}{K_i}\right)}$$

7. For noncompetitive inhibition:

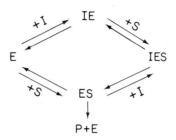

the rate law is

$$r_p = \frac{V_{max}S}{(S + K_m)\left(1 + \dfrac{I}{K_i}\right)}$$

8. An enzyme can be reduced during the course of one reaction and then be reoxidized by a second substrate during the course of a second reaction

$$E_o + S_r \rightleftharpoons E_o \cdot S_r \rightleftharpoons E_r \cdot S_o \rightleftharpoons E_r + S_o$$
$$E_r + O_2 \longrightarrow E_o + H_2O_2$$

9. A *holoenzyme* is formed from the reaction of an enzyme and its cofactor

$$\text{Enzyme} + \text{Cofactor} \rightleftharpoons \text{Enzyme} \cdot \text{Cofactor}$$
(INACTIVE)   (ACTIVE)

The holoenzyme is usually reduced or oxidized during the course of its reaction with the substrate.

QUESTIONS AND PROBLEMS

P7-1. The pyrolysis of acetaldehyde is believed to take place according to the sequence

$$CH_3CHO \xrightarrow{k_1} CH_3\cdot + CHO\cdot$$
$$CH_3\cdot + CH_3CHO \xrightarrow{k_2} CH_3\cdot + CO + CH_4$$
$$CHO\cdot + CH_3CHO \xrightarrow{k_3} CH_3\cdot + 2CO + H_2$$
$$2CH_3\cdot \xrightarrow{k_4} C_2H_6$$

Derive the rate expression for the rate of disappearance of acetaldehyde, $-r_{Ac}$. Under what conditions does it reduce to Equation (7-2)?

P7-2. The following observations have been made about the reaction

$$2GH_2 \longrightarrow 2GH + H_2$$

1. The reaction rate is independent of hydrogen concentration.

2. The reaction rate decreases with increasing concentration of GH.

3. The initial rate appears to be first order w.r.t. GH_2.

4. When there is an extremely high concentration of GH and a low concentration of GH$_2$, the reaction appears to be second order w.r.t. GH$_2$.
   a. Deduce a form of the rate law from the above information.
   b. Suggest a mechanism.

P7-3. For the decomposition of ozone in an inert gas M, the rate expression is

$$-r_{O_3} = \frac{k(O_3)^2(M)}{(O_2)(M) + k'(O_3)}$$

Suggest a mechanism.

P7-4. The reaction of hypochloride and iodide ions in *aqueous* solution,

$$I^- + OCl^- \longrightarrow OI^- + Cl^-$$

occurs rapidly. The rate law for this reaction is reported to be

$$r_{OI^-} = \frac{k(I^-)(OCl^-)}{(OH^-)}$$

Suggest a mechanism. (*Hint*: It is believed that either HOCl or HOI or both may be active intermediates.)

P7-5. The gas phase homogeneous oxidation of nitrogen monoxide (NO) to the dioxide (NO$_2$)

$$2NO + O_2 \xrightarrow{k} 2NO_2,$$

is known to have a form of third-order kinetics that suggests that the reaction is elementary as written, at least for the low partial pressures of the nitrogen oxides. However, the rate constant k actually *decreases* with increasing absolute temperature, indicating an apparently *negative* activation energy. Since the activation energy of any elementary reaction must be positive, some explanation is in order.

Provide an explanation, starting from the fact that an active intermediate species, NO$_3$, is a participant in some other known reactions that involve oxides of nitrogen.

P7-6. The thermal decomposition of diethyl ether is believed to proceed by the mechanism

$$C_2H_5OC_2H_5 \underset{k_{-1}}{\overset{k_1}{\rightleftharpoons}} CH_3\cdot + \cdot CH_2OC_2H_5$$

$$CH_3\cdot + C_2H_5OC_2H_5 \xrightarrow{k_2} C_2H_6 + \cdot CH_2OC_2H_5$$

$$\cdot CH_2OC_2H_5 \xrightarrow{k_3} CH_3\cdot + CH_3CHO$$

Show that this mechanism can lead to the rate law

$$-r_{(C_2H_5)_2O} = k[C_2H_5OC_2H_5]$$

where $k = k_2(k_1k_3/k_{-1}k_2)^{1/2}$.

P7-7. The thermal decomposition of ethane to ethylene and hydrogen is believed to follow the sequence

$$C_2H_6 \xrightarrow{k_1} 2CH_3\cdot$$

$$CH_3\cdot + C_2H_6 \xrightarrow{k_2} CH_4 + C_2H_5\cdot$$

$$C_2H_5\cdot \xrightarrow{k_3} C_2H_4 + H\cdot$$

$$H\cdot + C_2H_6 \xrightarrow{k_4} H_2 + C_2H_5\cdot$$

$$2C_2H_5\cdot \xrightarrow{k_5} C_4H_{10}$$

Derive a rate law for the rate of formation of ethylene.

P7-8. Derive a rate law for the enzyme catalyzed reaction sequence

$$E + S \underset{k_2}{\overset{k_1}{\rightleftharpoons}} E\cdot S \underset{k_4}{\overset{k_3}{\rightleftharpoons}} P + E$$

in terms of the substrate concentration, the total enzyme concentration, and the specific reaction rates k_1, k_2, k_3, and k_4.

P7-9. The use of beef liver catalase has been used to accelerate the decomposition of hydrogen peroxide to yield water and oxygen. (*J. Chemical Engineering Education*, 5, (1971) 141.) The concentration of hydrogen peroxide is given as a function of time for a reaction mixture with a pH of 6.76 and maintained at 30°C.

t (min)	0	10	20	50	100
$C_{H_2O_2}$ (mole/l)	.02	.01775	.0158	.0106	.005

    a. Determine the Michaelis-Menten parameters V_{max} and K_m.

    b. If the total enzyme concentration is tripled, what will the substrate concentration be after 20 min?

P7-10. At high substrate concentrations, the substrate itself may tie up the enzyme substrate complex by forming the complex $S \cdot E \cdot S$:

$$E \cdot S + S \rightleftharpoons S \cdot E \cdot S \qquad \text{(P7-10-1)}$$

Derive the rate law for the reaction

$$E + S \rightleftharpoons E \cdot S \longrightarrow E + P \qquad \text{(P7-10-2)}$$

in which substrate inhibition is also taking place.

P7-11. The enzymatic reaction

$$E + S \rightleftharpoons E \cdot S \qquad \text{(P7-11-1)}$$
$$E \cdot S + A \longrightarrow P + E \qquad \text{(P7-11-2)}$$

is carried out in the liquid phase. Derive the rate laws for this reaction for:

    a. No inhibitor is present.

    b. Product, P, inhibition of the competitive type.

    c. Product inhibition of the noncompetitive type.

    d. Product inhibition of the uncompetitive type.

P7-12. Consider the application of the pseudo-steady-state hypothesis to epidemiology. We shall treat each of the following steps as elementary in that the rate will be proportional to the number of people in a particular state of health. A healthy person, H, can become ill, I, spontaneously,

$$H \longrightarrow I \qquad \text{(P7-12-1)}$$

or he may become ill through contact with another ill person

$$I + H \longrightarrow 2I \qquad \text{(P7-12-2)}$$

The ill person may become well,

$$I \longrightarrow H \qquad \text{(P7-12-3)}$$

or he may expire

$$I \longrightarrow D \qquad \text{(P7-12-4)}$$

The reaction given in Equation (P7-12-4) is normally considered completely irreversible, although the reverse reaction has been reported to occur.

    a. Derive an equation for the death rate.

    b. At what concentration of healthy people does the death rate become critical?

    c. Comment on the validity of the pseudo-steady-state hypothesis under the conditions of part b.

P7-13. It has been observed that substrate inhibition occurs in the enzymatic reaction

$$E + S \rightleftharpoons E \cdot S$$
$$E \cdot S \longrightarrow P + E$$
$$E \cdot S + S \rightleftharpoons S \cdot E \cdot S$$

   a. Show that the rate law for the above sequence is consistent with the plot in Figure P7-13 of $-r_s$ (mmoles/l/min) vs. the substrate concentration, S, (mmoles/l).

   b. If this reaction is carried out in a CSTR that has a volume of 1000 l to which the volumetric flow rate is 3.2 l/min, determine the three possible steady states, noting, if possible, which are stable. The entrance concentration of the substrate is 50 mmoles/l.

   c. What is the highest conversion possible for this CSTR when it is operated at conditions specified above?

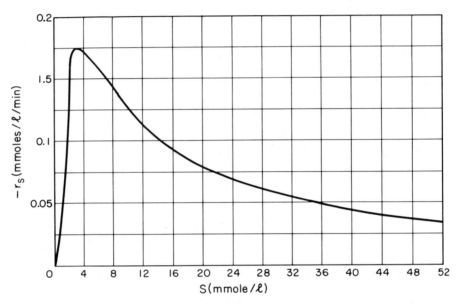

Figure P7–13

P7-14. Phthalic acid (OPA) is an intermediate which has many uses in the chemical process industries. It can be obtained by oxidizing *o*-toluic acid (OTA) using sodium dichromate,

[structural formula: NaOTA] + $\tfrac{3}{2}$ Na$_2$Cr$_2$O$_7$ ⟶ [structural formula: NaOPA] + Cr$_2$O$_3$ + Na$_2$CrO$_4$ + $\tfrac{3}{2}$ H$_2$O

(NaOTA)   (NaOPA)

where NaOTA and NaOPA are the sodium salts of OTA and OPA. The following reaction mechanism has been proposed [*I.E.C. Process Design and Development* 5, (1966), 376]:

Reaction 1 $(Cr_2O_7)^{-2} + H_2O \longrightarrow 2(HCrO_4)^-$ (P7-14-1)

Reaction 2 $(HCrO_4)^- \longrightarrow CrO_4^{-2} + H^+$ (P7-14-2)

Reaction 3 $H^+ + (Cr_2O_7)^{-2} \longrightarrow (HCr_2O_7)^-$ (P7-14-3)

Reaction 4   [structural formula with CH$_3$, COONa] + $(HCr_2O_7)^- \longrightarrow (H_2Cr_2O_7)^- +$ [structural formula with COONa, COONa]

It is believed that Reaction 4 is the rate-limiting step. Assuming that this is true, determine if the rate law

$$-r_{\text{NaOTA}} = \frac{k(\text{NaOTA})\text{Cr}_2\text{O}_7^{-2})^{3/2}(\text{H}_2\text{O})^{1/2}}{(\text{CrO}_4^{-2})}$$

is consistent with the above mechanism.

P7-15. At 200°C the rate expression for the thermal reaction between H_2 and Cl_2 in the presence of oxygen is

$$r_{\text{HCl}} = \frac{k(H_2)(Cl_2)^2}{(Cl_2) + k'(O_2)[(H_2) + k''(Cl_2)]}$$

Suggest a mechanism. (*Hint*: Conceptually, it is possible to consider the overall reaction to be

$$H_2 + Cl_2 + 2O_2 \longrightarrow HCl + HO_2 + ClO_2$$

and also assume that a scavanger is present that could remove a chlorine free radical.)

P7-16. As an example of the application of the pseudo-steady-state hypothesis to preditor-prey. relationships, consider the following sequence:

1. Animal A eats grass G (which is kept in constant supply) and reproduces:

$$A + G \xrightarrow{k_1} 2A \qquad (\text{P7-16-1})$$

2. Animal M devours animal A and reproduces:

$$M + A \xrightarrow{k_2} 2M \qquad (\text{P7-16-2})$$

3. Finally animal M dies:

$$M \xrightarrow{k_3} D \qquad (\text{P7-16-3})$$

a. From a population balance (analogous to a mole balance) derive differential equations describing the rate of change of the number of animals A and M.

b. Apply the pseudo-steady-state hypothesis to animals A and M to obtain the death rate of animal M.

c. Solve the equations obtained in part a, i.e.,

$$\frac{dx}{dt} = k_1 x - k_2(x)(y)$$

$$\frac{dy}{dt} = k_2(x)(y) - k_3 y,$$

to obtain the number of animals A and M as a function of time. From the resulting equation, make a sketch of the number of animals A and M as a function of time. [J. Higgens, *Applied Kinetics and Reaction Engineering*, ed. R. L. Gorring and V. W. Weekman, (Washington: ACS, 1967) or M. E. Gilpin, *Science* 177, (1972) 902.]

P7-17. The following data on Bakers Yeast in a particular medium at 23.4°C and various oxygen partial pressures was obtained.

P_{O_2}	Q_{O_2} (no sulfanilamide)	Q_{O_2} (20 mg sulfanilamide/ml added to medium)
0.0	0.0	0.0
0.5	23.5	17.4
1.0	33.0	25.6
1.5	37.5	30.8
2.5	42.0	36.4
3.5	43.0	39.6
5.0	43.0	40.0

P_{O_2} = oxygen partial pressure in mm. of Hg.
Q_{O_2} = oxygen uptake rate in μ liters of O_2 per hour per mgm of cells.

(a) Calculate the Q_{o_2} maximum (V_{max}) and the Michaelis-Menten constant K_m.

(b) By use of the Lineweaver-Burk plot, determine whether sulfanilamide is a competitive or non-competitive inhibitor to O_2 uptake.   [U. of Penn]

P7-18. Determine if the noncompetitive inhibition scheme

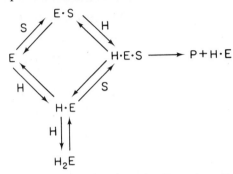

can explain the Michaelis pH function as given by Equation (7-46). Consider the complex H_2E to be inactive, and that

$$K_a = \frac{(H \cdot E)(H)}{(H_2E)}$$

$$K_b = \frac{(H)(E)}{(H \cdot E)}$$

$(H) \equiv (H^+) \equiv C_{H^+}$

SUPPLEMENTARY READING

1. A discussion of complex reactions involving active intermediates is found in Chapters 3 and 4 of

LAIDLER, K. J., *Chemical Kinetics* Vol. I, New York: Pergamon Press, 1963;

Chapter 10 of

FROST, A. A., and R. G. PEARSON, *Kinetics and Mechanism*, 2nd ed., New York: John Wiley and Sons, 1961;

Chapter 2 of

LEVENSPIEL, O., *Chemical Reaction Engineering*, 2nd ed., New York: John Wiley and Sons, 1972;

Chapter 6 of

PANNETIER, G. and P. SOUCHAY, *Chemical Kinetics*, New York: American Elsevier Publishing Co., 1967;

and Chapter 13 of

BENSON, S. W., *Foundations of Chemical Kinetics*, New York: McGraw-Hill Book Co., 1960.

2. Further discussion of enzymatic reactions is presented in the following books:

DIXON, M. and E. C. WEBB, *Enzymes*, New York: Academic Press, 1958.

MAHLER, H. R. and E. H. CORDES, *Basic Biological Chemistry*, New York: Harper & Row, 1968.

CHRISTENSEN, H. N. and G. A. PALMER, *Enzyme Kinetics*, Philadelphia: W. B. Saunders Co., 1967.

WEBB, J. L., *Enzyme and Metabolic Inhibitors*, Vol. 1, New York: Academic Press, 1963.

REINER, J. M., *Behavior of Enzyme Systems*, Minneapolis: Burgess Publishing Company, 1959.

GUTFRIEND, H., *Enzymes: Physical Principles*, New York: Wiley-Interscience, 1972.

WINGARD, L. B., *Enzyme Engineering*, New York: Wiley-Interscience, 1972.

Some aspects of enzyme reactions in heterogeneous systems are presented in

MCLAREN, A. D. and L. Packer, *Advances in Enzymology*, Vol. 33, ed. R. R. Nord, New York: Wiley-Interscience, 1970, p. 245.

3. A recent and thorough review of enzyme kinetics modeling in biological systems is

HUMPHREY, A. E., "Kinetics of Biological Systems," *Advances in Chem. No. 109*, 630, 1972.

8
Chemical Reactor Energy Balances

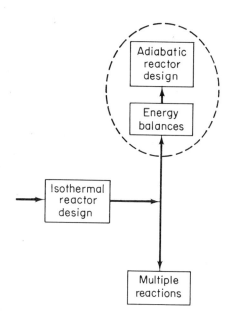

We shall now focus our attention on heat effects in chemical reactions with emphasis on adiabatic reactions. Section 8.1 involves the derivation and manipulation of energy balance. In Sections 8.2 through 8.5, the reactor energy balance is coupled with the various forms of the mole balances and reaction rate laws.

8.1–Energy Balances

We shall begin with the application of the first law of thermodynamics first to a closed system and then to an open system. A system is any bounded portion of the universe, moving or stationary, which is chosen for the application of the various thermodynamic equations. For a closed system in which no mass crosses the system boundaries, the change in total energy, dE, of the system is equal to the heat flow, δQ, *to* the system less the work, δW, done *by* the system *on* the

surroundings. For a closed system the energy balance is

$$dE = \delta Q - \delta W \tag{8-1}$$

The δs signify that δQ and δW are not exact differentials of a state function.

For an open system in which some of the energy exchange is brought about by the flow of mass across the system boundaries, the energy balance becomes

$$\begin{Bmatrix} \text{Rate of} \\ \text{accumulation} \\ \text{of energy} \\ \text{within the} \\ \text{system} \end{Bmatrix} = \begin{Bmatrix} \text{Rate of flow} \\ \text{of heat to} \\ \text{the system} \\ \text{from the} \\ \text{surroundings} \end{Bmatrix} - \begin{Bmatrix} \text{Rate of work} \\ \text{done by} \\ \text{the system} \\ \text{on the} \\ \text{surroundings} \end{Bmatrix} + \begin{Bmatrix} \text{Rate of energy} \\ \text{added to the} \\ \text{system by mass} \\ \text{flow } into \text{ the} \\ \text{system} \end{Bmatrix} - \begin{Bmatrix} \text{Rate of energy} \\ \text{leaving the} \\ \text{system by mass} \\ \text{flow } out \text{ of} \\ \text{the system} \end{Bmatrix} \tag{8-2}$$

The unsteady energy balance for an open system which has n species, each entering and leaving the system at their respective rates F_i (moles of i per time) and with their respective energies E_i (calories per mole of i), is

$$\frac{dE}{dt} = \dot{Q} - \dot{W} + \sum_{i=1}^{n} E_i F_i \bigg|_{\text{in}} - \sum_{i=1}^{n} E_i F_i \bigg|_{\text{out}} \tag{8-3}$$

It is customary to separate the work terms into *flow work* and *other work*, $\dot{W}_s$. Flow work is that work which is necessary to get the mass into and out of the system. For example, when shear stresses are absent, we write

$$\dot{W} = \overbrace{-\sum_{i=1}^{n} F_i \Pi \underset{\sim}{V}_i \bigg|_{\text{in}} + \sum_{i=1}^{n} F_i \Pi \underset{\sim}{V}_i \bigg|_{\text{out}}}^{[\text{Rate of Flow Work}]} + \dot{W}_s \tag{8-4}$$

where Π is the pressure and $\underset{\sim}{V}_i$ the specific volume.

In most instances, the flow work term is combined with those terms in the energy balance which represent the energy exchange by mass flow across the system boundaries. Substituting Equation (8-4) into (8-3) and grouping terms:

$$\frac{dE_{\text{sys}}}{dt} = \dot{Q} - \dot{W}_s + \sum_{i=1}^{n} F_i(E_i + \Pi \underset{\sim}{V}_i) \bigg|_{\text{in}} - \sum_{i=1}^{n} F_i(E_i + \Pi \underset{\sim}{V}_i) \bigg|_{\text{out}} \tag{8-5}$$

The energy E_i is the sum of the internal energy U_i, the kinetic energy, $u_i^2/2$, and the potential energy, gz_i:

$$E_i = U_i + \frac{u_i^2}{2} + gz_i \tag{8-6}$$

We recall that the enthalpy, H_i, is defined in terms of the internal energy U_i and the product $\Pi \underset{\sim}{V}_i$:

$$H_i = U_i + \Pi \underset{\sim}{V}_i \tag{8-7}$$

Typical units of H_i are

$$[H_i] = \frac{\text{Btu}}{\text{lb-mole of } i}, \quad \text{or} \quad \frac{\text{calories}}{\text{gmole } i}$$

Enthalpy carried into (or out of) the system can be expressed as the sum of the net internal energy carried into (or out of) the system by mass flow plus the flow work, i.e.,

$$H_i F_i = F_i(U_i + \Pi \underset{\sim}{V}_i)$$

Combining Equations (8-5), (8-6) and (8-7), we can now write the energy balance in the form

$$\frac{dE}{dt}\bigg|_{\text{sys}} = \dot{Q} - \dot{W}_s + \sum_{i=1}^{n} \left(H_i + \frac{u_i^2}{2} + gz_i\right) F_i \bigg|_{\text{in}} - \sum_{i=1}^{n} \left(H_i + \frac{u_i^2}{2} + gz_i\right) F_i \bigg|_{\text{out}} \tag{8-8}$$

The energy of the system at any instant in time, E_{sys}, is the sum of the products of the number of moles of each species in the system times their respective energies, i.e.,

$$E_{\text{sys}} = \sum_{i=1}^{n} E_i N_i \tag{8-9}$$

Write the energy balance for a system in which there is no *other* work being done on or by the system; kinetic and potential energy changes are neglected. Refer to Equation (8-8) if necessary.

(H1) ___

___

For liquids, the time rate of change of the product ΠV is negligible w.r.t. the other terms. It is also negligible in many gas phase reactions; consequently we shall neglect it in our analysis; the energy balance then reduces to

$$\boxed{\dot{Q} + \sum_{i=1}^{n} F_i H_i \bigg|_{\text{in}} - \sum_{i=1}^{n} F_i H_i \bigg|_{\text{out}} = \frac{d\left(\sum_{i=1}^{n} H_i N_i\right)}{dt}} \tag{8-10}$$

We are neglecting any mixing effects so that the partial molal enthalpies are equal to the molal enthalpies of the pure components. The molal enthalpy of species i, H_i, at a particular temperature and pressure, is usually expressed in terms of an enthalpy of formation of species i, $H_i^\circ(T_R)$, at some reference temperature, T_R, plus the change in enthalpy, ΔH_{Qi}, that is caused by going from the reference temperature to some temperature, T.

$$H_i = H_i^\circ(T_R) + \Delta H_{Qi} \tag{8-11}$$

For an inert gas i, that is being heated from T_1 to T_2,

$$\Delta H_{Qi} = \int_{T_1}^{T_2} C_{p_i} \, dT \tag{8-12}$$

Typical units of C_{p_i} are

$$[C_{p_i}] = \frac{\text{Btu}}{(\text{lb-mole of } i)(^\circ\text{R})}, \quad \frac{\text{calories}}{(\text{gmole } i)(^\circ\text{K})}$$

Consider a substance in the gaseous state at temperature T. The enthalpy of formation is given at temperature T_R, which is below the freezing point, T_s, of the substance. Write an expression for the molal enthalpy of this substance in terms of the heat capacities of the solid, liquid, and gas phases, and the heats of formation, fusion, and vaporization. The boiling point of the substance is T_B.

(H2) ___

___

In the absence of phase changes, what is the equation for molal enthalpy of a gas at temperature T as a function of the enthalpy of formation at $T_R(\text{gas})$ and the heat capacity C_{p_i}.

(H1)

From the problem statement $\dot{W}_s = u_i = z_i = 0$

$$\dot{Q} + \sum_{i=1}^{n} F_i H_i \bigg|_{\text{in}} - \sum_{i=1}^{n} F_i H_i \bigg|_{\text{out}} = \frac{d\left(\sum_{i=1}^{n} U_i N_i\right)}{dt} \tag{H1-1}$$

Substituting for U_i in terms of enthalpy and noting

$$\sum_{i=1}^{n} \mathbf{V}_i N_i \equiv V \tag{H1-2}$$

we find

$$\dot{Q} + \sum_{i=1}^{n} F_i H_i \bigg|_{\text{in}} - \sum_{i=1}^{n} F_i H_i \bigg|_{\text{out}} = \frac{d\left(\sum_{i=1}^{n} N_i H_i - \Pi V\right)}{dt} \tag{H1-3}$$

(H2)

$$\begin{bmatrix} \text{Enthalpy} \\ \text{of Species} \\ i \text{ at } T \end{bmatrix} = \begin{bmatrix} \text{Enthalpy of} \\ \text{Formation of} \\ \text{Species } i \text{ at } T_R \end{bmatrix} + \begin{bmatrix} \Delta H_{Qi} \text{ in Heating} \\ \text{Solid From } T_R \\ \text{to } T_S \end{bmatrix} + \begin{bmatrix} \text{Heat of} \\ \text{Fusion} \\ \text{at } T_S \end{bmatrix} + \begin{bmatrix} H_{Qi} \text{ in Heating} \\ \text{Liquid from} \\ T_S \text{ to } T_B \end{bmatrix}$$

$$+ \begin{bmatrix} \text{Heat of} \\ \text{Vaporization} \\ \text{at } T_B \end{bmatrix} + \begin{bmatrix} \Delta H_{Qi} \text{ in} \\ \text{Heating Gas} \\ \text{from } T_B \text{ to } T \end{bmatrix}$$

$$H_i(T) = H_i^\circ(T_R) + \int_{T_R}^{T_S} C_{psi}\, dT + \Delta H_{si}(T_s) + \int_{T_S}^{T_B} C_{pli}\, dT + \Delta H_{\text{vap}i}(T_B) + \int_{T_B}^{T} C_{pi}\, dT$$

(H3)

For the generalized reaction,
$$aA + bB \rightleftharpoons cC + dD \quad (2\text{-}1)$$
it is usually most convenient to develop the energy balance equations on the basis of 1 mole of a specific reactant, such as A, in which case we write the above reaction in the form
$$A + \frac{b}{a}B \rightleftharpoons \frac{c}{a}C + \frac{d}{a}D \quad (2\text{-}2)$$

Consider the steady-state system shown in Figure 8-1, in which material enters and leaves the system volume.

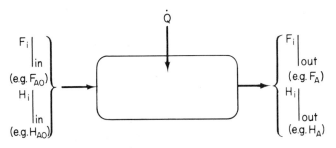

Figure 8–1

What form does Equation (8-10) take for a continuous flow system operated at steady state?

(H4)

In the material that follows we shall primarily be concerned with heat effects in CSTRs and tubular reactors. In particular, we shall be interested in developing the steady-state energy balance, Equation (H4-1). In addition to compounds A, B, C, and D, there will sometimes be nonreacting inerts, I, entering and leaving the system. Recall that F_{A0} is the flow rate of A entering the system volume (e.g., moles of A/hr) and X is the number of moles of A reacted per mole of A fed; then the molar flow rate of A leaving is
$$F_A = F_{A0} - XF_{A0}$$
Write similar expressions for F_B, F_C, F_D, F_I, and F_{total} for the generalized reaction given by Equation (2-2), remembering to use only species A as your basis.

(H5)

313

(H3)

$$H_i(T) = H_i^\circ(T_R) + \Delta H_{Qi} \tag{8-11}$$

For an ideal gas being heated from T_R to T,

$$\Delta H_{Qi} = \int_{T_R}^{T} C_{p_i} \, dT \tag{H3-1}$$

$$H_i(T) = H_i^\circ(T_R) + \int_{T_R}^{T} C_{p_i} \, dT \tag{H3-2}$$

(H4)

At steady state, conditions at a point do not change with time. Therefore,

$$\frac{d\left(\sum_{i=1}^{n} N_i H_i\right)}{dt} = 0$$

The energy balance on the system volume for a continuous flow system is

$$\dot{Q} + \sum_{i=1}^{n} F_i H_i \bigg|_{\text{in}} - \sum_{i=1}^{n} F_i H_i \bigg|_{\text{out}} = 0 \tag{H4-1}$$

In a number of situations $\dot{Q}$ takes the form

$$\dot{Q} = UA_s(T_A - T) \tag{H4-2}$$

where
- U = Overall heat transfer coefficient
- A_s = External surface area
- T_A = Ambient temperature
- T = Temperature of system volume.

(H5)

$$F_A = F_{A0} - F_{A0} X \tag{H5-1}$$

$$F_B = F_{B0} - \frac{b}{a} X F_{A0} \tag{H5-2}$$

$$F_C = F_{C0} + \frac{c}{a} X F_{A0} \tag{H5-3}$$

$$F_D = F_{D0} + \frac{d}{a} X F_{A0} \tag{H5-4}$$

$$F_I = F_{I0} \tag{H5-5}$$

$$F_t = F_{t0} + F_{A0} X \delta \tag{H5-6}$$

In Frame (H6) expand the summations $\sum_1^n F_i H_i$ in the equation

$$\dot{Q} + \sum_{i=1}^n F_i H_i \bigg|_{in} - \sum_{i=1}^n F_i H_i \bigg|_{out} = 0 \qquad (H4\text{-}1)$$

for the generalized reaction given by Equation (2-2) in order to write Equation (H4-1) in terms of F_{A0}, H_{A0}, F_A, H_A, H_B, etc. (i.e., let species 1 be A, 2 be B, 3 be C, 4 be D, and 5 be I, so that $F_1 H_1 |_{in} = F_{A0} H_{A0}$, $\quad F_2 H_2 |_{in} = F_{B0} H_{B0}$, etc., $\quad$ and $F_1 H_1 |_{out} = F_A H_A$, etc.).

(H6)
_____

_____

Now substitute Equations (H5-1)–(H5-5) for F_A, F_B, etc. in terms of F_{A0}, X, and the stoichiometric coefficients, into Equation (H6-1), grouping the coefficients of $F_{A0} X$.

(H7)
_____

_____

The coefficient of $F_{A0} X$ in Equation (H7-1) is called the *heat of reaction*.

$$\Delta H_{\text{Rxn,A}}(T) = \left[\frac{c}{a} H_C(T) + \frac{d}{a} H_D(T) - \frac{b}{a} H_B(T) - H_A(T) \right] \equiv \Delta H_{\text{Rxn}} \qquad (8\text{-}13)$$

Note that all the enthalpies in this expression are given at the temperature of the system volume.

How would you express completely the symbol $\Delta H_{\text{Rxn}}(T)$ in words?

(H8)
_____

_____

In a discussion of the heat of reaction for a specific reaction, it should be remembered that ΔH_R refers to the heat of reaction of a particular species involved in the reaction.

For the reaction

$$2A + B \longrightarrow C$$

circle the statements in Frame (H9) that are true.

(H9)
_____

1. The reaction is exothermic.
2. The units of the heat of reaction are cal./gmole reacted.
3. The units of the heat of reaction are cal./gmole of A reacted.
4. $\Delta H_{\text{Rxn,A}} = \frac{1}{2} \Delta H_{\text{Rxn,B}}$.
5. $\Delta H_{\text{Rxn,A}} = 2 \Delta H_{\text{Rxn,B}}$.

(H6)

After transposing $\sum_{i=1}^{n} F_i H_i|_{\text{out}}$ to the right-hand side of the equal sign, we expand each summation to obtain

$$\dot{Q} + F_{A0}H_{A0} + F_{B0}H_{B0} + F_{C0}H_{C0} + F_{D0}H_{D0} + F_{I0}H_{I0}$$
$$= F_A H_A + F_B H_B + F_C H_C + F_D H_D + F_I H_I \quad \text{(H6-1)}$$

In this expression F_I represents the molar flow rate of inerts into and out of the system.

(H7)

$$F_{A0}H_{A0} + F_{B0}H_{B0} + F_{C0}H_{C0} + F_{D0}H_{D0} + F_{I0}H_{I0} + \dot{Q}$$
$$= F_{A0}H_A + F_{B0}H_B + F_{C0}H_C + F_{D0}H_D + F_{I0}H_I + F_{A0}X\left[\frac{c}{a}H_C + \frac{d}{a}H_D - \frac{b}{a}H_B - H_A\right] \quad \text{(H7-1)}$$

(H8)

It is the heat of reaction *per mole of A reacted at temperature T.*

(H9)

Ans. 3: You must refer to a specific species when discussing the heat of reaction:

$$[\Delta H_{\text{Rxn, A}}] = \frac{\text{calories}}{\text{gmole of A reacted}}$$

$$[\Delta H_{\text{Rxn, B}}] = \frac{\text{calories}}{\text{gmole of B reacted}}$$

$$\Delta H_{\text{Rxn, A}} = \frac{\text{calories}}{\text{gmole of B reacted}} \cdot \frac{\text{moles of B reacted}}{\text{moles of A reacted}} = \Delta H_{\text{Rxn, B}} \cdot \frac{\text{moles of B reacted}}{\text{moles of A reacted}}$$

Ans. 4: For every mole of A that reacts, $\frac{1}{2}$ mole of B reacts:

$$\Delta H_{\text{Rxn, A}} = \tfrac{1}{2} \Delta H_{\text{Rxn, B}}$$

Unless otherwise stated, we shall let ΔH_R represent the heat of reaction per mole of A reacted, i.e.,

$$\Delta H_{Rxn} \equiv \Delta H_R \equiv \Delta H_{Rxn,A}$$

We now rearrange Equation (H7-1) to collect the coefficients of F_{A0}, F_{B0}, etc., and write the energy balance in the form,

$$\dot{Q} = F_{A0}(H_A - H_{A0}) + F_{B0}(H_B - H_{B0}) + F_{C0}(H_C - H_{C0})$$
$$+ F_{D0}(H_D - H_{D0}) + F_{I0}(H_I - H_{I0}) + XF_{A0}\,\Delta H_{Rxn}(T) \qquad (8\text{-}14)$$

where $H_i \equiv H_i(T)$ and $H_{i0} \equiv H_i(T_0)$.

Equation (8-14) can be written a little more compactly using the summation convention:

$$\boxed{\dot{Q} = \sum_{i=1}^{n} F_{i0}(H_i - H_{i0}) + F_{A0} X\,\Delta H_{Rxn}(T)} \qquad (8\text{-}15)$$

For a reaction in which there is no phase change, the enthalpy of component i at temperature T, H_i, can be expressed in terms of a standard enthalpy of formation of compound i, $H_i^\circ(T_R)$, at some reference temperature T_R, plus the change in enthalpy that occurs going from T_R to temperature T. Express these conditions in a mathematical statement.

(H10) _____

_____

Using Equation (H10-1), express the differences between the enthalpies of streams entering and leaving the system in terms of the heat capacity and any other appropriate quantities, assuming that no change of phase takes place [e.g., $(H_A - H_{A0}) = $?].

(H11) _____

_____

If none of the species entering the reactor undergo a phase change, the energy balance is expressed in the form

$$\dot{Q} = F_{A0}\int_{T_{A0}}^{T} C_{pA}\,dT + F_{B0}\int_{T_{B0}}^{T} C_{pB}\,dT + F_{C0}\int_{T_{C0}}^{T} C_{pC}\,dT + F_{D0}\int_{T_{D0}}^{T} C_{pD}\,dT$$
$$+ F_{I0}\int_{T_{I0}}^{T} C_{pI}\,dT + XF_{A0}\,\Delta H_{Rxn}(T) \qquad (8\text{-}16)$$

In Frame (H12) replace the heat capacity integrals in Equation (8-16) by mean heat capacities between T_{i0} and T, for each species, e.g.,

$$\tilde{C}_{pA} = \frac{\int_{T_{A0}}^{T} C_{pA}\,dT}{T - T_{A0}} \qquad (8\text{-}17)$$

and write the energy balance in terms of $\tilde{C}_{pi}$ ($i = $ A, B, etc.).

(H10)

If there is no phase change in going from T_R to T, the enthalpy of species i at temperature T is

$$H_i(T) = H_i^\circ(T_R) + \int_{T_R}^{T} C_{p_i} dT \qquad \text{(H10-1)}$$

The temperature dependence of the heat capacity of a species in the gas phase can be approximated by the quadratic

$$C_{p_i} = \alpha_i + \beta_i T + \gamma_i T^2 \qquad \text{(H10-2)}$$

(H11)

T_{A0} is the temperature of species A entering the system volume (i.e., reactor).

$$(H_A - H_{A0}) = \left[H_A^\circ(T_R) + \int_{T_R}^{T} C_{pA} dT \right] - \left[H_A^\circ(T_R) + \int_{T_R}^{T_{A0}} C_{pA} dT \right] = \int_{T_R}^{T} C_{pA} dT - \int_{T_R}^{T_{A0}} C_{pA} dT$$

$$= \int_{T_R}^{T} C_{pA} dT + \int_{T_{A0}}^{T_R} C_{pA} dT = \int_{T_{A0}}^{T} C_{pA} dT \qquad \text{(H11-1)}$$

At any specified point in the reactor all species are at the same temperature.

(H12)

When one of the species undergoes a change of phase, Equation (H12-1) will not apply and one must use the energy balance in the form given by Equation (8-15).

Will the exit temperature of a reacting mixture increase or decrease as the amount of inerts at the temperature of the feed stream is increased in an endothermic reaction? Assume that the conversion that is exiting the adiabatic reactor somehow remains unchanged.

(H13)

Consider a reaction that is being carried out isothermally in a reactor in which 70% conversion is achieved for an entering flow rate of A of 2 lb-moles/min. All species enter at the same temperature, T_0. The reaction is exothermic, with a heat of reaction at the exit temperature of 600 Btu/lb-mole of A. What rate of heat addition or withdrawal is necessary to maintain this *isothermal* condition in the reactor?

(H14)

$\dot{Q}$ is positive when heat is added to the system. Conversely, when $\dot{Q}$ is negative, heat is being given off by the system to the surroundings. Therefore, in the previous example eight hundred and forty Btu must be absorbed by the surroundings each minute to maintain the system in an isothermal condition.

Next, consider the case in which the reactor volume is insulated from the surroundings, and the reaction

$$A + B \rightleftharpoons C + D$$

takes place adiabatically ($\dot{Q} = 0$).

An equimolar mixture of A and B enters a reactor at a rate of 4 lb-moles/min; 70% conversion is to be achieved. The mean heat capacities of A, B, C, and D are 7, 5, 5, and 7 Btu/lb-mole °F, respectively. If both the reactants enter at 77°F, calculate the exit temperature when the exothermic heat of reaction is constant at 600 Btu/lb-mole of A reacted.

(H15)

Notice that in calculating the exit temperature for the adiabatic case $\dot{Q} = 0$, it is not neces-

(H12)

$$\theta_i = \frac{F_{i0}}{F_{A0}}$$

$$\dot{Q} = F_{A0}[\tilde{C}_{pA}(T - T_{A0}) + \theta_B \tilde{C}_{pB}(T - T_{B0}) + \theta_C \tilde{C}_{pC}(T - T_{C0}) + \theta_D \tilde{C}_{pD}(T - T_{D0})$$
$$+ \theta_I \tilde{C}_{pI}(T - T_{I0})] + F_{A0} X \Delta H_R(T) \tag{H12-1}$$

(H13)

The temperature will increase when the flow rate of inerts is increased. The energy absorbed by the endothermic reaction will be drawn from a greater number of moles; consequently, there will be a smaller temperature drop per mole of inerts entering the reactor.

(H14)

For isothermal operation, $T = T_0$, Equation (H12-1) reduces to

$$\dot{Q} = \Delta H_{Rxn} F_{A0} X = (-600 \text{ Btu/lb-mole of A})(2 \text{ lb-mole of A/min})(.7)$$
$$= -840 \text{ Btu/min}$$

Heat must be withdrawn from the system at a rate of 840 Btu/min.

(H15)

$$\dot{Q} = F_{I0} = F_{D0} = F_{C0} = 0$$

Consequently, the energy balance, Equation (H12-1), becomes

$$-\Delta H_{Rxn} X F_{A0} = (F_{A0} \tilde{C}_{pA} + F_{B0} \tilde{C}_{pB})(T - T_0) \tag{H15-1}$$

Solving for T

$$T = T_0 - \frac{\Delta H_{Rxn} X F_{A0}}{\left(\tilde{C}_{pA} + \frac{F_{B0}}{F_{A0}} \tilde{C}_{pB}\right) F_{A0}} \tag{H15-2}$$

$$= T_0 - \frac{\Delta H_{Rxn} X}{\tilde{C}_{pA} + \theta_B \tilde{C}_{pB}} \tag{H15-3}$$

$$\theta_B = 1$$

$$T = 77°F - \frac{(-600)(.7)}{7 + 5} \left(\frac{\text{Btu/lb-mole}}{\text{Btu/lb-mole/°F}}\right)$$

$$= 77 + 35 = 112°F$$

sary to know the magnitude of the molar flow rates of the components entering the reactor, but only the ratio of the flow rate of each species to our basis A, θ_i.

Next, calculate the exit temperature for the same adiabatic reaction,

$$A + B \rightleftharpoons C + D$$

when the entering feed is 30% A, 30% B, and 40% inerts, and the heat capacity of the inert substance is 10 Btu/lb-mole °F. A 70% conversion of A is realized in this reaction.

(H16) _____

_____

In the previous examples we stated that the heat of reaction was a constant. We shall now determine under what conditions this is true.

The heat of reaction per mole of A reacted at temperature T is

$$\Delta H_{\text{Rxn}} = \left[\frac{c}{a} H_C + \frac{d}{a} H_D - \frac{b}{a} H_B - H_A\right] \tag{8-13}$$

Replace the enthalpies, $H_i(T)$, of each of the species at temperature T in Equation (8-13) with their respective enthalpy of formation, $H_i^\circ(T_R)$, at a reference temperature T_R and with an expression containing the mean heat capacities.

(H17) _____

_____

The symbol $\hat{C}_{p_i}$ represents the mean heat capacity of species i between temperatures T_R and T, and $\tilde{C}_{p_i}$ represents the mean heat capacity of species i between temperature T_0 and T, i.e.,

$$\tilde{C}_{p_i} = \frac{\int_{T_0}^{T} C_{p_i} dT}{T - T_0}, \qquad \hat{C}_{p_i} = \frac{\int_{T_R}^{T} C_{p_i} dT}{T - T_R}$$

In most situations $\tilde{C}_{p_i}$ and $\hat{C}_{p_i}$ will be essentially equal, and will only differ significantly when C_{p_i} is a strong function of temperature and when T_0 and T_R differ significantly.

The enthalpies of formation of various compounds, $H_i^\circ(T_R)$, are usually tabulated at 25°C and can readily be found in the *Handbook of Chemistry and Physics* and similar handbooks. From these values of the standard heat of formation, $H_i^\circ(T_R)$, we can calculate the heat of reaction at the reference temperature T_R from the equation

$$\boxed{\Delta H_{\text{Rxn}}^\circ(T_R) = \left[\frac{c}{a} H_C^\circ(T_R) + \frac{d}{a} H_D^\circ(T_R) - \frac{b}{a} H_B^\circ(T_R) - H_A^\circ(T_R)\right]} \tag{8-18}$$

We define $\Delta \tilde{C}_p$ as

$$\boxed{\Delta \tilde{C}_p = \left[\frac{c}{a} \hat{C}_{p_C} + \frac{d}{a} \hat{C}_{p_D} - \frac{b}{a} \hat{C}_{p_B} - \hat{C}_{p_A}\right]} \tag{8-19}$$

(H16)

For the conditions given, Equation (H12-1) becomes

$$-\Delta H_R X F_{A0} = (F_{A0}\tilde{C}_{p_A} + F_{B0}\tilde{C}_{p_B} + F_{I0}\tilde{C}_{p_I})(T - T_0)$$

Rearranging,

$$T = T_0 - \frac{\Delta H_R X}{\tilde{C}_{p_A} + \frac{F_{B0}}{F_{A0}}\tilde{C}_{p_B} + \frac{F_{I0}}{F_{A0}}\tilde{C}_{p_I}}$$

$$= 77 - \frac{(-600)(.7)}{\left(7 + \frac{.3}{.3}5 + \frac{.4}{.3}10\right)}$$

$$= 77 + \frac{420}{7 + 5 + 13.3} = 77 + 16.6$$

$$= 93.6°F$$

(H17)

Assuming that there is no phase change going from T_R to T, we replace the enthalpies $H_i(T)$ in Equation (8-14) by

$$H_i(T) = H_i°(T_R) + \int_{T_R}^{T} C_{p_i} \, dT \tag{H17-1}$$

$$= H_i°(T_R) + \hat{C}_{p_i}(T - T_R) \tag{H17-2}$$

to obtain

$$\Delta H_R(T) = \left[\frac{c}{a} H_C°(T_R) + \frac{d}{a} H_D°(T_R) - \frac{b}{a} H_B°(T_R) - H_A°(T_R)\right]$$

$$+ \left(\frac{c}{a}\hat{C}_{p_C} + \frac{d}{a}\hat{C}_{p_D} - \frac{b}{a}\hat{C}_{p_B} - \hat{C}_{p_A}\right)(T - T_R) \tag{H17-3}$$

Then

$$\Delta H_R(T) = \Delta H_R(T_R) + \left(\frac{c}{a}\hat{C}_{p_C} + \frac{d}{a}\hat{C}_{p_D} - \frac{b}{a}\hat{C}_{p_B} - \hat{C}_{p_A}\right)(T - T_R) \tag{H17-4}$$

If the heat capacity of each species is a polynomial function of temperature of the form given by Equation (H10-2),

$$C_{pi} = \alpha_i + \beta_i T + \gamma_i T^2 \tag{H10-2}$$

$\Delta \tilde{C}_p$ can be written in the form

$$\Delta \tilde{C}_p = \frac{\left[\int_{T_R}^{T} (\Delta\alpha + \Delta\beta T + \Delta\gamma T^2)dT\right]}{(T - T_R)} \tag{8-20}$$

where

$$\Delta\alpha = \frac{c}{a}\alpha_C + \frac{d}{a}\alpha_D - \frac{b}{a}\alpha_B - \alpha_A$$

$$\Delta\beta = \frac{c}{a}\beta_C + \frac{d}{a}\beta_D - \frac{b}{a}\beta_B - \beta_A$$

$$\Delta\gamma = \frac{c}{a}\gamma_C + \frac{d}{a}\gamma_D - \frac{b}{a}\gamma_B - \gamma_A$$

We can now put the heat of reaction at temperature T in a more compact form than that given by Equation (H17-4)

$$\boxed{\Delta H_{\text{Rxn}}(T) = \Delta H^\circ_{\text{Rxn}}(T_R) + \Delta \tilde{C}_p (T - T_R)} \tag{8-21}$$

In the previous examples of the reaction $A + B \rightleftharpoons C + D$ we said ΔH_{Rxn} was constant at 600 Btu/lb-mole of A. The heat capacities of A, B, C, and D were assumed to be independent of temperature, with values of 7, 5, 5, and 7 Btu/lb-mole/°F, respectively. Determine whether the assumption of constant heat of reaction was justified.

(H18)

As in Frame (H18), we can readily see that whenever $\Delta \tilde{C}_p$ is zero the heat of reaction is constant over the particular temperature range for which the mean heat capacities are given. If the heat capacities of each of the reactants and products are all equal, will the heat of reaction always be constant?

(H19)

If one mole of product is formed from one mole of reactant (e.g., $A \rightarrow B$, or $A + B \rightleftharpoons C + D$) will the heat of reaction always be independent of temperature?

(H20)

(H18)

For this example, a = b = c = d = 1:
$$\Delta \tilde{C}_p = \hat{C}_{p_C} + \hat{C}_{p_D} - \hat{C}_{p_B} - \hat{C}_{p_A}$$
$$= 5 + 7 - 5 - 7 = 0$$
$$= 0$$

Thus,
$$\Delta H_{\text{Rxn}}(T) = \Delta H_{\text{Rxn}}(T_R)$$

and the heat of reaction is indeed independent of temperature.

(H19)

No! Suppose that A $\longrightarrow$ 3B:
$$\hat{C}_{p_A} = \hat{C}_{p_B}$$
$$\Delta \tilde{C}_p = 3\hat{C}_{p_B} - \hat{C}_{p_A} = 2\hat{C}_{p_B} \neq 0$$

(H20)

No! Suppose that A + B $\rightleftharpoons$ C + D, (i.e., a = b = c = d = 1) and that
$$\hat{C}_{p_A} = 10, \quad \hat{C}_{p_B} = 5, \quad \hat{C}_{p_C} = 7, \quad \hat{C}_{p_D} = 6$$
$$\Delta \tilde{C}_p = 7 + 6 - 10 - 5 = -2 \neq 0$$

If $\Delta \tilde{C}_p$ is positive the heat of reaction increases with increasing temperature, and if $\Delta \tilde{C}_p$ is negative the heat of reaction decreases with increasing temperature.

The energy balance over the system volume can now be written in the form which is most readily used in the solution of problems involving reactions with heat effects:

$$\frac{\dot{Q}}{F_{A0}} - X[\Delta H_{Rxn}(T_R) + \Delta \tilde{C}_p(T - T_R)] = \sum_{i=1}^{n} \theta_i \tilde{C}_{pi} \Delta T_i \qquad (8\text{-}22)$$

where n is the number of species entering the reactor,

$$\theta_i = \frac{F_{i0}}{F_{A0}} \quad , \quad \Delta T_i = T - T_{i0}$$

and the summation is

$$\sum_{i=1}^{n} \theta_i \tilde{C}_{pi} \Delta T_i = \tilde{C}_{pA}(T - T_{A0}) + \frac{F_{B0}}{F_{A0}} \tilde{C}_{pB}(T - T_{B0}) + \frac{F_{C0}}{F_{A0}} \tilde{C}_{pC}(T - T_{C0})$$
$$+ \frac{F_{D0}}{F_{A0}} \tilde{C}_{pD}(T - T_{D0}) + \frac{F_{I0}}{F_{A0}} \tilde{C}_{pI}(T - T_{I0}) \qquad (8\text{-}23)$$

If all the species enter at the same temperature, $T_0 = T_{i0}$, the energy balance takes the form

$$\frac{\dot{Q}}{F_{A0}} - X[\Delta H_{Rxn}(T_R) + \Delta \tilde{C}_p(T - T_R)] = \left(\sum_{i=1}^{n} \theta_i \tilde{C}_{pi}\right)(T - T_0) \qquad (8\text{-}24)$$

Sometimes the summation term on the r.h.s. of Equation (8-24) is referred to as the heat capacity of the mixture or solution C_{PS}, i.e.,

$$\sum_{i=1}^{n} \theta_i \tilde{C}_{pi} = C_{PS}$$

If T_0 and T_R are roughly the same (within 100°K for gases), or if the heat capacity of species i is not a strong function of temperature, we can make the approximation,

$$\hat{C}_{pi} \doteq \tilde{C}_{pi} \qquad (8\text{-}25)$$

In the remaining examples in this chapter, with the exception of the SO_2 oxidation example, we shall use the approximation given by Equation (8-25).

In dealing with very viscous liquids at high flow rates, can $\dot{Q}$ be used to represent the heat generated from viscous dissipation?

(H21) ───

───

The dependence of the specific reaction rate k on temperature is given by the Arrhenius equation

$$k = Ae^{-E/RT} \qquad (3\text{-}2)$$

where E is the activation energy, A the frequency factor, R the ideal gas constant, and T the absolute temperature.

Equation (8-24) shows a unique relationship between temperature and conversion. When this equation is coupled with the reactor mole balance equation, it may be used to predict the reactor volume necessary to achieve a given conversion X.

(H21)
No. The viscous dissipation effects were neglected in this equation as stated below Equation (8-3) where shear stresses were ignored. These viscous effects are usually accounted for by incorporating the rate shear work in the other work, $\dot{W}_S$. $\dot{Q}$ is the exchange of heat from (to) the surroundings to (from) the system and can therefore never account for an internal effect *in* the system.

8.2–The Adiabatic Operation of a CSTR

The design equation for the backmix reactor or CSTR that gives the reactor volume necessary to achieve a conversion X is

$$V = \frac{F_{A0}X}{-r_A} \qquad (2\text{-}13)$$

As mentioned above, the temperature dependence of the reaction rate is given by the Arrhenius equation, (3-2). The coupling of the energy balance and the reactor mole balance can best be illustrated by considering the following example.

Example: A conversion of 90% is desired in a CSTR in which the liquid feed enters at 27°C at a flow rate of 2 cu ft/min. The heat of reaction at 273°K is -600 Btu/lb-mole of A. The decomposition of A is an elementary irreversible reaction

$$A \longrightarrow B + C$$

Additional data.

$$\tilde{C}_{p_A} = 7 \text{ Btu/lb-mole-°C}, \qquad \tilde{C}_{p_C} = \tilde{C}_{p_B} = 4 \text{ Btu/lb-mole-°C}$$
$$E = 25{,}000 \text{ cal./gmole}, \qquad k_1 = .12 \text{ hr}^{-1} \text{ at } 25°\text{C}$$

The overall objective is to calculate the CSTR volume necessary to achieve 90% conversion. First, replace $-r_A$ in the CSTR design equation by the appropriate kinetic expression in terms of k, C_{A0}, and X.

(H22) _____

_____

The specific reaction rate at 25°C is .12 hr^{-1}. The ratio of specific reaction rates at two different temperatures, T and T_1, is

$$\boxed{\begin{aligned}\frac{k(T)}{k_1(T_1)} &= \frac{e^{-E/RT}}{e^{-E/RT_1}} = e^{-E/R(1/T - 1/T_1)} \\ k(T) &= k_1 e^{(-E/R)\left[\frac{T_1-T}{T_1 T}\right]}\end{aligned}} \qquad (8\text{-}26)$$

Once the specific reaction rate constant is known at some temperature T_1, it is easy to calculate k at any other temperature T, provided the activation energy has been specified. Now replace all letters and symbols in Equation (H22-3) by their appropriate numerical values, if known.

(H23) _____

_____

We find that after substitution of the appropriate numbers we still are unable to determine the reactor volume, since we do not know the temperature of the reacting mixture. Assuming that the reaction vessel is completely insulated, calculate the temperature in the CSTR.

(H22)

$$V = \frac{F_{A0}X}{-r_A} \tag{2-13}$$

$$-r_A = k_1 C_A = k_1 C_{A0}(1-X) \tag{H22-1}$$

$$V = \frac{F_{A0}X}{k_1 C_{A0}(1-X)} = \frac{C_{A0}v_0 X}{k_1 C_{A0}(1-X)} \tag{H22-2}$$

$$= \frac{v_0 X}{k_1(1-X)} \tag{H22-3}$$

(H23)

$$V_R = \frac{v_0 X}{k_1 \exp\left[\frac{-E}{R} \cdot \frac{(T_1 - T)}{TT_1}\right](1-X)} \tag{H23-1}$$

$$= \frac{2 \,\text{cu ft/min} \times 60 \,\text{min/hr} \times .9}{(.12/\text{hr}) \exp\left[\left(-\frac{25000}{1.98}\right)\left(\frac{298 - T}{298T}\right)\right](1-.9)} \tag{H23-2}$$

$$= \frac{9000}{\exp\left[12{,}550\left(\frac{T-298}{298T}\right)\right]} \,\text{cu ft}$$

In Frame (H24) evaluate $\dot{Q}$ and $\Delta \tilde{C}_p$.

(H24)

Next, determine the temperature of the reaction mixture for a conversion of 90%.

(H25)

Note: There are three different base temperatures given in this example.
T_R = Reference temperature at which the heat of reaction is given; $T_R = 273°K$.
T_0 = Temperature at which the feed enters; $T_0 = 300°K$.
T_1 = Temperature at which the specific reaction rate constant, k_1, is evaluated;
$k_1 = .12 \text{ hr}^{-1}$ at 25°C., (i.e., $T_1 = 298°K$.)

Assuming a feed of 66.6% A and 33.3% inerts, calculate the reactor volume for 90% conversion for this liquid phase reaction.

(H26)

8.3.–Adiabatic Operation of a Plug Flow Reactor

In cases where the reaction rate can be expressed as a single conversion variable X, the design equation giving the reactor volume necessary to achieve a conversion X in a tubular reaction is

$$V = F_{A0} \int_0^X \frac{dX}{-r_A} \tag{2-17}$$

Numerical evaluation of the integral is necessary to evaluate V when the reaction takes place under nonisothermal conditions.

Example: The gas phase reaction
$$A \longrightarrow 2B,$$
is to be carried out in a tubular flow reactor. The rate feed is 300 lb-moles/hr of A and 300 lb-moles/hr of inert materials. The feed enters the reactor at a temperature of 27°C and a pressure of 10 atm. Find the required exit temperature and reactor volume necessary to achieve 80% conversion. The reactor is operated adiabatically. Data:

(H24)

Since the vessel is well insulated, we may assume that the reaction takes place under adiabatic conditions; hence, $\dot{Q} = 0$:

$$\Delta \tilde{C}_p = [\hat{C}_{p\text{B}} + \hat{C}_{p\text{C}} - \hat{C}_{p\text{A}}] = 4 + 4 - 7 = 1.0$$

$$\Delta \tilde{C}_p = 1.0 \, \frac{\text{Btu}}{\text{lb-mole of A-°K}}$$

(H25)

$$-X[\Delta H_{\text{Rxn}}(T_R) + \Delta \tilde{C}_p(T - T_R)] = \left(\tilde{C}_{p\text{A}} + \frac{F_{\text{B0}}}{F_{\text{A0}}} \tilde{C}_{p\text{B}} + \frac{F_{\text{C0}} \tilde{C}_{p\text{C}}}{F_{\text{A0}}} + \frac{F_{\text{I}}}{F_{\text{A0}}} \tilde{C}_{p\text{I}} \right)(T - T_0) \quad \text{(H25-1)}$$

$$- .9[(-600) + 1.0 \times (T - 273)] = [7 + ? + ? + ?][T - 300] \quad \text{(H25-2)}$$

We cannot proceed further because we have not been given F_{B0}, F_{C0}, and F_{I}.

(H26)

$$\theta_{\text{I}} = \frac{F_{\text{I}}}{F_{\text{A0}}} = \frac{.333}{.666} = \tfrac{1}{2}, \qquad F_{\text{B0}} = F_{\text{C0}} = 0$$

Substituting these values into Equation (H25-1),

$$540 - .9(T - 273) = (7 + \tfrac{1}{2} \times 8)(T - 300) \quad \text{(H26-1)}$$

Solving Equation (H26-1) for T,

$$T = 343.3°\text{K}$$

and substituting for T in Equation (H23-2),

$$V_R = \frac{9000}{\exp\left[12{,}550 \left(\dfrac{343.3 - 298}{(343.3)(298)} \right) \right]} = \frac{9000}{259} \quad \text{(H26-2)}$$

we find the CSTR volume to be

$$V_R = 34.75 \text{ cu ft}$$

We also note that the "8" in Equation (H26-1) is the heat capacity of the inert.

$$\Delta H_R \text{ (at } 300°\text{K)} = \frac{-2000 \text{ Btu}}{\text{lb-mole of A}}, \quad \tilde{C}_{p_A} = \frac{30 \text{ Btu}}{\text{lb-mole of A-°C}}$$

$$\tilde{C}_{p_B} = \frac{20 \text{ Btu}}{\text{lb-mole B-°C}}, \quad \tilde{C}_{p_I} = \frac{25 \text{ Btu}}{\text{lb-mole of inerts-°C}}$$

$$k = k_1 e^{[E/R(1/T_1 - 1/T)]} \quad E = 5500 \text{ cal./gmole}, \quad R = 1.98 \text{ cal/gmole/°K}$$

$$k_1 = .217 \text{ min}^{-1}, \text{ at } T_1 = 300°\text{K}$$

8.3A Rate Laws

The reaction is first order w.r.t. A and is irreversible; the rate law is

$$-r_A = kC_A$$

The concentration of species A at any position in the reactor is

$$C_A = \frac{F_A}{v}$$

In Frame (H27) write the rate law in terms of X, $k(T_1)$, E, C_{A0}, ϵ, T_0 and T.

(H27) ───

───

Now substitute all known numerical quantities for their appropriate symbols in the design equation for a plug flow reactor.

(H28) ───

───

We see we cannot proceed further until we determine the temperature as a function of conversion. The reaction is carried out adiabatically, and the reference temperature is the same as the temperature of the entering gas streams. Before substituting any numerical values, write the energy balance, deleting all symbols and letters that are unnecessary to this specific example.

(H29) ───

───

(H27)

In this example, $T_1 = T_0$:
$$k = k_1 \exp\left[\frac{E}{R}\left(\frac{1}{T_1} - \frac{1}{T}\right)\right]$$

$$C_A = \frac{F_A}{v} = \frac{F_{A0}(1-X)}{v_0(1+\epsilon X)f} = \frac{C_{A0}(1-X)}{(1+\epsilon X)f}$$

where $f = T/T_0$:
$$-r_A = C_{A0}k_1 \exp\left[\frac{E}{R}\left(\frac{1}{T_1} - \frac{1}{T}\right)\right]\frac{(1-X)}{(1+\epsilon X)f}$$

(H28)

$$V_R = \frac{F_{A0}}{C_{A0}} \int_0^X \frac{(1+\epsilon X)f\, dX}{k_1 \exp\left[\frac{E}{R}\left(\frac{1}{T_1} - \frac{1}{T}\right)\right](1-X)} \tag{H28-1}$$

$$v_0 = \frac{F_{A0}}{C_{A0}} = \frac{F_{t0}}{C_{t0}} = 600 \text{ lb-mole/hr} \times 1 \text{ hr/60 min} \times [359 \text{ cu ft/lb-mole}$$
$$\times 1 \text{ atm/10 atm} \times 300°K/273°K]$$

$v_0 = 395$ cu ft/min

The units of time in v_0 and k must be consistent.

$\epsilon = y_{A0}\delta = (.5)(2-1) = .5$

$V_R = [395 \text{ cu ft/min}] \times$ [integral from 0 to .8]

$$= 395 \int_0^{.8} \frac{(1+.5X)\left(\frac{T}{300}\right)dX}{.217 \exp\left[2777\left(\frac{T-300}{300T}\right)\right](1-X)} \tag{H28-2}$$

(H29)

Applying the energy balance in the form given by Equation (8-24) we obtain

$$-X[\Delta H(T_0) + \Delta \tilde{C}_p(T-T_0)] = \left[\tilde{C}_{p_A} + \frac{F_{I0}}{F_{A0}}\tilde{C}_{p_I}\right][T-T_0] \tag{H29-1}$$

Solve for the temperature as a function of the conversion X, substituting numerical values of all known quantities.

(H30)

We now have all the information needed to evaluate the integral that gives the plug flow reactor volume in Equation (2-17).

8.3B Numerical and Graphical Techniques

The evaluation of this integral can sometimes be carried out systematically by preparing a table:

X	T	f	k	$(1 + \epsilon X)f$	$k(1 - X)$	$\dfrac{C_{A0}}{-r_A} = \dfrac{f(1 + \epsilon X)}{k(1 - X)}$

For this example problem

$$f = \frac{T}{300}$$

$$k = 0.217 \exp\left[27.7\left(\frac{T - 300}{3T}\right)\right], \text{ min}^{-1}$$

$$\epsilon = 0.5$$

$$T = 300 + \frac{200X}{5.5 + X} \,°\text{K}$$

We adopt the procedure in Table 8-1 to determine the plug flow reactor volume:

TABLE 8–1

1. Choose $X = 0$.
2. Calculate T.
3. Calculate f and k.
4. Calculate $(1 + \epsilon X)f$ and $k(1 - X)$.
5. Calculate $C_{A0}/-r_A$.
6. Increment X and go to 2.
7. Plot $C_{A0}/-r_A$ vs. a function of X.
8. Determine the area under the curve.

Note that in this and similar examples, whenever $C_{A0}/-r_A$ changes slowly with respect to X, it is possible to use relatively large increments of X in order to evaluate the integral. To do this, plot $C_{A0}/-r_A$ vs. X and measure the area under the curve (see Figure 8-2).

(H30)

$$\Delta \tilde{C}_p = 2\tilde{C}_{p_B} - \tilde{C}_{p_A}$$

$$\Delta \tilde{C}_p = [(2)(20) - 30] = 10$$

Solving Equation (H29-1) for T,

$$T = T_0 - \frac{\Delta H_R \cdot X}{X\Delta \tilde{C}_p + \tilde{C}_{p_A} + \frac{F_{I0}}{F_{A0}}\tilde{C}_{p_I}} \tag{H30-1}$$

Recalling $F_{I0} = F_{A0}$

$$T = 300 + \frac{2000X}{30 + 25 + 10X} = 300 + \frac{2000X}{55 + 10X} \tag{H30-2}$$

$$= 300 + \frac{200X}{5.5 + X} \; °K$$

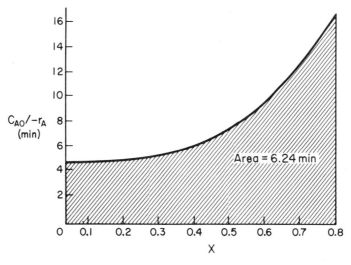

Figure 8-2

$$V = v_0 \times \text{Area} = 395 \text{ cu ft/min} \times 6.24 \text{ min}$$
$$V = 2465 \text{ cu ft}$$

From the data in Table (8-2) we see in this example that the temperature of the gas exiting from the reactor is 325.5°K and from Figure 8-2 we find the reactor volume of 2465 cu ft.

TABLE 8-2

X	T (°K)	k (min^{-1})	f	$f(1 + \epsilon X)$	$k(1 - X)$	$C_{A0}/-r_A$
0	300.0	.217	1.000	1.00	.217	4.61
.2	307.0	.268	1.023	1.12	.214	5.25
.4	313.5	.324	1.045	1.25	.195	6.40
.6	319.6	.384	1.065	1.38	.154	9.00
.7	322.6	.415	1.075	1.46	.125	11.70
.8	325.5	.450	1.085	1.52	.090	16.90

8.4 OXIDATION OF SULFUR DIOXIDE EXAMPLE

As the last example of this chapter we shall look at the oxidation of sulfur dioxide in a sulfuric acid plant. The reaction

$$SO_2 + \tfrac{1}{2}O_2 \rightleftharpoons SO_3$$

is catalyzed by V_2O_5 deposited on a pumice carrier. The oxidation is to be carried out adiabatically in a tubular packed bed reactor filled with cylindrical pellets (8 mm in diameter and 25 mm long) carrying the V_2O_5-promoted catalyst. The rate law for the SO_2 oxidation on this particular catalyst is[†]

$$-r'_{SO_2} = k\sqrt{\frac{P_{SO_2}}{P_{SO_3}}}\left(P_{O_2} - \frac{P_{SO_3}^2}{K_p P_{SO_2}^2}\right) \quad (8\text{-}27)$$

where P_i is the partial pressure of species i. This equation can be used when the conversion is

[†] S. M. Walas, *Reaction Kinetics for Chemical Engineers*, (New York: McGraw-Hill, Inc., 1959), p. 248. From a dissertation by Eklund of the Royal Institute of Technology, Stockholm, 1956.

greater than 5%. Below 5% conversion, the rate is assumed constant at the rate given by Equation (8-27) for 5% conversion.

The feed stream enters a packed bed reactor at 910°F, 1 atm, and at a flow rate of 100 lb-moles/hr. The composition of the feed (mole %) is 10% O_2, 79% inerts (considered as N_2), and 11% SO_2. Write the design equation relating catalyst weight to conversion in terms of the conversion of SO_2, numerically evaluating all symbols possible in the rate law.

(H31)

(H31) The general mole balance equations (design equations) based on the weight of catalyst were given in their differential and integral forms by Equations (6-43) and (6-44):

$$F_{A0} \frac{dX}{dW} = -r'_A$$

Integrating,

$$W = F_{A0} \int_0^X \frac{dX}{-r'_A} \tag{6-44}$$

$$SO_2 + \frac{1}{2} O_2 \longrightarrow SO_3$$

Letting $A \equiv SO_2$

$$-r'_{SO_2} = k \sqrt{\frac{P_{SO_2}}{P_{SO_3}}} \left(P_{O_2} - \frac{P_{SO_3}^2}{K_p P_{SO_2}^2} \right) \tag{8-27}$$

$$\epsilon = y_{A0}\delta = .11(1 - \tfrac{1}{2} - 1) = -.055$$

$$\theta_{SO_2} = 1.0, \quad \theta_{O_2} = \frac{.10}{.11} = .91$$

$$\theta_{SO_3} = 0.0, \quad \theta_{N_2} = \frac{.79}{.11} = 7.17$$

$$P_{SO_2} = C_{SO_2} RT = (C_{SO_2,0}) \frac{1-X}{(1+\epsilon X)\left(\frac{T}{T_0}\right)} RT = C_{SO_2,0} \frac{1-X}{1-.055X} RT_0$$

$$C_{SO_2,0} RT_0 = P_{SO_2,0}$$

$$P_{SO_3} = C_{SO_3} RT = C_{SO_2,0} \frac{XRT_0}{1+\epsilon X} = (P_{SO_2,0}) \frac{X}{1-.055X}$$

$$P_{O_2} = C_{O_2} RT = \frac{P_{SO_2,0}\left(\theta_{O_2} - \frac{X}{2}\right)}{1+\epsilon X} = (P_{SO_2,0}) \frac{0.91 - .5X}{1 - .055X}$$

$$-r_{SO_2} = k \sqrt{\frac{1-X}{X}} \left[P_{SO_2,0} \frac{(0.91 - 0.5X)}{1 - .055X} - \frac{X^2}{(1-X)^2 K_p} \right] \tag{H31-1}$$

$$P_{SO_2,0} = y_{SO_2,0} \Pi = .11 \text{ atm}$$

Rearranging the first term in the brackets of Equation (H31-1), we note

$$P_{A0} \frac{(.91 - .5X)}{(1 - .055X)} = \frac{(.91 - .5X)}{\left(\frac{1 - .055X}{P_{A0}}\right)} = \left(\frac{.91 - .5X}{9.1 - .5X}\right)$$

$$F_{A0} = (.11) F_{T_0} = 11 \text{ lb-moles of } A/\text{hr} = .00306 \text{ lb-moles of } A/\text{sec}$$

By substituting for F_{A0}, $-r_{SO_2}$, and P_{A0} in Equations (8-27) and (6-44) and rearranging, the design equation becomes

$$W = .00306 \int_0^X \frac{dX}{k \sqrt{\frac{1-X}{X}} \left(\frac{.91 - .5X}{9.1 - .5X} - \frac{X^2}{(1-X)^2 K_p} \right)} \tag{H31-2}$$

Since the reaction is carried out adiabatically, the temperature (and thus the specific reaction rate) will be related to conversion. This relationship between T and X can be obtained from an energy balance.

As mentioned earlier, this exothermic reaction is to be carried out adiabatically. Keeping in mind the general mole balances and any other necessary balances, list all the additional data or information needed to solve this problem.

(H32)

To evaluate the equilibrium constant we can utilize Van't Hoff's equation (see Appendix C) to determine $K_p(T)$:

$$\frac{d \ln K_p}{dT} = \frac{\Delta H_R(T)}{RT^2} \tag{8-28}$$

The temperature dependence of ΔH_R is given by Equations (8-20) and (8-21). Therefore, if K_p at a given temperature and $\Delta H_R(T)$ are both known, the equilibrium constant at any other temperature can be found by evaluating the integral in Equation (8-28). The equilibrium constant at a given temperature, T_R, can be determined from the Gibbs free energies of formation:

$$RT_R \ln K(T_R) = -\Delta G°(T_R) = -\sum_{i=1}^{n} v_i G_i° \tag{8-29}$$

where $G_i°$ is the Gibbs free energy of species i at T_R and v_i the stoichiometric coefficient of species i.

(H32)

The only remaining information we need to know to completely determine the mole balance on SO_2, Equation (H31-2), is

(1) the specific reaction rate k, as a function of temperature, and
(2) the equilibrium constant as a function of temperature.

Since the reaction is carried out adiabatically, the temperature T and conversion X are related through the energy balance. For an adiabatic reaction, Equation (8-17) can be written in the form

$$-X \Delta H_R(T) = \int_{T_0}^{T} \sum_{i=1}^{n} \theta_i C_{p_i}(dT) \tag{H32-1}$$

From Equation (H32-1) it is apparent that the following information is also necessary to solve the problem. (The temperature is in degrees Rankine and units of C_p are Btu/lb-mole/°R):

(3) the heat capacity of SO_2,

$$C_{P_{SO_2}} = 6.95 + .0055T - 1.17 \times 10^{-6}T^2 \tag{H32-2}$$

(4) the heat capacity of SO_3,

$$C_{P_{SO_3}} = 7.45 + .0105T - 2.04 \times 10^{-6}T^2 \tag{H32-3}$$

(5) the heat capacity of O_2,

$$C_{P_{O_2}} = 6.12 + .00176T - .31 \times 10^{-6}T^2 \tag{H32-4}$$

(6) the heat capacity of N_2,

$$C_{P_{N_2}} = 6.46 + .00077T - 2.13 \times 10^{-8}T^2 \tag{H32-5}$$

and

(7) the heat of reaction as a function of temperature T. To determine $\Delta H_R(T)$ we need to know the heats of formation of SO_2 and SO_3 at a reference temperature (77°F = 25°C), along with the above heat capacities. These heats of formation at 77°F are

$$H^{\circ}_{SO_2}(77°F) = -127{,}700 \text{ Btu/lb-mole}$$

$$H^{\circ}_{SO_3}(77°F) = -170{,}000 \text{ Btu/lb-mole}$$

The heats of formation, Gibbs free energies, and heat capacities of various compounds can be found in a number of sources.[†] The enthalpies and Gibbs Free Energies of the elements in their standard stable states (e.g., O_2 at 25°C and 1 atm) are, by convention, zero. ($H°_{O_2}(77°F) = 0$.)

In the temperature range of 850–1050°F, the equilibrium constant and specific reaction rate for the SO_2 oxidation can be calculated from

$$\ln K_p = \frac{42{,}300}{T} - 24.2 + .170 \ln T \tag{8-30}$$

$$\ln k = -\frac{149{,}750}{T} + 92.5 \tag{8-31}$$

The temperature T is given in degrees Rankine, K_p, in atm^{-1} and k in lb-moles/sec/lb-cat./atm.

As an exercise, you may want to make the appropriate substitutions for the SO_2 oxidation into Equation (8-28) and carry out the integration to determine whether K_p is indeed given by Equation (8-30).

In Frame (H33) we will apply the energy balance, Equation (H32-1), more specifically to the SO_2 oxidation. Carry this out by expanding the right- and left-hand sides, replacing the C_{p_i}'s by the appropriate heat capacities of SO_2, SO_3, O_2, and N_2. Also note that the limits of integration of the temperature on the two sides of the equation will be different.

(H33)

By specifying the inlet temperature, we have fixed the maximum conversion that we can achieve for a reversible exothermic reactor carried out adiabatically. It can be shown that if the inlet temperature for this reaction falls much below 850°F, the reaction essentially will not take place because an unrealistically large amount of catalyst would be required to obtain any measurable conversion. This points out the necessity of preheating feed streams. However, the higher the inlet temperature, the lower the maximum attainable conversion will be. Consequently, there is usually some optimum inlet temperature in reversible adiabatic reactions. (See Problem P8-8).

[†]*Handbook of Chemistry and Physics*, 48th ed., R. C. Weast, ed., (Cleveland, Ohio: Chemical Rubber Co., 1967).

Perry's Chemical Engineering Handbook, 4th ed., (New York: McGraw-Hill Book Co., 1963).

O. A. Hougen, K. M. Watson and R. A. Ragatz, *Chemical Process Principle*, 2nd ed., Part 1, (New York: John Wiley & Sons, 1959).

R. C. Reid and T. K. Sherwood, *The Properties of Gases and Liquids*, 2nd ed., (New York: McGraw-Hill Book Co., 1966).

(H33)

$$-X\Delta H_R(T) = \int_{T_0}^{T} \sum \theta_i C_{p_i} dT \tag{H32-1}$$

Applying Equation (H32-1) to the SO_2 oxidation, and expanding the right- and left-hand sides, we obtain

$$-X\left[\Delta H_R(T_R) + \int_{T_R}^{T}(C_{p_{SO_3}} - \tfrac{1}{2}C_{p_{O_2}} - C_{p_{SO_2}})dT\right] =$$

$$\int_{T_0}^{T}(\theta_{SO_2}C_{p_{SO_3}} + \theta_{SO_2}C_{p_{SO_2}} + \theta_{O_2}C_{p_{O_2}} + \theta_{N_2}C_{p_{N_2}})dT \tag{H33-1}$$

$$T_R = 537°R, \qquad T_0 = 1370°R$$

Equation (H33-1) is the solution requested in Frame (H33). The heat capacities were given by Equations (H32-2) through (H32-5), and

$$\Delta\tilde{C}_p(T - T_R) = \int_{T_R}^{T}(C_{p_{SO_3}} - \tfrac{1}{2}C_{p_{O_2}} - C_{p_{SO_2}})dT \tag{H33-2}$$

Substituting Equations (H32-2), (H32-3), and (H32-4) into (H33-2) and assuming that T is roughly 1500°R, we make the approximation,

$$\int_{537}^{T}(-2.56 + .004T - .715 \times 10^{-6}T^2)dT \simeq 1.8(T - 537) \tag{H33-3}$$

i.e.,
$$\Delta\tilde{C}_p \doteq 1.8\left(\frac{\text{Btu}}{\text{lb-mole } SO_2\,°R}\right)$$

In this example we wish to determine the conversion and temperature as a function of catalyst weight and of distance along a reactor, which is 6 in. in diameter. Since the cross-sectional area is constant along the length of the reactor, the catalyst weight and reactor length are related to the equation

$$W = \rho_B A_c z \tag{8-32}$$

For ease of calculation we shall make the following approximations:

1. If the temperature T is approximately 1050°F, it can be shown that

$$\int_{T_R}^{T} (C_{P_{SO_3}} - C_{P_{SO_2}} - \tfrac{1}{2} C_{P_{O_2}}) dT \doteq 1.8(T - T_R)$$

2. Between T_0 and T the heat capacities of SO_3, SO_2, O_2, and N_2 are assumed to be constant at 20, 13, 8.4, and 7.9, respectively.

Using these approximations, write the energy balance in terms of temperature and conversion, evaluating numerically all symbols possible.

(H34)

With the aid of a flow chart, or a table similar to Table 8–1, explain how you would proceed to calculate X and T as functions of catalyst weight W and distance z along the reactor.

(H35)

Without carrying out any calculations, in Figure 8-3 sketch qualitatively the equilibrium conversion X_e, the temperature, the reaction rate $-r'_{SO_2}$, and the conversion X, along the length of a packed bed reactor 160 ft long. Assume that the equilibrium is closely approached at a point approximately 100 ft from the entrance to the reactor.

(H34)

Rewriting Equation (H33-1) for the approximation in the problem statement:

$$-X[\Delta H_R(T_R) + \Delta \tilde{C}_P(T - T_R)] = \sum \theta_i \tilde{C}_{P_i}(T - T_0) \tag{H33-1}$$

Solving for T,

$$T = \frac{-X[\Delta H_R(T_R) - \Delta \tilde{C}_P T_R] + \sum \theta_i \tilde{C}_{P_i} T_0}{X \Delta \tilde{C}_P + \sum \theta_i \tilde{C}_{P_i}} \tag{H34-1}$$

$$\Delta H_R(T_R) = \Delta H_R(77°F) = H_{SO_3}^\circ(77°F) - \tfrac{1}{2}H_{O_2}^\circ(77°F) - H_{SO_2}^\circ(77°F)$$

$$\Delta H_R(T_R) = [-170{,}000 - \tfrac{1}{2} \times 0 - (-127{,}700)]$$

$$= -42{,}300 \text{ Btu/lb-mole } SO_2$$

$$\Delta \tilde{C}_P = 1.8 \text{ Btu/lb-mole } SO_2/°R$$

$$\sum \theta_i C_{P_i} = \theta_{SO_3} C_{P_{SO_3}} + \theta_{SO_2} C_{P_{SO_2}} + \theta_{O_2} C_{P_{O_2}} + \theta_{N_2} C_{P_{N_2}}$$

$$= (0)(20) + (1)(13) + (.91)(8.4) + (7.17)(7.9)$$

$$\sum \theta_i C_{P_i} = 77.3, \quad T_R = 77°F = 537°R, \quad T_0 = 910°F = 1370°R$$

$$\sum \theta_i C_{P_i} T_0 = (77.3)(1370) = 105{,}900$$

$$T = \frac{43260X + 105{,}900}{1.8X + 77.3} \tag{H34-2}$$

(H35)

$$-r'_{SO_2} = k\sqrt{\frac{1-X}{X}}\left(\frac{.91 - .5X}{9.1 - .5X} - \frac{X^2}{(1-X)^2 K_p}\right) \tag{H31-1}$$

After substituting $X = .05$ into Equation (H31-1), we find that the rate of reaction between $X = 0.0$ and $X = .05$ is given by

$$-r'_{SO_2} = k\left(.4249 - \frac{.01176}{K_p}\right) \tag{H35-1}$$

1. Set $X = 0.005$.
2. Calculate T from Equation (H34-2).
3. Calculate k from Equation (8-31).
4. Calculate K_p from Equation (8-30).
5. If $X \leq 0.05$, calculate $-r'_{SO_2}$ from Equation (H35-1).
   If $X > 0.05$, calculate $-r'_{SO_2}$ from Equation (H31-1).
6. Calculate W from Equation (H35-2), using some numerical integration technique (e.g., Simpson's rule). One of the simplest is Euler's scheme:

$$W_{i+1} = W_i + \frac{X_{i+1} - X_i}{-r'_{SO_2}} F_{A0} \tag{H35-2}$$

   The reaction rate $-r'_{SO_2}$ in Equation (H35-2) is evaluated at the conversion X_{i+1}.

7. Calculate the equilibrium conversion by solving Equation (H35-3) for X_e:

$$K_p(T) = \frac{P_{SO_3}^2}{P_{O_2} P_{SO_2}^2} = \frac{X_e^2}{(1 - X_e)^2 \left(\frac{.91 - .5X_e}{9.1 - .5X_e}\right)} \tag{H35-3}$$

8. Check to determine if equilibrium has been reached:

   If $X > X_e$ (i.e., if $-r'_{SO_2} < 0$), GO TO STEP 10.

   If $X < X_e$ (i.e., if $-r'_{SO_2} > 0$), CONTINUE.

9. Increment X:

   $$X = X + \Delta X; \text{ GO TO STEP 2.}$$

10. STOP.

(H36)

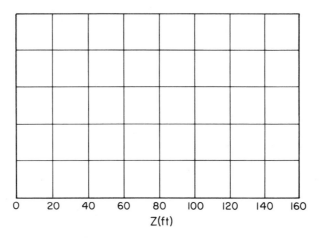

Figure 8-3

The computer program in Figure 8-5 in FORTRAN IV, is used to solve Equations (H31-2) and (H34-2) simultaneously to obtain temperature and conversion profiles within the reactor. This program or parts of it may be useful in the solution to Problems P8-7 through P8-10.

```
>    1     C     W=CATALYST WEIGHT            AC=CROSS SECTIONAL AREA
>    2     C     R=REACTION RATE              AK1=SPECIFIC REACTION RATE
>    3     C     T=TEMPERATURE                AKE=EQUILIBRIUM CONSTANT
>    4     C     X=CONVERSION                 DX=INCREMENT IN CONVERSION
>    5     C     Z=DISTANCE DOWN THE REACTOR  RHO=BULK DENSITY
>    6     C     FAO=ENTERING MOLAR FLOW RATE OF A
>    7           TO=1370.
>    8           FAO=0.00306
>    9           AC=0.196
>   10           RHO=150.
>   11           DX=0.001
>   12           WRITE(6,10)TO,FAO,AC,RHO,DX
>   13        10 FORMAT(15X,5F15.5)
>   14           X1=0.0
>   15           X=0.0
>   16           W=0.0
>   17        20 CONTINUE
>   18     C     CALCULATE ADIABATIC TEMPERATURE
>   19           T=(43260.*X +77.3*TO)/(1.8*X +77.3)
>   20     C     CALC SPECIFIC RXN. RATE & EQUILIBRIUM CONSTANT AT THIS TEMPERATURE
>   21           AK1=EXP(92.5-149750./T)
>   22           AKE=EXP(42300./T -24.2 +0.17*ALOG(T))
>   23     C     CALCULATE RATE OF REACTION
>   24           IF(X .GT. 0.05) GO TO 30
>   25           R=AK1*(0.4292 - 0.01176/AKE)
>   26           GO TO 40
>   27        30 R=AK1*SQRT((1.-X)/X)*((.91- .5*X)/(9.1-.5*X)-(X/(1.-X))**2./AKE)
>   28     C     CALCULATE THE CATALYST WEIGHT AND REACTOR LENGTH
>   29     C     NECESSARY TO ACHIEVE THIS CONVERSION
>   30        40 W=W+FAO*(X-X1)/R
>   31           Z=W/RHO/AC
>   32           WRITE(6,50)X,T,AK1,AKE,R,W,Z
>   33        50 FORMAT(1X,1P7E15.5)
>   34     C     CHECK TO SEE IF EQUILIBRIUM HAS BEEN REACHED
>   35           IF(R .LE. 0.0)GO TO 60
>   36     C     INCREMENT X AND THEN REPEAT THE CALCULATIONS
>   37           X1=X
>   38           X=X+DX
>   39           GO TO 20
>   40        60 CONTINUE
>   41           RETURN
```

Figure 8-5

(H36)

For constant pressure, the equilibrium conversion X_e is solely a function of the gas temperature; for exothermic reactions, it decreases with increasing temperature. As the SO_2 oxidation takes place adiabatically down the reactor, the conversion, temperature, and reaction rate continue to increase until chemical equilibrium is closely approached. At this point the net rate of reaction goes to zero, and, as a result, no further change in temperature or conversion will occur down the insulated reactor. The temperature and conversion profiles are shown in Figure 8-4.

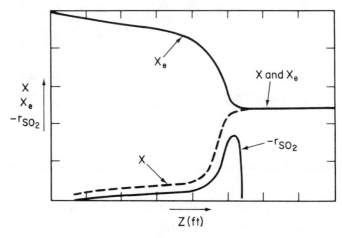

Figure 8–4

The shape of the temperature profile is nearly identical to the shape of the conversion profile shown in Figure 8-4.

The rate of reaction $-r'_{SO_2}$ was determined from Equation (H31-1), and the total catalyst weight W up to any distance z from the reactor entrance is given by Equation (8-33). The bulk density of the catalyst is 150 lb/cu ft, and the reactor cross-sectional area is 0.196 ft. Selected values of X, T, $-r'_{SO_2}$, W, and z from the computer output are listed in Table 8-3.

TABLE 8-3

X	T (°R)	$-r'_{SO_2}\left(\dfrac{\text{lb-moles } SO_2}{\text{lb cat.-sec}}\right)$	W (lb)	Z (ft)
.00	1370	2.16×10^{-8}	.0	.0
.01	1375	3.28×10^{-8}	1135	38.6
.02	1380	4.97×10^{-8}	1882	64.0
.05	1396	1.69×10^{-7}	2921	99.3
.10	1422	8.18×10^{-7}	3377	114.9
.15	1449	4.25×10^{-6}	3468	117.9
.20	1475	2.17×10^{-5}	3485	118.5
.35	1553	2.13×10^{-3}	3489	118.7
.50	1631	4.82×10^{-2}	3490	118.7
.522	1642	7.62×10^{-3}	3490	118.7
.523	1643	$\sim .0$	3490	118.7

The temperature and conversion profiles are shown in Figure 8-6. Note that the equilibrium conversion is closely approached at a distance of 118 ft from the reactor entrance, and that the conversion and temperature increase very rapidly at this point, resulting in a reaction front. The reaction front is a condition produced at a point or location in the reactor where values of dX/dz and dT/dz are very large compared to values of these derivatives at preceding points.

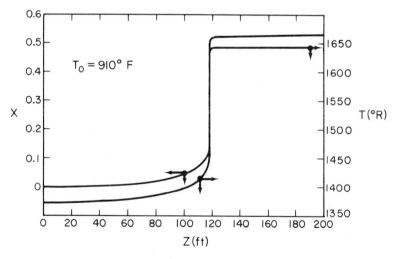

Figure 8-6

Once the reaction has reached equilibrium, which in this case is a distance 118 ft down the reactor for an inlet temperature of 910°F, what would be the effect of removing the insulation from the pipe or of cooling the reacting mixture with a heat exchange system at distances greater than 118 feet from the inlet?

(H37)

8.5–Adiabatic Operation of a Batch Reactor

Up to now we have focused on the steady state operation of non-isothermal flow reactors. However, the development of the unsteady energy balance coupled with the batch reactor design equation is analogous to the steady state development. For a batch system containing n species, Equation (8-10) becomes

$$\dot{Q} = \frac{d\left(\sum_{i=1}^{n} N_i H_i\right)}{dt} = \sum_{i=1}^{n} N_i \frac{dH_i}{dt} + \sum_{i=1}^{n} H_i \frac{dN_i}{dt} \qquad (8\text{-}33)$$

A mole balance on species i gives

$$\frac{dN_i}{dt} = r_i V \qquad (8\text{-}34)$$

The rates of reaction of the various species, r_i, are related to the rate of reaction of species A through the stoichiometric coefficients, v_i,

$$r_i = v_i(-r_A) \qquad (8\text{-}35)$$

Substituting Equations (8-34) and (8-35) into (8-33) we find

$$\dot{Q} = \sum_{i=n}^{n} N_i \frac{dH_i}{dt} + \Delta H_R(T)(V)(-r_A) \qquad (8\text{-}36)$$

We now substitute for dH_i in terms of $\tilde{C}_{pi}$ and dT, and use $N_i = N_{A0}(\theta_i + v_i X)$ to obtain

$$\dot{Q} = N_{A0}\left(\sum_{i=1}^{n} \theta_i \tilde{C}_{pi} + \Delta \tilde{C}_p X\right)\frac{dT}{dt} + \Delta H_R V(-r_A) \qquad (8\text{-}37)$$

where

$$\sum_{i=1}^{n} v_i \tilde{C}_{pi} = \Delta \tilde{C}_p \text{ and } \sum_{i=1}^{n} v_i H_i = \Delta H_R$$

Substitute Equation (B10-4)

$$N_{A0}\frac{dx}{dt} = -r_A V \qquad (B10\text{-}4)$$

into Equation (8-37) and integrate the resulting expression for the case of adiabatic operation of a batch reactor to obtain an algebraic equation between T and X.

(H38)

Once we have T as a function of X for our batch reactor, we can construct a table similar to 8-1 (Page 333) and use analogous techniques to those discussed in Section 8.3B to evaluate the design equation in order to determine

$$t = N_{A0} \int_0^X \frac{dX}{-r_A V} \qquad (2\text{-}14)$$

the time necessary to achieve a specified conversion. (See Problems 8-16 and 8-17).

(H37)

If the reacting mixture is cooled after chemical equilibrium is achieved, the equilibrium conversion X_e will increase. If the mixture is cooled slowly enough, the conversion will also increase. However, if the mixture is cooled very rapidly, producing a large decrease in temperature within a very short (space) time, the specific reaction rate will decrease sharply. In the latter case the reaction will be *frozen* at the previous conversion, $X_e = .522$.

(H38)

Substituting Equation (B10-4) into (8-37) we obtain

$$\dot{Q} = N_{A0}(\sum \theta_i \tilde{C}_{pi} + \Delta \tilde{C}_p X)\frac{dT}{dt} + \Delta H_R N_{A0}\frac{dX}{dt} \tag{H38-1}$$

Expanding ΔH_R and letting $\left(\sum_{r=1}^{n} \theta_i \tilde{C}_{pi}\right) = C_{PS}$ = heat capacity of the mixture

$$\dot{Q} = N_{A0}\left\{[C_{PS} + \Delta \tilde{C}_p X]\frac{dT}{dt} + [\Delta H_R(T_R) + \Delta \tilde{C}_p(T - T_R)]\frac{dX}{dt}\right\} \tag{H38-2}$$

For adiabatic operation, we set $\dot{Q} = 0$, and rearrange to obtain

$$-[\Delta H_R(T_R) - \Delta \tilde{C}_p T_R + \Delta \tilde{C}_p T]\frac{dX}{dt} = [C_{PS} + \Delta \tilde{C}_p X]\frac{dT}{dt} \tag{H38-3}$$

Further rearrangement gives

$$\frac{dX}{C_{PS} + \Delta \tilde{C}_p X} = \frac{-dT}{[\Delta H_R(T_R) - \Delta \tilde{C}_p T_R + \Delta \tilde{C}_p T]} \tag{H38-4}$$

Integrating between the initial conditions, $X = 0$ and $T = T_0$, and later conditions at which $X = X$ and $T = T$:

$$\frac{1}{\Delta \tilde{C}_p} \ln\left[\frac{C_{PS} + \Delta \tilde{C}_p X}{C_{PS}}\right] = \frac{1}{\Delta \tilde{C}_p} \ln\left[\frac{\Delta H_R(T_R) + \Delta \tilde{C}_p(T_0 - T_R)}{\Delta H_R(T_R) + \Delta \tilde{C}_p(T - T_R)}\right] \tag{H38-5}$$

$$= \frac{1}{\Delta \tilde{C}_p} \ln\left[\frac{\Delta H_R(T_0)}{\Delta H_R(T)}\right]$$

Taking the antilog:

$$\frac{C_{PS} + \Delta \tilde{C}_p X}{C_{PS}} = \frac{\Delta H_R(T_0)}{\Delta H_R(T)} \tag{H38-6}$$

Solving for X

$$\boxed{X = \left[\left(\frac{C_{PS}}{\Delta \tilde{C}_p}\right)\frac{\Delta H_R(T_0)}{\Delta H_R(T)} - \frac{C_{PS}}{\Delta \tilde{C}_p}\right]} \tag{H38-7}$$

For the case $\Delta \tilde{C}_p = 0$, Equation (H38-4) integrates to

$$\boxed{T = T_0 - \frac{\Delta H_R X}{C_{PS}} \equiv T_0 - \frac{\Delta H_R X}{\sum\limits_{i=1}^{n} \theta_i \tilde{C}_{pi}}} \tag{H38-8}$$

SUMMARY

For the reaction

$$A + \frac{b}{a}B \rightleftharpoons \frac{c}{a}C + \frac{d}{a}D$$

1. The heat of reaction at temperature T, *per mole of* A, is

$$\Delta H_{Rxn}(T) = \left[\frac{c}{a}H_C + \frac{d}{a}H_D - \frac{b}{a}H_B - H_A\right] \quad \text{(S8-1)}$$

2. The standard heat of reaction *per mole of* A at reference temperature T_R is given in terms of the heats of formation of each species:

$$\Delta H_{Rxn}^\circ(T_R) = \left[\frac{c}{a}H_C^\circ(T_R) + \frac{d}{a}H_D^\circ(T_R) - \frac{b}{a}H_B^\circ(T_R) - H_A^\circ(T_R)\right] \quad \text{(S82-2)}$$

3. The mean heat capacity difference, $\Delta \tilde{C}_p$, per mole of A is

$$\Delta \tilde{C}_p = \left[\frac{c}{a}\hat{C}_{p_C} + \frac{d}{a}\hat{C}_{p_D} - \frac{b}{a}\hat{C}_{p_B} - \hat{C}_{p_A}\right] \quad \text{(S8-3)}$$

4. When there are no phase changes the heat of reaction at temperature T is related to the standard reference heat of reaction by

$$\Delta H_{Rxn}(T) = \Delta H_{Rxn}^\circ(T_R) + \Delta \tilde{C}_p(T - T_R) \quad \text{(S8-4)}$$

$\hat{C}_{pi}$ is the mean heat capacity of species i between temperatures T_R and T. $\tilde{C}_{pi}$ is the mean heat capacity of species i between temperatures T_0 and T.

5. Neglecting changes in potential energy, kinetic energy, and viscous dissipation, the steady-state energy balance is

$$\frac{\dot{Q}}{F_{A0}} - \frac{\dot{W}_s}{F_{A0}} - X[\Delta H_{Rxn}^\circ(T_R) + \Delta \tilde{C}_p(T - T_R)] = \sum_{i=1}^{n}\theta_i \tilde{C}_{pi}\Delta T_i \quad \text{(S8-5)}$$

where n is the number of species entering the reactor:

$$\theta_i = \frac{F_{i0}}{F_{A0}}, \quad \Delta T_i = T - T_{i0}$$

6. If all species enter at the same temperature, $T_0 = T_{i0}$, and no work is done on the system, the energy balance reduces to

$$\frac{\dot{Q}}{F_{A0}} - X[\Delta H_{Rxn}^\circ(T_R) + \Delta \tilde{C}_p(T - T_R)] = \left(\sum_{i=1}^{n}\theta_i \tilde{C}_{pi}\right)(T - T_0) \quad \text{(S8-6)}$$

7. The temperature dependence of the specific reaction rate is given in the form

$$k(T) = k_1(T_1)\exp\left[-\left(\frac{T_1 - T}{T_1 T}\right)\frac{E}{R}\right] \quad \text{(S8-7)}$$

8. The temperature dependence of the equilibrium constant is given by Van't Hoff's equation:

$$\frac{d \ln K_p}{dT} = \frac{\Delta H_{Rxn}(T)}{RT^2} \quad \text{(S8-8)}$$

QUESTIONS AND PROBLEMS

P8-1. The compounds A, B, and C are associated with a particular exothermic reaction. If the heat of reaction for each of the compounds is 7 kcal/gmole of A reacted, -14 kcal/gmole of B reacted, and 21 kcal/gmole of C reacted, deduce the stoichiometry of the reaction. Write the equation so that all coefficients are integers (e.g., $x + z \longrightarrow 2w$).

P8-2. Explain whether or not it is possible to find a unique temperature which will maximize conversion in a stirred tank reactor of fixed volume for:

   a. An irreversible reaction.

   b. An elementary exothermic reversible reaction.

   c. An elementary endothermic reversible reaction.

P8-3. The endothermic liquid phase elementary reaction

$$A + B \longrightarrow 2C$$

proceeds, substantially, to completion in a single steam-jacketed, continuous stirred reactor. From the following data, calculate the steady-state reactor temperature:

$$\text{Reactor volume} = 125 \text{ gal}$$
$$\text{Steam jacket area} = 10 \text{ sq ft}$$
$$\text{Jacket steam} = 150 \text{ psig } (365.9°F \text{ saturation temperature})$$
$$\text{Overall heat transfer coefficient jacket} = U = 150 \text{ Btu/hr sq ft}°F$$
$$\text{Agitator shaft horsepower} = 25 \text{ HP}$$
$$\text{Heat of reaction} = \Delta H_R = +20{,}000 \text{ Btu/lb-mole of A}$$
$$\text{(independent of temperature)}$$

	Component		
	A	B	C
Feed (moles/hr)	10.0	10.0	0
Feed temperature (°F)	80	80	—
Specific heat, Btu/lb-mole/°F†	51.0	44.0	47.5
Molecular weight	128	94	—
Density (lb/cu ft)	63.0	67.2	65.0

†Independent of temperature.

(Ans.: T=196°F)

[Calif PECEE, August 1967]

P8-4. The reversible first-order *exothermic* reaction

$$A \underset{k_1'}{\overset{k_1}{\rightleftharpoons}} B$$

is carried out in an adiabatic CSTR. The feed is at 70°F and is 10,000 lb/hr of a solution containing only A at a concentration of .1 lb-moles/cu ft. Find the volume necessary for 90% conversion of A.

Data:

$$\text{Solution density} = 56 \text{ lb/cu ft (constant)}$$
$$\text{Heat capacity of solution} = 0.7 \text{ Btu/lb/°F (constant)}$$
$$\text{Forward rate constant} = k_1 = 1.2 \text{ hr}^{-1} \text{ at } 25°C$$
$$\text{Activation energy} = E_1 = 25{,}000 \text{ cal./gmole}$$
$$\text{Heat of reaction} = \Delta H_R = -8000 \text{ Btu/lb-mole (constant)}$$
$$\text{Equilibrium constant} = K_e = 12.2 \text{ at } 25°C$$
$$\text{Gas constant} = R = 1.987 \text{ cal/gmole/°C}$$

(Ans.: 4166 ft.³)

[Exam, U. of M.]

P8-5. The elementary gas phase reaction (with kinetics implied as written)

$$A \underset{k_2}{\overset{k_1}{\rightleftharpoons}} 2B$$

is to be carried out in the various reactors discussed below. The feed, which is at a temperature of 27°C, consists of 30 mole % of A and the remainder inerts. The volumetric flow rate entering the reactor at this temperature is 1000 cu ft/min. The concentration of A in the feed at 27°C is 0.5 lb-moles. For 40% conversion,

    a. Calculate the volume of a plug flow reactor when the reaction is carried out adiabatically.

    b. Calculate the reactor volume for a backmix reactor which is operated adiabatically.

Data:

$$C_{p_A} = 25 \text{ Btu/lb-mole/}°C$$
$$C_{p_B} = 20 \text{ Btu/lb-mole/}°C$$
$$C_{p_I} = 30 \text{ Btu/lb-mole/}°C$$

The heat of reaction is a function of temperature, and its value at 300°K is $-15{,}000$ Btu/lb-mole of A.   At 300°K,

$$k_1 = .217 \text{ min}^{-1}$$
$$K_c = 10 \text{ moles/cu ft}$$

k_1 varies with temperature as follows:

T (°K)	300	340
k_1	.217	.324

(Ans.: a. 2890 ft.³, b. 5930 ft.³)

P8-6. The second-order irreversible gas phase reaction

$$A + B \longrightarrow C$$

is carried out isothermally at 1500°R and 5 atm in a 4-in. diameter tubular plug flow reactor. The feed rate is 20 lb-moles/hr, consisting of stoichiometric amounts of the reactants A and B is and 20 mole % inerts, I.

    a. At what distance along the reactor will the conversion be 30%?

    b. What will the wall heat flux (Btu/hr-sq ft) be at this position if the endothermic heat of the reaction is 23,000 Btu/lb-mole at 1500°R? The specific heats are:

$$k = .3 \times 10^6 (\text{hr}^{-1})(\text{cu ft/lb-mole})$$
$$C_{p_A} = .25 \text{ Btu/lb-mole/}°F$$
$$C_{p_B} = .30 \text{ Btu/lb-mole/}°F$$
$$C_{p_C} = .20 \text{ Btu/lb-mole/}°F$$
$$C_{p_I} = .40 \text{ Btu/lb mole/}°F$$

P8-7. *Position of reaction front.* Using the computer program in Figure 8-5 (or a slight variation thereof), determine the effect on the temperature and conversion profiles of varying the inlet temperature. In particular, study the effect of the inlet temperature on the position of the reaction front. In the example problem, the reaction front occurred at a distance of 118 feet from the reactor entrance for an inlet temperature of 1370°R.

P8-8. *Optimum inlet temperature for sulfur dioxide oxidation.* Using the same data and feed flow rate and composition as given in the SO₂ example problem, determine the inlet temperature which will give the maximum conversion of SO_2 in an existing tubular reactor containing 2500 lbs of V_2O_5. *Hint*: Only a slight modification of the computer program shown in Figure 8-5 may be necessary to find this optimum in T_0.

P8-9. *Sulfur dioxide oxidation in a tubular reactor with heat losses.* Using the same data, feed temperature, flow rate, and composition given in the example problem beginning on p. 335, reconsider the SO_2 oxidation, taking into account heat losses along the length of the reactor. The heat flux at any location along the reactor is given by Newton's law of cooling:

$$q = U(T_A - T)$$

where
U = overall heat transfer coefficient based on the external surface area, $A = \pi DL$
D = pipe diameter
T_A = ambient temperature

a. Perform an energy balance on a differential section of the reactor, Δz. Take the limit as $\Delta z \to 0$ and arrange the resulting equation in the form

$$\frac{dT}{dz} = \frac{U\pi D(T_A - T) - F_{A0}[\Delta H_R(T_R) + \Delta \tilde{C}_P(T - T_R)]\frac{dX}{dz}}{F_{A0}(\sum \theta_i \Delta \tilde{C}_{P_i} + \tilde{C}_P X)} \quad \text{(P8-9-1)}$$

b. The weight of catalyst up to any length z in the reactor has been given by

$$W = \left(\frac{\pi D^2}{4}\right)\rho_B z \quad \text{(P8-9-2)}$$

Show how the differential equation resulting from a mole balance on A can be written in the form

$$\frac{dX}{dz} = 29.5 \frac{k_1(T)}{F_{A0}} \sqrt{\frac{1-X}{X}} \left[\frac{.91 - .5X}{9.1 - .5X} - \left(\frac{X}{1-X}\right)^2 \frac{1}{K_P}\right] \quad \text{(P8-9-3)}$$

c. Explain how you would solve Equations (P8-9-1) and (P8-9-3) simultaneously to obtain the temperature and conversion profiles down the reactor.

d. Develop and run a computer program to integrate numerically Equations (P8-9-1) and (P8-9-2) to obtain the temperature and conversion profiles down a reactor that is 120 feet long. Assume that $U = 5$ Btu/hr/sq ft/°F.

e. Determine the effect of changing various parameter values (e.g. U, F_{A0}) on the conversion and temperature profiles.

P8-10. *Interstage Cooling.* In the reactor system shown in Figure P8-10, reactor 1 is operated at conditions identical to those given in the example problem on page 335. The stream leaving reactor 1 is cooled to T_2 in a double-pass tube and shell heat exchanger before entering reactor 2, which is operated adiabatically.

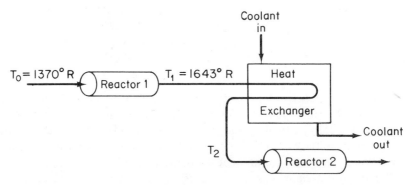

Figure P8–10

a. Assuming that reactor 2 contains the same weight of catalyst as reactor 1, determine the optimum feed temperature to reactor 2 along with the conversion associated with this feed temperature.

b. Determine the minimum cost of the combination of the heat exchanger and reactor 2 to achieve an overall conversion of 95% (based on the feed to reactor 1). Air is to be used as the coolant and is available at 77°F. and a rate of 100 moles/hr. The cost of the reactor is

$$\$ = 7W + 347W^{.367} \quad (W \text{ is in lb of catalyst.}) \quad (P8\text{-}10\text{-}1)$$

This cost includes both the catalyst and the reaction container costs. The cost of the heat exchanger is

$$\$ = 127A^{.59} \quad (A \text{ is in sq ft}) \quad (P8\text{-}10\text{-}2)$$

c. Rather than using air as a coolant in the heat exchanger, the feed stream to reactor 1 is used. The feed stream enters the exchanger at 600°F, is preheated to temperature T_0, and is fed directly to reactor 1. The stream then leaves reactor 1 at temperature T_1, again enters the air heat exchanger (this time as the hot stream), exits the exchanger at T_2, and enters reactor 2. The weight of catalyst in reactor 1 is equal to the weight of catalyst in reactor 2. Determine the minimum cost of this reactor system (HE + reactor 1 + reactor 2) necessary to achieve an overall conversion of 92%. The cost functions are given by Equations (P8-10-1) and (8-10-2). Reasonable first estimates of the catalyst weight and T_0 in each reactor are 1000 lb and 910°F, respectively.

P8-11. *Multiplicity of the Steady State.* A first-order exothermic reaction is carried out in a CSTR. Reactant A enters the CSTR at concentration C_{A0} and temperature T_0 and leaves at concentration C_A and temperature T.

a. For this first-order reaction in a CSTR show that the rate of reaction can be written in the form

$$-r_A = \frac{kC_{A0}}{1 + \tau k} \quad (P8\text{-}11\text{-}1)$$

b. Show that the energy balance can be written in the form

$$-F_{A0}X\Delta H_R(T) = (F_{A0}\sum \theta_i C_{p_i} + UA)(T - T_c) \quad (P8\text{-}11\text{-}2)$$

where

$$T_c = \frac{T_0 F_{A0}\sum \theta_i C_{p_i} + UAT_A}{F_{A0}\sum \theta_i C_{p_i} + UA}$$

A = external surface area of the reactor
T_A = ambient temperature
U = overall heat transfer coefficient

c. After expressing the specific reaction rate in the form

$$k = k_0 e^{-E/RT} \quad (P8\text{-}11\text{-}3)$$

combine Equations (P8-11-1) and (P8-11-2) to obtain

$$\frac{k_0 F_{A0}\tau[-\Delta H_R(T)]}{\tau k_0 + e^{E/RT}} = (F_{A0}\sum \theta_i C_{p_i} + UA)(T - T_c) \quad (P8\text{-}11\text{-}4)$$

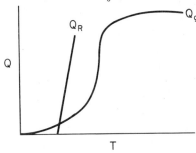

Figure P8–11d

d. The left-hand side of Equation (P8-11-4) will be referred to as heat generated,

$$Q_g = \frac{k_0 F_{A0}[-\Delta H_R(T)]}{\tau k_0 + e^{E/RT}} \quad (P8\text{-}11\text{-}5)$$

and the right-hand side as heat released:

$$Q_R = (F_{A0}\sum \theta_i C_{p_i} + UA)(T - T_c) \quad (P8\text{-}11\text{-}6)$$

Determine whether the plots of Q_g and Q_R in Figure P8-11d are qualitatively correct.

e. At the point where the Q_g and the Q_R lines intersect the steady-state energy balance is satisfied. Is it possible for the Q_R line to intersect the Q_g line three times to produce a multiplicity of the steady state? What parameters (and how do they) affect the slope of the Q_R-vs.-T line that would bring this multiplicity? Explain!

Further discussion of the multiplicity of the steady state may be found in item 2 of Supplementary Reading at the end of this chapter.

P8-12. An endothermic liquid phase reaction,

$$A + B \longrightarrow 2E$$

is to be carried out in a heated tubular flow reactor. For certain reasons it is desirable to operate the reactor so that the rate of reaction remains constant all along the reactor and equal to the inlet reaction rate. The reaction is known to be elementary and irreversible;

$$r_A = -kC_A C_B$$

where the rate constant k is equal to $.200$ l/mole-sec at $60°C$ and also exhibits Arrhenius temperature dependence, with an activation energy $E = 4{,}000$ cal/gmole of A.

    a. Will this mode of reactor operation require a temperature profile that has increasing or decreasing temperatures along the reactor? Explain briefly. (To prevent boiling of the reaction mixture the maximum allowable temperature in the reactor is $T = 227°C$.)

    b. Under these circumstances, what is the greatest possible outlet conversion of A for a feed at $60°C$, consisting of an equimolar concentration of 5 gmoles/l of reactants A and B in an inert solvent I? The inert is present in 50 mole %.

    c. What reactor volume is necessary for the conversion of the feed specified in part b, assuming a feed rate of 1 l/sec? (If you are unable to obtain conversion for part b, assume 50% conversion here and in part d.)

    d. What total rate of heat supply is necessary for the reactor?

Further data: The heat of the (endothermic) reaction is $\Delta H_R = 1{,}000$ cal/gmole of A at $60°C$ and the molar heat capacities may be taken as constants:

$$\tilde{C}_{p_A} = \tilde{C}_{p_B} = \tilde{C}_{p_E} = \tilde{C}_{p_I} = 1.0 \text{ cal/gmole-}°C$$

Also, the solution density is constant and equal to 1.2 g/ml, and the gas constant is $R = 1.98$ cal/gmole-°K. The inert solvent, along with A and B, enters the reactor at $60°C$. [Exam, U. of M.]

P8-13. You have just been hired as an engineer for a chemical plant. Your first job is to explain why the production system used by your late predecessor exploded. First of all, he ran the liquid-phase reaction

$$A \longrightarrow B$$

in a pilot plant reactor. He had pretty strong evidence from the chemistry lab that it was a first-order, irreversible reaction.

The pilot plant reactor was a jacketed kettle with a stirring device, operated as a CSTR. The kettle was not insulated, and saturated steam was run through the jacket. This run was successful, producing a 30% conversion of A. In addition, it was noticed that 2.47 lb/hr of steam were condensed.

After this run in the pilot plant reactor, your late predecessor decided that he could run the same reaction in a production reactor. The production reactor was also a jacketed kettle with a stirring device, operated as a CSTR, and it also was not insulated. Saturated steam was also run through the jacket. Since the production reactor was very similar to the pilot plant reactor (except for size, of course), your late predecessor thought there would be no trouble operating it. However, there was trouble. Very shortly after going on steam, the reactor exploded.

Why did it explode? Justify any qualitative discussion with numbers.

Data:

Feed: feed is pure A.
temperature: 80°F
molecular weight: 100
density: 50 lb/cu ft
heat capacity: 0.5 Btu/lb-°F

Pilot plant reactor: diameter: 1.29 ft
height: 1.29 ft
area for heat transfer: 6.48 sq ft
volume for reaction: 6.67 cu ft

Production reactor: diameter: 3.17 ft
height: 3.17 ft
area for heat transfer: 39.5 sq ft
volume for reaction: 100 cu ft

Pilot plant run: volumetric feed rate: 6.67 cu ft/hr
molar feed rate: 3.33 lb-moles of A/hr
conversion: 30% of A converted
steam condensed: 2.47 lb/hr

Production run: volumetric feed rate: 100 cu ft/hr
molar feed rate: 50 lb-moles A/hr
conversion: analytical instrument destroyed in explosion

Reaction: Reaction is first order and irreversible;
activation energy seems to be somewhere in the neighborhood of 54,000 Btu/lb-mole.

Heat transfer coefficients: stirred liquid—condensing steam: 500 Btu/hr-sq ft-°F
stirred liquid—gas under forced convection: 50 Btu/hr-sq ft-°F
condensing steam—gas under natural convection: 10 Btu/hr-sq ft-°F
gas under forced convection—gas under natural convection: 5 Btu/hr-sq ft-°F

Miscellaneous: Density of reaction mixture: 50 lb/cu ft
Heat capacity of reaction mixture: 0.5 Btu/lb-°F
Temperature of steam: 212°F
Temperature of air outside reactor: 75°F
Heat of condensation of steam: 971 Btu/lb
R: 1.99 Btu/lb-mole-°F
Boiling point of reaction mixture: 325°F
Area of Rocky Mountain National Park: 395 sq miles

[CU]

P8-14. The endothermic elementary liquid phase reaction

$$A + B \xrightarrow{k} C$$

is carried out in a CSTR, which is heated by a steam jacket on the exterior of the reactor. The reactants, consisting only of A and B, are fed in equimolar concentrations (2 gmole/l) at 27°C to the reactor at a volumetric feed rate of 30 l/min. The total reactor volume is 1200 l, and it is desired to achieve 60% conversion of A to product C.

a. Find the reaction temperature necessary to achieve this conversion.

b. What temperature of the steam in the surrounding jacket is necessary to operate the reactor at the conditions specified?

c. What is the corresponding saturated steam pressure?

d. If the steam enters at these conditions, how many lb/hr will condense?

Additional information:

Gas constant R: 2.00 cal/gmole/°K
Activation energy E: 2400 cal/gmole
Rate constant k: 0.01725 l/gmole/min, at 27°C
Heat of reaction ΔH_R: 10,000 cal/gmole of A, at 27°C
Specific heats: $C_{p_A} = C_{p_B} = 20$ cal/gmole/°K
$C_{p_C} = 40$ cal/gmole/°K
Heat transfer area A': 6 sq m
Heat transfer coefficient (overall between reactor and steam) U: 1,000 cal/min/sq m

[Exam, U. of M.]

P8-15. You are operating a batch reactor and the reaction is first order, liquid phase, and exothermic. An inert coolant is added to the reaction mixture to control the temperature. The temperature is kept constant by varying the flow rate of the coolant. See Figure P8-15. Calculate the flow rate of the coolant 2 hr after the start of the reaction.

$$A \longrightarrow B, \qquad r_A = -kC_A$$

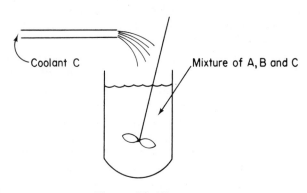

Figure P8-15

Data:

Temperature of reaction: 100°F
Value of k at 100°F: 1.2×10^{-4} sec^{-1}
Temperature of coolant: 80°F
Heat capacity of all components: 0.5 Btu/lb-°F
Density of all components: 50 lb/cu ft
ΔH_R: -25000 Btu/lb-mole
Initially: the vessel contains only A (no B or C present):
$C_{A0} = 0.5$ lb-moles/cu ft. The initial volume of reacting mixture is 50 cu ft.

[CU]

P8-16. Calculate the time necessary to achieve 90% conversion of A in the reaction described on Page 327 when this reaction is carried out adiabatically in a batch reactor. The 50 ft³ reactor is charged (filled) with 900 lb-moles which is 66.67% A and 33.33% inerts.

P8-17. Rework Problem 8-16 for the case when $C_{p_B} = C_{p_C} = 3.5$ Btu/lb-mole/°C. All other parameters and conditions remain the same.

SUPPLEMENTARY READING

1. An excellent development of the energy balance is presented in Chapters 3 and 7 of

    ARIS, R., *Elementary Chemical Reactor Analysis.* Englewood Cliffs, N. J.: Prentice-Hall, Inc., 1969.

and Chapters 4 and 6 of

    HIMMELBLAU, D. M., *Basic Principles and Calculations in Chemical Engineering*, 2nd ed. Englewood Cliffs, N. J.: Prentice-Hall, Inc, 1967.

A number of example problems dealing with nonisothermal reactors can be found in Chapter 5 of

    SMITH, J. M., *Chemical Engineering Kinetics*, 2nd ed. New York: McGraw-Hill Book Co., 1970.

and in Chapter 3 of

    WALAS, S. M., *Reaction Kinetics for Chemical Engineers.* New York: McGraw-Hill Book Co., 1959.

For a thorough discussion on the heat of reaction and equilibrium constant one might also consult

    DENBIGH, K. G., *Principles of Chemical Equilibrium*, 3rd ed. London: Cambridge University Press, 1971.

2. Multiplicity of the steady state is discussed in a recent series of articles:

    ROOT, R. B. and R. A. SCHMITZ, *AIChE J.* 15 (1969), 670.

    ———, *AICHE J.* 16, (1970), 356.

    VEJTASA, S. A. and R. A. SCHMITZ, *AIChE J.* 16 (1970), 410.

and also in Chapter 7 of

    ARIS, R., *Op. Cit.* There is an extensive set of pertinent references at the end of this chapter on nonisothermal operation of reactors.

In addition, see

    PERLMUTTER, D. D., *The Stability of Chemical Reactors.* Englewood Cliffs, N. J.: Prentice-Hall, Inc., 1972.

3. Optimization and interstage cooling of reactors are discussed in

    ARIS, R., *The Optional Design of Chemical Reactors.* New York: Academic Press, 1961.

The optimum temperature progression is discussed in Chapter 8 of

    LEVENSPIEL, O., *Chemical Reaction Engineering*, 2nd ed. New York: John Wiley and Sons, 1972.

4. A brief discussion of the thermal characteristics of chemical reactors (including radial temperature variations in tubular reactors, heat transfer to catalyst particles, temperature effects on diffusion limitations, and multiplicity of the steady state) can be found in Chapter 8 of

    DENBIGH, K. G. and J. C. R. TURNER, *Chemical Reactor Theory*, 2nd ed. London: Cambridge University Press, 1971.

5. Partial differential equations describing axial and radial variations in temperature and concentration in chemical reactors are developed in Chapter 8 of

    WALAS, S. M., *Reaction Kinetics for Chemical Engineers.* New York: McGraw-Hill Book Co., 1959.

6. The heats of formation, $H_i^\circ(T)$, Gibbs free energies, $G_i^\circ(T_R)$, and the heat capacities of various compounds can be found in

   Handbook of Chemistry and Physics, 48 Edition, R. C. WEAST, Editor, Chemical Rubber Co., Cleveland, Ohio (1967).

   Perry's Chemical Engineering Handbook, 4th Edition, McGraw-Hill Book Co., New York (1963).

   Chemical Process Principle, 2nd Edition, Part 1, O. A. HOUGEN, K. M. WATSON and R. A. RAGATZ, John Wiley & Sons, New York (1959).

   The Properties of Gases and Liquids, 2nd Edition, R. C. REID and T. K. SHERWOOD, McGraw-Hill Book Co., New York (1966).

9
Competing Reactions

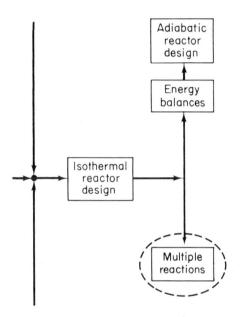

In this chapter we shall discuss reactor selection and general mole balances for simultaneous reactions. There are two basic types of multiple reactions: series and parallel. In *parallel reactions* the reactant is consumed by two different reactions to form different products. An example of a parallel reaction is

$$A \underset{k_2}{\overset{k_1}{\rightleftarrows}} \begin{matrix} B \\ C \end{matrix}$$

In a *series reaction* the reactant forms an intermediate product, which reacts further to form another product. One example is

$$A \xrightarrow{k_1} B \xrightarrow{k_2} C$$

A number of multiple reactions involve a combination of both series and parallel reaction,

such as

$$A + B \longrightarrow C + D$$
$$A + C \longrightarrow E$$

In the first part of this chapter we shall primarily be concerned with parallel reactions. Of particular interest are reactants that are consumed in the formation of a *desired product*, D, as well as in the formation of an *undesired product*, U, in a competing or side reaction. In the reaction sequence

$$A \xrightarrow{k_D} D$$
$$A \xrightarrow{k_U} U,$$

we want to minimize the formation of U and maximize the formation of D, since the greater the amount of undesired product formed, the greater will be the cost of separating the undesired product U from the desired product D. (See Figure 9-1.)

In a highly efficient and costly reactor scheme in which very little of the undesired product U is formed in the reactor, the cost of the separation process could be quite low. On the other hand, if the reactor scheme is inexpensive and inefficient, resulting in the formation of substantial amounts of U, the cost of the separation system could be quite high. Normally, as the cost of the reactor system increases in an attempt to minimize U, the cost of separating species U from D decreases (see Figure 9-2).

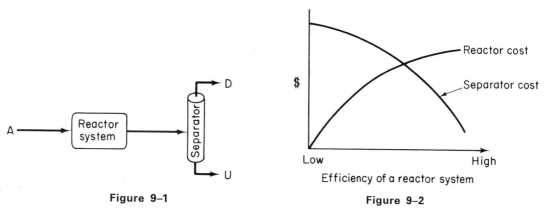

Figure 9–1   Figure 9–2

9.1–Schemes for Minimizing the Undesired Product

First we shall discuss various means of minimizing the undesired product, U, through the selection of reactor type and conditions, and the development of efficient reactor schemes.

For the competing reactions

$$A \xrightarrow{k_D} D \quad \text{(Desired)} \qquad (9\text{-}1)$$

$$A \xrightarrow{k_U} U, \quad \text{(Undesired)} \qquad (9\text{-}2)$$

the rate laws are

$$r_D = k_D C_A^{\alpha_1} \qquad (9\text{-}3)$$
$$r_U = k_U C_A^{\alpha_2} \qquad (9\text{-}4)$$

The rate of disappearance of A for this reaction sequence is the sum of the rates of formation of U and D.

$$-r_A = r_D + r_U \qquad (9\text{-}5)$$
$$-r_A = k_D C_A^{\alpha_1} + k_U C_A^{\alpha_2} \qquad (9\text{-}6)$$

where α_1 and α_2 are positive constants. We want the rate of formation of D, r_D, to be high with respect to the rate of formation of U, r_U. Taking the ratio of these rates we obtain

$$\frac{r_D}{r_U} = \frac{k_D}{k_U} C_A^{\alpha_1 - \alpha_2} \tag{9-7}$$

We shall let this ratio be a selectivity parameter, **S**, which is to be maximized

$$\mathbf{S} = \frac{r_D}{r_U} \tag{9-8}$$

9.1A Maximizing S for One Reactant

Case 1: $\alpha_1 > \alpha_2$. For the case where the reaction order of the desired product is greater than the reaction order of the undesired product, let a be a positive number that is the difference in these reaction orders, i.e.,

$$\alpha_1 - \alpha_2 = a$$

Then

$$\mathbf{S} = \frac{r_D}{r_U} = \frac{k_D}{k_U} C_A^a \tag{9-9}$$

To make this ratio as large as possible we want to carry out the reaction in a manner that will keep the concentration of reactant A as high as possible during the reaction, since, if the concentration of A is low, we see from Equation (9-9) that the ratio of the desired product to the undesired product will also be low. If the reaction is carried out in the gas phase, we shall wish to run it without inerts and at high pressures in order to keep C_A high. If the reaction is in the liquid phase, the use of diluents should be kept at a minimum.†

A batch or plug flow reactor should be used in this case, since in these two reactors the concentration of A starts at a high value and drops progressively during the course of the reaction. In the *perfectly mixed* CSTR model, the concentration of the reactant within the reactor is always at its lowest value; i.e., that of the outlet concentration.

To keep C_A high throughout a plug flow reactor we might, at first glance, believe that we should feed A into the reactor at various points along the reactor, as shown in Figure 9-3. However, further examination of the situation shows that although the addition of pure A (stream 2) to stream 1 increases the concentration of A in stream 1, the stream of pure A is diluted by this addition. Consequently, instead of diluting stream 2 (pure A) by feeding it into the side of the reactor, it is preferable to feed it to the entrance of either another reactor or of the reactor in Figure 9-3, along with stream 1. This scheme, shown in Figure 9-3, will maintain C_A at a low value.

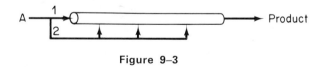

Figure 9-3

Case 2: $\alpha_2 > \alpha_1$. Assume that the reaction order of the undesired product is greater than that of the desired product and select the conditions and reactors you would use to maximize the selectivity parameter **S**. Assume that the reaction occurs in the gas phase.

---

†For a number of liquid phase reactions, the proper choice of a solvent can enhance the selectivity. See, for example, K. F. Wong and C. A. Eckert, *I.E.C.* 62, (1970), 16.

(J1)

1. Feed pure A.
2. Dilute feed stream with inerts.
3. High pressure.
4. Low pressure.
5. Low temperature.
6. High temperature.
7. Tubular reactor.
8. Batch reactor.
9. CSTR.

Since the activation energies of the two reactions in case 1 and case 2 are not known, it cannot be determined whether the reaction should be run at high or low temperatures. The sensitivity of the selectivity parameter to temperature can be determined from the ratio of the specific reaction rates,

$$\frac{k_D}{k_U} = \frac{A_D}{A_U} e^{-[1/RT(E_D - E_U)]} \tag{9-10}$$

where A = frequency factor
E = activation energy

and the subscripts D and U refer to reactions (9-1) and (9-2), respectively.

Case 3: $E_D > E_U$. In this case the specific reaction rate of the desired reaction k_D (and therefore the overall rate r_D) increases more rapidly with increasing temperature than does the specific rate of the undesired reaction k_U. Consequently, the reaction system should be operated at the highest possible temperature in order to maximize S. When $E_U > E_D$, the reaction should be carried out at a low temperature to maximize S, but not so low that the desired reaction does not proceed to any significant extent.

In the following example, reactant A decomposes by three simultaneous reactions to form three products: one which is desired, D, and two which are undesired, T and U. These gas phase reactions, along with their corresponding rate laws, are:

Desired product:

$$A \longrightarrow D$$

$$r_D = \left\{.0012 \exp\left[26{,}000 \left(\frac{1}{300} - \frac{1}{T}\right)\right]\right\} C_A \tag{9-11}$$

Unwanted product:

$$A \longrightarrow U$$

$$r_U = \left\{.0018 \exp\left[25{,}500 \left(\frac{1}{300} - \frac{1}{T}\right)\right]\right\} C_A^{1.5} \tag{9-12}$$

Unwanted product:

$$A \longrightarrow T$$

$$r_T = \left\{.00452 \exp\left[5000 \left(\frac{1}{300} - \frac{1}{T}\right)\right]\right\} C_A^{.5} \tag{9-13}$$

Explain fully how and under what conditions (e.g., reactor type, pressure, temperature, etc.) you should carry out the above reactions in order to minimize the concentrations of the unwanted products U and T.

(J1)

$$A \xrightarrow{k_D} D$$

$$A \xrightarrow{k_U} U$$

$$\alpha_2 > \alpha_1$$

Let $a = \alpha_2 - \alpha_1$, where a is a positive number:

$$S = \frac{r_D}{r_U} = \frac{k_D C_A^{\alpha_1}}{k_U C_A^{\alpha_2}} = \frac{k_D}{k_U C_A^{(\alpha_2-\alpha_1)}} = \frac{k_D}{k_U C_A^a}$$

For the ratio r_D/r_U to be high, the concentration of A should be as low as possible. The means of accomplishing this are:

2. Dilute the feed with inerts.
4. Run at low pressures (i.e., low concentrations of A), if the reaction is in the gas phase.
9. Use a CSTR, since the reactor concentrations of the reactant are maintained at a low level (the exit concentration).

We cannot determine whether to run the reaction at high or low temperatures since we do not know the magnitude of the temperature dependence of each of the specific reaction rates.

9.1B Maximizing S for Two Reactants

We shall next consider two simultaneous reactions in which there are two reactants, A and B, being consumed to produce a desired product, D, along with an unwanted product, U, resulting from a side reaction. The rate laws for the reactions

$$A + B \xrightarrow{k_1} D \tag{9-14}$$

$$A + B \xrightarrow{k_2} U \tag{9-15}$$

are

$$r_D = k_1 C_A^{\alpha_1} C_B^{\beta_1} \tag{9-16}$$

$$r_U = k_2 C_A^{\alpha_2} C_B^{\beta_2} \tag{9-17}$$

The ratio of the rate of formation of D to that of U

$$\boxed{S = \frac{r_D}{r_U} = \frac{k_1}{k_2} C_A^{(\alpha_1-\alpha_2)} C_B^{(\beta_1-\beta_2)}} \tag{9-18}$$

(J2) Since the activation energies of reactions (9-11) and (9-12) are much greater than the activation energy of reaction (9-13), the rate of formation of T will be negligible w.r.t. the rates of formation of D and U at high temperatures, i.e.,

$$S_{DT} = \frac{r_D}{r_T} \simeq \text{very large}$$

Now we need only consider the relative rates of formation of D and U at high temperatures.

$$S_{DU} = \frac{r_D}{r_U} = \frac{.66 e^{500[(1/300)-(1/T)]}}{C_A^{.5}} \tag{J2-1}$$

Since there are three reactions occurring simultaneously (two of which are undesired), it is necessary to subscript the selectivity parameter, S, in order to be certain to which undesired product it refers. From Equation (J2-1) we observe that the amount of undesired product, U, can be minimized by carrying out the reaction at low concentrations. Therefore, to maximize the conversion of A to D we would want to operate our reactor at high temperatures (to minimize the amount of T formed) and low concentrations of A (to minimize the amount of U formed). That is, carry out the reaction at

1. High temperatures.
2. Low concentrations of A, which may be accomplished by
   a. adding inerts,
   b. using low pressures (if gas phase),
   c. using a CSTR.

is to be maximized. Shown in Figure 9-4 are various reactor schemes and conditions which might be used to maximize **S**. From Figure 9-4, a to l, select the schemes and conditions for cases 4–7 [Frames (J3)–(J6)] that might be used to maximize **S** in each case.

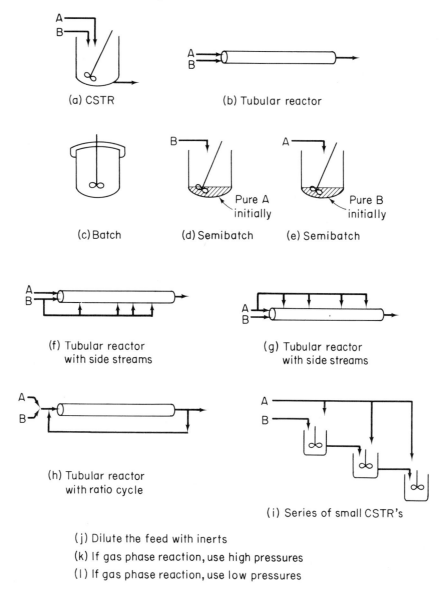

Figure 9–4

Case 4: $\alpha_1 > \alpha_2$, $\beta_1 > \beta_2$. Explain why you selected the particular reaction schemes you did.

(J3)

(J3)
$$\alpha_1 > \alpha_2, \quad \beta_1 > \beta_2$$

Let $a = \alpha_1 - \alpha_2$ and $b = \beta_1 - \beta_2$, where a and b are positive constants. Using these definitions we can write Equation (9-18) in the form

$$S = \frac{r_D}{r_U} = \frac{k_1}{k_2} C_A^a C_B^b$$

To maximize the ratio r_D/r_U, maintain the concentrations of both A and B as high as possible. To do this, use

- (b) a tubular reactor,
- (c) a batch reactor, and
- (k) high pressures (if gas phase).

Case 5: $\alpha_1 > \alpha_2, \beta_2 > \beta_1$. Again explain your reasoning for choosing the schemes which would maximize (r_D/r_U).

(J4)

Case 6: $\alpha_2 > \alpha_1, \beta_2 > \beta_1$. Explain your choices for this case.

(J5)

Case 7: $\alpha_2 > \alpha_1, \beta_1 > \beta_2$. Explain your choices for this case.

(J6)

(J4)

$$\alpha_1 > \alpha_2, \quad \beta_2 > \beta_1$$

Let
$$a = \alpha_1 - \alpha_2, \quad \text{a positive constant}$$
$$b = \beta_2 - \beta_1, \quad \text{a positive constant}$$

Substituting a and b into Equation (9-18), we get

$$S = \frac{r_D}{r_U} = \frac{k_1}{k_2} \frac{C_A^a}{C_B^b} \qquad (J4\text{-}1)$$

To make S as large as possible we want to make the concentration of A high and the concentration of B low. To achieve this, use:

(d) a semibatch reactor in which B is fed slowly into a large amount of A and

(f) a tubular reactor with side streams of B continually fed to the reactor.

(J5)

$$\alpha_2 > \alpha_1, \quad \beta_2 > \beta_1$$

Let
$$a = \alpha_2 - \alpha_1, \quad \text{a positive constant}$$
$$b = \beta_2 - \beta_1, \quad \text{a positive constant}$$

Substituting for a and b into Equation (9-18), we get

$$S = \frac{r_D}{r_U} = \frac{k_1}{k_2 C_A^a C_B^b}$$

To make S as large as possible the reaction should be carried out at low concentrations of A and of B. Use:

(a) A CSTR.

(h) A tubular reactor in which there is a large recycle ratio. The product stream will then act as a dilutent and maintain the entering concentrations of A and B at a low value.

(j) Dilute the feed with inerts.

(l) Use low pressures (if gas phase).

(J6)

$$\alpha_2 > \alpha_1, \quad \beta_1 > \beta_2$$

Let
$$a = \alpha_2 - \alpha_1, \quad \text{a positive constant}$$
$$b = \beta_1 - \beta_2, \quad \text{a positive constant}$$

Then

$$S = \frac{r_D}{r_U} = \frac{k_1}{k_2} \frac{C_B^b}{C_A^a}$$

To maximize S, run the reaction at high concentrations of B and low concentrations of A. Use:

(e) A semibatch reactor in which A is slowly fed to a large amount of B.

(g) A tubular reactor with side streams of A.

(i) A series of small CSTRs with fresh A fed to each reactor.

An alternate definition of selectivity found in current literature is given in terms of the flow rates leaving the reactor:

$$S_E = \text{Selectivity} = \frac{F_D}{F_U} = \frac{\text{Exit molar flow rate of desired product}}{\text{Exit molar flow rate of undesired product}}$$

For a batch reactor the selectivity is given in terms of the number of moles of D and U at the end of the reaction time:

$$S_E = \frac{N_D}{N_U}$$

One also finds that the reaction yield, like the selectivity, has two definitions—one based on the ratio of reaction rates and one based on the ratio of molar flow rates. In the first case, the yield at a point can be defined as the ratio of the reaction rate of a given product to the reaction rate of the key reactant A:[†]

$$Y_C = \frac{r_D}{-r_A}$$

In the case of reaction yield based on molar flow rates, the yield is defined in terms of the ratio of the moles of product formed at the end of the reaction to the number of moles of the key reactant, A, that have been consumed.

For a batch system,

$$Y_E = \frac{N_D}{N_{A0} - N_A}$$

For a flow system,

$$Y_E = \frac{F_D}{F_{A0} - F_A}$$

As a consequence of the different definitions for selectivity and yield, one should check carefully, when reading the literature dealing with multiple reactions, to ascertain the definition intended by the author.

9.2–The Stoichiometric Table for Multiple Reactions

Now we will develop stoichiometric relationships for simultaneous reactions. To generalize our discussion we shall revert to the general reaction,

$$aA + bB \xrightarrow{k_1} cC + dD \tag{2-1}$$

which is now accompanied by a side reaction,

$$gA + fC \xrightarrow{k_2} eE \tag{9-19}$$

The capital letters again denote the species, and the lower case letters denote the stoichiometric coefficients. Since species A appears as the reactant in both reactions, we shall choose it as our basis of calculation and divide Equations (2-1) and (9-19) through by their respective stoichiometric coefficients of A:

Reaction 1:
$$A + \frac{b}{a}B \xrightarrow{k_{A1}} \frac{c}{a}C + \frac{d}{a}D \tag{2-2}$$

Reaction 2:
$$A + \frac{f}{g}C \xrightarrow{k_{A2}} \frac{e}{g}E \tag{9-20}$$

[†] J. J. Carberry, in *Applied Kinetics and Chemical Reaction Engineering*, ed. R. L. Gorring and V. W. Weekman, (Washington, D.C.: ACS Publications, 1967), p. 89.

We now define X_{A1} as the conversion of A to form C and D in reaction (Rxn) 1, i.e.,

$$X_{A1} = \frac{\text{moles of A consumed by reaction 1}}{\text{moles of A initially}}$$

and we define X_{A2} as the conversion of A to form E in reaction 2, i.e.,

$$X_{A2} = \frac{\text{moles of A consumed by reaction 2}}{\text{moles of A initially}}$$

The number of moles of A remaining at any time t will be

$$\begin{Bmatrix}\text{moles of A}\\\text{remaining}\end{Bmatrix} = \begin{Bmatrix}\text{moles of A}\\\text{initially}\end{Bmatrix} - \begin{Bmatrix}\text{moles of A}\\\text{consumed by}\\\text{Rxn 1}\end{Bmatrix} - \begin{Bmatrix}\text{moles of A}\\\text{consumed by}\\\text{Rxn 2}\end{Bmatrix} \quad (9\text{-}21)$$

$$N_A = N_{A0} - N_{A0}X_{A1} - N_{A0}X_{A2} \quad (9\text{-}22)$$

The number of moles of species C remaining after a time t is

$$\begin{Bmatrix}\text{moles of C}\\\text{remaining}\end{Bmatrix} = \begin{Bmatrix}\text{moles of C}\\\text{initially}\end{Bmatrix} + \begin{Bmatrix}\text{moles of C}\\\text{produced by}\\\text{Rxn 1}\end{Bmatrix} - \begin{Bmatrix}\text{moles of C}\\\text{consumed by}\\\text{Rxn 2}\end{Bmatrix} \quad (9\text{-}23)$$

$$\begin{pmatrix}\text{moles of C}\\\text{produced by}\\\text{Rxn 1}\end{pmatrix} = \begin{bmatrix}\dfrac{\text{moles C produced in Rxn 1}}{\text{moles A consumed in Rxn 1}}\end{bmatrix}\begin{pmatrix}\text{moles A}\\\text{consumed in}\\\text{Rxn 1}\end{pmatrix} = \frac{c}{a}(N_{A0}X_{A1}) \quad (9\text{-}24)$$

After writing an expression similar to Equation (9-24) for the moles of C consumed by Rxn 2, write an equation giving the number of moles of C remaining, N_C in terms of N_{A0}, X_{A1}, X_{A2}, etc.

(J7)

(J7) Species C is produced in the system by reaction 1 and deleted from the system by reaction 2.

$$\begin{pmatrix} \text{moles of C} \\ \text{consumed by} \\ \text{reaction 2} \end{pmatrix} = \begin{bmatrix} \text{moles C consumed in Rxn 2} \\ \text{moles A consumed in Rxn 2} \end{bmatrix} \begin{pmatrix} \text{moles of A} \\ \text{consumed in} \\ \text{reaction 2} \end{pmatrix} = \frac{f}{g} N_{A0} X_{A2} \qquad \text{(J7-1)}$$

Substituting Equations (J7-1) and (9-24) into (9-23), we obtain

$$N_C = N_{C0} + \frac{c}{a} N_{A0} X_{A1} - \frac{f}{g} N_{A0} X_{A2} \qquad \text{(J7-2)}$$

$$N_C = N_{A0} \left(\theta_C + \frac{c}{a} X_{A1} - \frac{f}{g} X_{A2} \right) \qquad \text{(J7-3)}$$

The complete stoichiometric table is listed in Table 9–1 for the reactions

$$A + \frac{b}{a}B \xrightarrow{k_{A1}} \frac{c}{a}C + \frac{d}{a}D \qquad (2\text{-}2)$$

$$A + \frac{f}{g}C \xrightarrow{k_{A2}} \frac{e}{g}E \qquad (9\text{-}20)$$

TABLE 9-1

Species	Initially (moles)	Change (moles)	Remaining (moles)
A	N_{A0}	$-(N_{A0}X_{A1} + N_{A0}X_{A2})$	$N_{A0}(1 - X_{A1} - X_{A2})$
B	N_{B0}	$-\frac{b}{a}N_{A0}X_{A1}$	$N_{A0}\left(\theta_B - \frac{b}{a}X_{A1}\right)$
C	N_{C0}	$+\frac{c}{a}N_{A0}X_{A1} - \frac{f}{g}N_{A0}X_{A2}$	$N_{A0}\left(\theta_C - \frac{f}{g}X_{A2} + \frac{c}{a}X_{A1}\right)$
D	N_{D0}	$+\frac{d}{a}N_{A0}X_{A1}$	$N_{A0}\left(\theta_D + \frac{d}{a}X_{A1}\right)$
E	N_{E0}	$+\frac{e}{g}N_{A0}X_{A2}$	$N_{A0}\left(\theta_E + \frac{e}{g}X_{A2}\right)$
Inerts, I	N_{I0}		N_I
	N_{t0}		$N_t = N_{t0} + \delta_1 N_{A0} X_{A1} + \delta_2 N_{A0} X_{A2}$

where

$$\delta_1 = \frac{c}{a} + \frac{d}{a} - \frac{b}{a} - 1 \qquad (9\text{-}25)$$

$$\delta_2 = \frac{e}{g} - \frac{f}{g} - 1 \qquad (9\text{-}26)$$

$$N_{t0} = N_{A0} + N_{B0} + N_{C0} + N_{D0} + N_{E0} + N_{I0} \qquad (9\text{-}27)$$

$$\frac{N_t}{N_{t0}} = 1 + \epsilon_1 X_{A1} + \epsilon_2 X_{A2} \qquad (9\text{-}28)$$

$$\epsilon_1 = y_{A0}\delta_1 \qquad (9\text{-}29)$$

$$\epsilon_2 = y_{A0}\delta_2 \qquad (9\text{-}30)$$

9.3–Applications of the Design Equations to Multiple Reactions

Express the volume as a function of the conversions X_{A1} and X_{A2} and any other appropriate terms when the reactions 1 and 2 are carried out under constant pressure. (Refer to Equation (3-9) if necessary.)

(J8) _____

_____

Write the concentrations of A and B in terms of θ_B, C_{A0}, X_{A1}, and X_{A2} when reactions 1 and 2 are carried out isothermally under constant pressure.

(J9) _____

_____

(J8)

If we neglect variations in the compressibility factor during the course of the reaction, Equation (3-9) reduces to

$$V = V_0 \left(\frac{N_t}{N_{t0}}\right) f \tag{J8-1}$$

for a constant pressure reactor. Replacing the ratio N_t/N_{t0} by Equation (9-28),

$$V = V_0(1 + \epsilon_1 X_{A1} + \epsilon_2 X_{A2}) f \tag{J8-2}$$

$$\epsilon_1 = y_{A0} \delta_1$$

$$\epsilon_2 = y_{A0} \delta_2$$

(J9)

$$C_A = \frac{N_A}{V} = \frac{N_{A0}(1 - X_{A1} - X_{A2})}{V_0(1 + \epsilon_1 X_{A1} + \epsilon_2 X_{A2})} \tag{J9-1}$$

$$= C_{A0} \frac{(1 - X_{A1} - X_A)}{(1 + \epsilon_1 X_{A1} + \epsilon_2 X_{A2})} \tag{J9-2}$$

B is consumed in reaction 1 only.

$$N_B = N_{B0} - \frac{b}{a} N_{A0} X_{A1}$$

$$= N_{A0} \frac{N_{B0}}{N_{A0}} - \frac{b}{a} N_{A0} X_{A1} = N_{A0} \left(\theta_B - \frac{b}{a} X_{A1}\right)$$

$$C_B = \frac{N_B}{V} = \frac{C_{A0}\left(\theta_B - \dfrac{b}{a} X_{A1}\right)}{(1 + \epsilon_1 X_{A1} + \epsilon_2 X_{A2})} \tag{J9-3}$$

Consider the rate of disappearance of A in reaction 1, Equation (2-2), to be first order in A and first order in B, and the side reaction, Equation (9-20), to be second order in C and zero order in A. Write the differential form of the mole balance equation for species B in a batch reactor in terms of the initial concentrations, the conversions X_{A1} and X_{A2}, ϵ_1, ϵ_2, and other appropriate quantities. The reaction is carried out isothermally and at constant pressure.

(J10)

Note that B is involved only in the first reaction. Next, write the differential form of the mole balance equation for species E, which is involved only in reaction 2, Equation (9-20), in a form similar to that presented in Frame (J10).

(J11)

Having seen how to relate and interrelate the number of moles of each species involved in the simultaneous reactions to the conversions X_{A1} and X_{A2}, we can proceed to determine these conversions as a function of time (space time). A method of solution to simultaneous reaction problems is given in Table 9–2.

(J10)

$$-\frac{1}{V}\frac{dN_B}{dt} = -r_B = -\frac{b}{a}r_{A1} = \frac{b}{a}k_1 C_A C_B \tag{J10-1}$$

$$N_B = N_{A0}\left(\theta_B - \frac{b}{a}X_{A1}\right) \tag{J10-2}$$

Substituting Equations (J10-2), (J9-2), and (J9-3) into Equation (J10-1),

$$\frac{dX_{A1}}{dt} = \frac{k_1 C_{A0}(1 - X_{A1} - X_{A2})\left(\theta_B - \frac{b}{a}X_{A1}\right)}{1 + \epsilon_1 X_{A1} + \epsilon_2 X_{A2}} \tag{J10-3}$$

The specific reaction rates are defined w.r.t. species A.

(J11)

Species E appears only in reaction 2, and a mole balance on E results in

$$\frac{1}{V}\frac{dN_E}{dt} = r_E = \frac{e}{g}(-r_{A2}) = \frac{e}{g}k_2 C_A^0 C_C^2 = \frac{e}{g}k_2 C_C^2 \tag{J11-1}$$

$$N_E = N_{A0}\left(\theta_E + \frac{e}{g}X_{A2}\right) \tag{J11-2}$$

For isothermal operation, $f = 1$ and

$$V = V_0(1 + \epsilon_1 X_{A1} + \epsilon_2 X_{A2}) \tag{J8-2}$$

The concentration of C is

$$C_C = C_{A0}\frac{\left(\theta_C - \frac{f}{g}X_{A2} + \frac{c}{a}X_{A1}\right)}{1 + \epsilon_1 X_{A1} + \epsilon_2 X_{A2}} \tag{J11-3}$$

Substituting Equations (J11-2), (J8-2), (J11-3), and (J9-2) into Equation (J11-1) yields

$$\frac{dX_{A2}}{dt} = \frac{k_2 C_{A0}\left(\theta_C - \frac{f}{g}X_{A2} + \frac{c}{a}X_{A1}\right)^2}{(1 + \epsilon_1 X_{A1} + \epsilon_2 X_{A2})} \tag{J11-4}$$

Referring to the mole balances on species B and E [Equations (J10-3) and (J11-4)] we see that we have two coupled nonlinear differential equations. Various methods of solving (numerical and analytical) equations of this type can be found in books on differential equations. Consequently, although steps 3 through 6 in Table 9–2 represent only one method of solving the equation, they have proven successful in the solutions of a number of coupled elementary equations of this type.

TABLE 9-2

Solution procedure:

1. Write the mole balance equation for a species which is involved only in reaction 1.
2. Write the mole balance equation for a species which is involved only in reaction 2.
3. Take the ratio of these two equations so that time (either space time or real time) is eliminated from the resulting equation.† Express all concentrations in terms of the conversions X_{A1} and X_{A2} so that your equation will be in the form

$$\frac{dX_{A1}}{dX_{A2}} = f(X_{A1}, X_{A2}) \quad (9\text{-}31)$$

4. Solve the differential equation (Equation 9-31) to obtain X_{A2} as a function of X_{A1} alone, or vice versa:

$$X_{A2} = g(X_{A1}) \quad (9\text{-}32)$$

Sometimes a closed form solution may not be possible.

5. Substitute either of the above mole balance equations for X_{A2} so that the resulting differential equation involves only X_{A1}:

$$\frac{dX_{A1}}{dt} = h(X_{A1}) \quad (9\text{-}33)$$

6. Solve Equation (9-33) for X_{A1} as a function time and then substitute $X_{A1}(t)$ into Equation(9-32) to obtain $X_{A2}(t)$. Again, closed form solution may not be possible. The time referred to in Equation 9-33 may be either space time or real time.

†If there are more than two simultaneous reactions, the mole balance equations should be solved simultaneously for the conversions X_1, X_2, X_3, etc., corresponding to each reaction, rather than proceeding through steps 3—6 above.

Example:

To illustrate this method of solution, consider the *elementary* simultaneous gas phase reactions:

$$\text{Reaction 1:} \quad A \xrightarrow{k_1} B + C \quad (9\text{-}34)$$

$$\text{Reaction 2:} \quad A \xrightarrow{k_2} D \quad (9\text{-}35)$$

with

$$k_1 = .01 \text{ min}^{-1}$$
$$k_2 = .0025 \text{ min}^{-1}$$

A mixture of 75% A and 25% inerts is fed to a tubular reactor at a volumetric flow rate of 200 l/min. Set up a stoichiometric table for this flow reaction system. (See Table 9-1 on p. 379, if necessary.)

(J12)

Species	Feed	Reacted	Leaving

(J12)

$$\text{Reaction 1:} \quad A \longrightarrow B + C$$
$$\text{Reaction 2:} \quad A \longrightarrow D$$

We can apply Table 9–1 to a flow system by replacing N_{j0} by F_{j0} and N_j by F_j, and defining the conversions

$$X_{A1} = \frac{\text{moles of A consumed by reaction 1}}{\text{moles of A fed to the reactor}}$$

$$X_{A2} = \frac{\text{moles of A consumed by reaction 2}}{\text{moles of A fed to the reactor}}$$

Species	Feed	Reacted	Leaving
A	F_{A0}	$-F_{A0}(X_{A1} + X_{A2})$	$F_A = F_{A0}(1 - X_{A1} - X_{A2})$
B	0	$F_{A0}X_{A1}$	$F_B = F_{A0}X_{A1}$
C	0	$F_{A0}X_{A1}$	$F_C = F_{A0}X_{A1}$
D	0	$F_{A0}X_{A2}$	$F_D = F_{A0}X_{A2}$
I	F_{I0}	—	$F_I = F_{I0}$
	F_{t0}		$F_t = F_{t0} + F_{A0}X_{A1}$

Apply the principles in steps 1–3 of Table 9–2 to this reaction to arrive at a differential equation similar to Equation (9-31).

(J13)

Both X_{A1} and X_{A2} are zero at the entrance of the reactor. Using this condition, we integrate Equation (J13-4) to obtain the relationship between X_{A1} and X_{A2}:

$$X_{A1} = \frac{k_1}{k_2} X_{A2} \tag{9-36}$$

Use Equations (J13-2), (J13-3), and (9-36) in carrying out steps 5 and 6 of Table 9–2 to obtain a solution for the space in terms of the conversion X_{A1}.

(J14)

9.4–Hydrodealkylation of Mesitylene

Consider a reaction sequence similar to but slightly less complicated than the generalized reactions (2-2) and (9-20), in which one of the reactants reacts simultaneously with one of the products. Specifically, we shall look at the production of meta-xylene by the hydrodealkylation of mesitylene over a Detol catalyst.[†]

Reaction 1:

$$\text{mesitylene} + H_2 \longrightarrow \text{meta-xylene} + CH_4 \tag{9-37}$$

Meta-xylene can also undergo hydrodealkylation to form toluene.

Reaction 2:

$$\text{meta-xylene} + H_2 \longrightarrow \text{toluene} + CH_4 \tag{9-38}$$

[†]*I.E.C. Process Design and Development* 4, (1965), 92.
Ibid., 5, (1966), 146.

(J13)

A mole balance on B yields

$$\frac{dF_B}{dV} = \frac{d[F_{A0}(\theta_B + X_{A1})]}{dV} = F_{A0}\frac{dX_{A1}}{dV} = r_B = -r_{A1} \qquad \text{(J13-1)}$$

$$v_0 C_{A0}\frac{dX_{A1}}{dV} = k_1 C_A = k_1 \frac{F_A}{v} = \frac{k_1 C_{A0}(1 - X_{A1} - X_{A2})}{(1 + \epsilon_1 X_{A1} + \epsilon_2 X_{A2})}$$

$$\frac{dX_{A1}}{d\tau} = \frac{k_1(1 - X_{A1} - X_{A2})}{1 + \epsilon_1 X_{A1} + \epsilon_2 X_{A2}} \qquad \text{(J13-2)}$$

A mole balance on D yields

$$\frac{dF_D}{dV} = F_{A0}\frac{dX_{A2}}{dV} = r_D = -r_{A2} = k_2 C_A$$

$$\frac{dX_{A2}}{d\tau} = \frac{k_2(1 - X_{A1} - X_{A2})}{(1 + \epsilon_1 X_{A1} + \epsilon_2 X_{A2})} \qquad \text{(J13-3)}$$

Taking the ratio of Equation (J13-2) and (J13-3), we obtain

$$\frac{dX_{A1}}{dX_{A2}} = \frac{k_1}{k_2} \qquad \text{(J13-4)}$$

(J14)

$$\epsilon_1 = y_{A0}\delta_1 = (.75)(1 + 1 - 1) = .75$$
$$\epsilon_2 = y_{A0}\delta_2 = (.75)(0) = 0$$

First we substitute the above values for ϵ_1 and ϵ_2 in Equation (J13-2),

$$\frac{dX_{A1}}{d\tau} = \frac{k_1(1 - X_{A1} - X_{A2})}{(1 + .75 X_{A1})} \qquad \text{(J14-1)}$$

and then use Equation (9-36) to substitute for X_{A2}:

$$\frac{dX_{A1}}{d\tau} = \frac{k_1\left[1 - \left(1 + \frac{k_2}{k_1}\right)X_{A1}\right]}{(1 + .75 X_{A1})} \qquad \text{(J14-2)}$$

With the aid of Equation (A2-10) in Appendix A2 we integrate Equation (J14-2)

$$\tau = \frac{1}{k_1}\left[\frac{1.75 + \frac{k_2}{k_1}}{\left(1 + \frac{k_2}{k_1}\right)^2}\ln\left(\frac{1}{1 - \left(1 + \frac{k_2}{k_1}\right)X_{A1}}\right) - \frac{.75 X_{A1}}{1 + \frac{k_2}{k_1}}\right] \qquad \text{(J14-3)}$$

$$\frac{k_2}{k_1} = .25, \qquad k_1 = .01 \text{ min}^{-1}$$

$$\tau = 128\left[\ln\left(\frac{1}{1 - 1.25 X_{A1}}\right)\right] - 60 X_{A1}, \text{ min} \qquad \text{(J14-4)}$$

Let's *calculate the space time to achieve 50% total conversion of A*:

$$X_{A1} + X_{A2} = .5 \qquad \text{(J14-5)}$$

$$X_{A1}\left(1 + \frac{k_2}{k_1}\right) = .5 \qquad \text{(J14-6)}$$

$$X_{A1} = \frac{.5}{1.25} = .4$$

$$X_{A2} = .1$$

$$\tau = 128\left(\ln\frac{1}{.5}\right) - 60(.4) = 64.7 \text{ min}$$

The second reaction is undesirable, since meta-xylene sells for about 105 cents a gallon and toluene for about 23 cents a gallon.† Thus we see that there is significant incentive to maximize the production of m-xylene.

The rate expressions of reactions 1 and 2 are, respectively,

$$-r_M = k_1 C_M C_H^{.5} \qquad (9\text{-}39)$$

$$r_T = k_2 C_X C_H^{.5} \qquad (9\text{-}40)$$

where the subscripts are M = mesitylene, X = m-xylene, T = toluene, H = hydrogen (H_2), and Me = methane.

At 1500°R the specific reaction rates are

Reaction 1: $k_1 = 55.20$ (cu ft/lb-mole)$^{.5}$/hr

Reaction 2: $k_2 = 30.16$ (cu ft/lb-mole)$^{.5}$/hr

The bulk density of the catalyst has been included in the specific reaction rate (i.e., $k_1 = k_1' \rho_B$).

9.4A Packed Bed Reactor Calculations

This reaction is to be carried out isothermally at 1500°R and 35 atm in a packed bed reactor in which the feed is 66.6 mole % hydrogen and 33.3 % mesitylene. Set up a stoichiometric table and then apply the principles in steps 1—3 of Table 9–2 to arrive at a differential equation similar to Equation (9-31).

(J15)

To solve Equation (J15-8), i.e.,

$$\frac{dX_{A1}}{dX_{A2}} = \frac{k_1}{k_2} \frac{(.5 - X_{A1})}{(X_{A1} - X_{A2})} \qquad (9\text{-}41)$$

for X_{A1} in terms of X_{A2}, we let

$$\frac{k_2}{k_1} = \gamma$$

$$y = .5 - X_{A1}$$

$$z = X_{A2} - .5$$

$$\frac{dy}{dz} = \frac{1}{\gamma} \frac{y}{z+y} \qquad (9\text{-}42)$$

After inverting,

$$\frac{dz}{dy} = \gamma + \gamma \frac{z}{y} \qquad (9\text{-}43)$$

we rearrange Equation (9-43) to obtain a form of Bernoulli's differential equation,

$$\frac{dz}{dy} - \frac{\gamma}{y} z = \gamma \qquad (9\text{-}44)$$

for which the integrating factor (i.f.) is

$$\text{i.f.} = \exp\left(\int f(y)\,dy\right) = \exp\left(-\gamma \int \frac{dy}{y}\right) = y^{-\gamma}$$

†April, 1973 prices from *Chemical Marketing Reporter: Oil Paint and Drug Reporter* Vol. 203, (*Nos.* 14–16) 1973. (Schnell Publishing Company).

(J15)

$$\text{Reaction 1:} \quad M + H \longrightarrow X + Me \quad (9\text{-}37)$$

$$\text{Reaction 2:} \quad X + H \longrightarrow T + Me \quad (9\text{-}38)$$

$$X_{A1} = \frac{\text{moles of H consumed in reaction 1}}{\text{mole of H fed}}$$

$$X_{A2} = \frac{\text{moles of H consumed in reaction 2}}{\text{mole of H fed}}$$

If only hydrogen and mesitylene are initially present (33.3% M and 66.6% H),

$$\theta_{Me} = \theta_X = \theta_T = 0, \quad \theta_M = \frac{.333}{.666} = .5$$

Hydrogen:	$F_A = F_{A0}(1 - X_{A1} - X_{A2})$	(J15-1)
Mesitylene:	$F_M = F_{A0}(\theta_M - X_{A1}) = F_{A0}(.5 - X_{A1})$	(J15-2)
Xylene:	$F_X = F_{A0}(\theta_X + X_{A1} - X_{A2}) = F_{A0}(X_{A1} - X_{A2})$	(J15-3)
Toluene:	$F_T = F_{A0}(\theta_T + X_{A2}) = F_{A0} X_{A2}$	(J15-4)
Methane:	$F_{Me} = F_{A0}(\theta_{Me} + X_{A1} + X_{A2}) = F_{A0}(X_{A1} + X_{A2})$	(J15-5)

Mole balances on mesitylene and toluene yield, respectively,

$$\frac{F_{A0}}{v_0} \frac{dX_{A1}}{d\tau} = k_1 C_H^{.5} C_M \quad (J15\text{-}6)$$

$$\frac{F_{A0}}{v_0} \frac{dX_{A2}}{d\tau} = k_2 C_H^{.5} C_X \quad (J15\text{-}7)$$

Dividing Equation (J15-6) by Equation (J15-7) and substituting for the concentration in terms of conversion, we obtain

$$\frac{dX_{A1}}{dX_{A2}} = \frac{k_1}{k_2}\frac{C_M}{C_X} = \frac{F_M}{F_X}\frac{k_1}{k_2} = \frac{(0.5 - X_{A1})}{(X_{A1} - X_{A2})}\frac{k_1}{k_2} \quad (J15\text{-}8)$$

Using the integrating factor, Equation (9-44) is rearranged

$$\frac{d}{dy}(y^{-\gamma}z) = \gamma y^{-\gamma} \tag{9-45}$$

The solution to Equation (9-45) for the initial conditions corresponding to $X_{A1} = X_{A2} = 0$ ($y = .5, z = -.5$) is

$$z = \frac{\gamma y - .5\left(\frac{y}{.5}\right)^\gamma}{1 - \gamma} \tag{9-46}$$

$$X_{A2} = \frac{\left(.5 - \frac{k_2}{k_1}X_{A1}\right) - .5(1 - 2X_{A1})^{\frac{k_2}{k_1}}}{\left(1 - \frac{k_2}{k_1}\right)} \tag{9-47}$$

Equation (9-47) gives the moles of toluene formed per mole of hydrogen fed, i.e., X_{A2}, as a function of the moles of mesitylene consumed per mole of hydrogen fed, i.e., X_{A1}. The molar feed rate of H_2 to the reactor is 10 lb-moles/hr. For a 60% *conversion of mesitylene*, i.e.,

$$\frac{\text{moles mesitylene consumed}}{\text{mole mesitylene fed}} = X'_{B1}$$

calculate:

1. X_{A1}, the conversion of mesitylene per mole of hydrogen fed.
2. The molar flow rate of mesitylene. (Ans. 2 lb mole/hr)
3. The molar flow rate of toluene.
4. The molar flow rate of xylene.
5. The molar flow rate of hydrogen.
6. The molar flow rate of methane.
7. The mole fraction of xylene.

[*Hint*: Refer to Equations (J15-1)—(J15-5), and (9-47).]

(J16)

(J16)

1. $X'_{B1} = \dfrac{\text{moles of mesitylene consumed}}{\text{moles of mesitylene fed}} = .6$

   $X_{A1} = \dfrac{\text{moles of mesitylene consumed}}{\text{moles of hydrogen fed}}$

   $$X_{A1} = \left(\dfrac{\text{moles of mesitylene fed}}{\text{moles of hydrogen fed}}\right)\left(\dfrac{\text{moles of mesitylene consumed}}{\text{moles of mesitylene fed}}\right) = \theta_M X'_{B1}$$

   $X_{A1} = \theta_M X'_{B1} = (.5)(.6) = .3$

   $$\boxed{X'_{B1} = 2X_{A1}} \qquad (J16\text{-}1)$$

2. From Equation (J15-2),

   $$F_M = F_{A0}(.5 - X_{A1}) = 10(.2) = 2 \text{ lb-moles of mesitylene/hr}$$

3. Equation (J15-4) gives the molar flow rate of toluene in terms of the conversion, X_{A2}:

   $$F_T = F_{A0} X_{A2} \qquad (J15\text{-}4)$$

   X_{A2} is obtained from Equation (9-47):

   $$X_{A2} = \dfrac{(.5 - \gamma X_{A1}) - .5(1 - 2X_{A1})^\gamma}{1 - \gamma} \qquad (9\text{-}47)$$

   $$\gamma = \dfrac{k_2}{k_1} = \dfrac{30.16}{55.2} = .546$$

   When $X_{A1} = .3$

   then $X_{A2} = \dfrac{[.5 - (.546)(.3)] - .5[1 - 2(.3)]^{.546}}{1 - .546} = .072$

   $F_T = (10)(.072) = .72 \text{ lb-moles/hr}$

4. The molar flow rate of xylene is

   $$F_X = F_{A0}(X_{A1} - X_{A2}) \qquad (J15\text{-}3)$$

   By substituting (9-47) into (J15-8) we can express the molar flow rate of xylene as a function of k_1, k_2 and X_{A1}:

   $$\dfrac{F_X}{F_{A0}} = X_{A1} - X_{A2} = X_{A1} - \dfrac{(.5 - \gamma X_{A1}) - .5(1 - 2X_{A1})^\gamma}{1 - \gamma}$$

   $$\boxed{X_{A1} - X_{A2} = \dfrac{.5(1 - 2X_{A1})^\gamma - (.5 - X_{A1})}{1 - \gamma}} \qquad (J16\text{-}2)$$

   $$X_{A1} - X_{A2} = \dfrac{.5[1 - 2(.3)]^{.546} - (.5 - .3)}{.454} = .228$$

   Also, $X_{A1} - X_{A2} = 0.3 - .072 = .228$

   $F_X = (10)(.228) = 2.28 \text{ lb-mole xylene/hr}$

5. $F_{Me} = F_{A0}(X_{A1} + X_{A2}) = 10(.3 + .072) = 3.72 \text{ lb-moles methane/hr} \qquad (J15\text{-}5)$

6. $F_t = F_X + F_M + F_{Me} + F_H + F_T \qquad (J16\text{-}3)$

   Substituting Equations (J15-6)—(J15-10) into (J16-3) and cancelling terms,

   $$F_t = 1.5 F_{A0} \qquad (J16\text{-}4)$$

7. The mole fraction of xylene is

   $$y_X = \dfrac{F_X}{F_t} = \dfrac{F_{A0}(X_{A1} - X_{A2})}{1.5 F_{A0}} = \dfrac{X_{A1} - X_{A2}}{1.5} = \dfrac{.228}{1.5} = .152$$

By continuing the calculations carried out in Frame (J16) for conversions of X_{A1} ranging from 0 to .5, (i.e., X'_{B1} ranging from 0 to 1) we can construct Figure 9-5, which shows the variation of the mole fraction of each of the species as a function of the mesitylene conversion (moles of M consumed/mole of M fed).

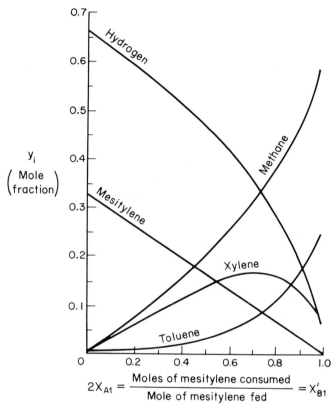

Figure 9–5

9.4B Optimization of Xylene Production in a Packed Bed Reactor

We note from Figure 9-5 that there is a point at which the mole fraction of xylene is at a maximum. We next wish to determine the distance down the reactor (reactor volume) at which the molar flow rate of xylene is at a maximum. To locate this optimum point we first determine the *conversion of mesitylene per mole of hydrogen fed*, X_{A1}, at which the molar flow of xylene will be at a maximum. After finding this value of the conversion, $X_{A1,\text{opt}}$, we can then rearrange and integrate the mesitylene mole balance,

$$C_{A0}\frac{dX_{A1}}{d\tau} = k_1 C_M C_H^{.5} \tag{J15-4}$$

to find the volume at which $X_{A1} = X_{A1,\text{opt}}$.

391

In frame (J17), determine the optimum conversion, X_{A1}, at which the molar flow rate of xylene will be at a maximum as a function of k_2 and k_1. [*Hint*: Apply elementary calculus principles to the appropriate equation, i.e., Equations (9-46), (J15-4), (J16-4) or (J16-7) in order to determine $X_{A1, opt} = f(k_1, k_2)$.]

(J17)

Next, write the reactor design equation that will give the volume in terms of k_1, k_2, X_{A1}, and $X_{A1, opt}$, at which F_X is at a maximum.

(J18)

If we continue along the same lines as the calculations in Frame (J16) we can derive Table 9–3.

TABLE 9–3

X_{A1}	X_{A2}	$(.5-X_{A1})$	$(1-X_{A1}-X_{A2})^{1/2}$	$-r_M$	$\dfrac{F_{A0}}{-r_M}$	$V = \int^{X_{A1}} \dfrac{F_{A0}}{-r_M} dX_{A1}$
0	0	.500	1.000	.0846	118	0
.1	.006	.400	.940	.0636	157	14
.2	.027	.300	.879	.0446	224	33
.3	.072	.200	.792	.0268	373	62
.368	.126	.132	.711	.0159	630	95
.40	.162	.100	.662	.0112	894	127
.49	.381	.010	.633	.0011	9347	—

Using Equations (J15-1)—(J15-5) in conjunction with the information given in Table 9–3, it is possible to construct Figure 9-6 showing the individual flow rates of hydrogen, mesitylene, xylene, toluene, and methane as a function of distance (volume) along the reactor. The tubular reactor volume for which the molar flow rate of xylene will be at a maximum is 95 cu ft.

(J17)

Equation (J16-2) gives the molar flow of xylene per mole of hydrogen fed as a function of the conversion of mesitylene per mole of hydrogen fed, X_{A1}. To find the maximum value of $X_{A1} - X_{A2}$ w.r.t. to X_{A1}, differentiate w.r.t. X_{A1}, set the derivative equal to zero, and solve for X_{A1}:

$$\frac{F_X}{F_{A0}} = X_{A1} - X_{A2} = \frac{.5(1 - 2X_{A1})^\gamma - (.5 - X_{A1})}{1 - \gamma}$$

$$\frac{d\frac{F_X}{F_{A0}}}{dX_{A1}} = \frac{d(X_{A1} - X_{A2})}{dX_{A1}} = 0 = \frac{d}{dX_{A1}}\left[\frac{.5(1 - 2X_{A1})^\gamma - (.5 - X_{A1})}{1 - \gamma}\right] \quad \text{(J17-1)}$$

$$0 = (.5)(\gamma)(1 - 2X_{A1})^{\gamma-1}(-2) + 1 \quad \text{(J17-2)}$$

$$X_{A1} = \tfrac{1}{2}\left[1 - \left(\frac{1}{\gamma}\right)^{\frac{1}{(\gamma-1)}}\right] = \tfrac{1}{2}\left[1 - \left(\frac{k_1}{k_2}\right)^{\frac{k_1}{(k_2-k_1)}}\right] \quad \text{(J17-3)}$$

$$\gamma = \frac{k_2}{k_1} = \frac{30.16}{55.2} = .546$$

$$X_{A1} = \tfrac{1}{2}\left[1 - \left(\frac{1}{.546}\right)^{\frac{1}{(.546-1)}}\right] = \tfrac{1}{2}(1 - .546^{2.2}) \quad \text{(J17-4)}$$

$$= .368$$

$$(X'_{B1} = X_{A1}/\theta_M = .368/.5 = .736)$$

(J18)

The mole balance on mesitylene, Equation (J15-4), is

$$C_{A0}\frac{dX_{A1}}{d\tau} = -r_M = k_1 C_M C_H^{.5} \quad \text{(J15-4)}$$

Substituting for C_M and C_H in terms of the conversions X_{A1} and X_{A2},

$$-r_M = k_1 C_{A0}^{1.5}(.5 - X_{A1})(1 - X_{A1} - X_{A2})^{.5} \quad \text{(J18-1)}$$

Recalling $\tau = V/v_0$, substitute Equation (J18-1) into (J15-4) and integrate to obtain the equation that gives the reaction volume V_{opt}, for which the molar flow rate of xylene will be at a maximum:

$$V_{opt} = F_{A0}\int_0^{X_{A1,opt}} \frac{dX_{A1}}{k_1 C_{A0}^{1.5}(.5 - X_{A1})(1 - X_{A1} - X_{A2})^{.5}} \quad \text{(J18-2)}$$

where

1. $X_{A2} = \dfrac{\left(.5 - \frac{k_2}{k_1}X_{A1}\right) - .5(1 - 2X_{A1})^{\frac{k_2}{k_1}}}{1 - \frac{k_2}{k_1}}$

2. $X_{A1,opt} = \tfrac{1}{2}\left[1 - \left(\frac{k_1}{k_2}\right)^{\frac{k_1}{(k_2-k_1)}}\right] = \tfrac{1}{2}\left[1 - \left(\frac{1}{\gamma}\right)^{\frac{1}{(\gamma-1)}}\right]$

3. $C_{A0} = \dfrac{y_{A0}\Pi_0}{RT} = \dfrac{(.66)(35)}{.73(1500)} = .021\,\dfrac{\text{lb-mole}}{\text{cu ft}}, \quad k_1 C_{A0}^{1.5} = .169\,\dfrac{\text{lb-mole}}{\text{cu ft/hr}}$

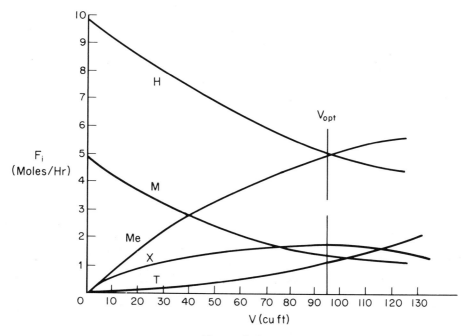

Figure 9-6

Using Table 9–3, calculate the selectivities (**S** and S_E) of xylene w.r.t. toluene, and the yields (Y_C and Y_E) of xylene at the point in the reactor where the molar flow rate of xylene is at a maximum. (*Hint*: Recall that **S** and Y_C are based on the ratio of reaction rates while S_E and Y_E are based on the ratio of molar flow rates.)

(J19)

9.4C Fluidized Bed Reactor Calculations

Now we shall consider a fluidized bed CSTR in which the same hydrodealkylation reactions,

$$H + M \xrightarrow{k_1} X + Me$$

$$H + X \xrightarrow{k_2} X + Me,$$

are taking place. For the purposes of this example, we shall consider the bulk density of the fluidized bed to be identical to the packed bed just discussed, in which case the specific reaction rates for the fluidized and fixed beds will be identical. As before, we would like to determine the volume of the CSTR, V_{opt}, at which the exit molar rate of xylene will be at a maximum.

(J19) The molar flow rate of xylene is a maximum at $X_{A1} = 0.368$. The corresponding value of X_{A2} is $.126$. The selectivity based on relative reaction rates of the desired to the undesired product is written as

$$S = \frac{r_X}{-r_T} = \frac{k_1 C_M C_H^{.5} - k_2 C_X C_H^{.5}}{k_2 C_X C_H^{.5}} \tag{J19-1}$$

$$= \frac{k_1 C_M}{k_2 C_X} - 1 = \frac{(.5 - X_{A1})}{\gamma(X_{A1} - X_{A2})} - 1 \tag{J19-2}$$

$$= \frac{1.83(.5 - .368)}{.368 - .126} - 1 = 0$$

$$= 0$$

This value of the selectivity at the optimum value of X_{A1} could have been deduced from Figure 9-5, showing the mole fraction of xylene as a function of $2X_{A1}$ (, i.e., X'_{B1}). At the maximum flow rate of xylene, the rate of formation of xylene from mesitylene is equal to the rate of decomposition of xylene to form toluene; consequently, the net rate of formation of xylene is zero (i.e., $r_X = 0$ at $X_{A1} = X_{A1,opt}$). A more meaningful definition under this circumstance is the exit selectivity, S_E.

$$S_E = \frac{F_X}{F_T} = \frac{F_{A0}(X_{A1} - X_{A2})}{F_{A0} X_{A2}} = \frac{X_{A1} - X_{A2}}{X_{A2}} \tag{J19-3}$$

$$S_E = \frac{.368 - .126}{.126} = 1.92$$

The yield at $X_{A1,opt}$, based on the ratio of reaction rates, is zero for the reasons discussed above:

$$Y_C = \frac{-r_X}{-r_M} = \frac{0}{-r_M} = 0 \tag{J19-4}$$

Again, the more meaningful definition is based on the ratio of molar flow rates. The key reactant on which the yield is based in this case is mesitylene.

$$Y_E = \frac{F_X}{F_{M0} - F_M} \tag{J19-5}$$

Substituting for the flow rates in terms of conversion,

$$Y_E = \frac{X_{A1} - X_{A2}}{X_{A1}} = \frac{.368 - .126}{.368} \tag{J19-6}$$

$$Y_E = 0.657$$

After writing mole balances for mesitylene and toluene for the CSTR, find the ratio of these two quantities, express their concentrations as functions of conversion, and then solve the resulting equation for X_{A2} as a function of X_{A1}, k_1, and k_2 ($\gamma = k_2/k_1$).

(J20)

9.4D Optimization of Xylene Production in a Fluidized Bed Reactor

The molar flow rate of xylene can be obtained by combining Equations (J15-3) and (J20-8):

$$F_X = F_{A0}(X_{A1} - X_{A2}) = F_{A0} \frac{(.5 - X_{A1})X_{A1}}{.5 + (\gamma - 1)X_{A1}} \qquad (9\text{-}48)$$

Determine the conversion X_{A1} at which the molar flow rate of xylene will be at a maximum.

(J21)

(J20) From a mole balance on mesitylene in a CSTR,

$$F_{M0} - F_M = F_{A0}\theta_M - F_{A0}(\theta_M - X_{A1}) = -r_M V \qquad (J20\text{-}1)$$

we obtain

$$\frac{V}{F_{A0}} = \frac{X_{A1}}{-r_M} = \frac{X_{A1}}{k_1 C_M C_H^{.5}} \qquad (J20\text{-}2)$$

From a mole balance on toluene,

$$F_T = F_{A0} X_{A2} = r_T V \qquad (J20\text{-}3)$$

we find that

$$\frac{V}{F_{A0}} = \frac{X_{A2}}{k_2 C_X C_H^{.5}} \qquad (J20\text{-}4)$$

Taking the ratio of (J20-4) and (J20-2) and substituting for the concentrations in terms of conversion we get

$$1 = \frac{k_1 C_M X_{A2}}{k_2 C_X X_{A1}} = \frac{(.5 - X_{A1}) X_{A2}}{\gamma (X_{A1} - X_{A2}) X_{A1}} \qquad (J20\text{-}5)$$

Recall $\gamma = k_2/k_1$. Solving for X_{A2},

$$X_{A2} = \frac{\gamma X_{A1}^2}{.5 + (\gamma - 1) X_{A1}} \qquad (J20\text{-}6)$$

To obtain the molar flow rate of xylene per mole of H_2 fed, subtract X_{A2} [Eqn. (J20-6)] from X_{A1}:

$$X_{A1} - X_{A2} = X_{A1} - \frac{\gamma X_{A1}^2}{.5 + (\gamma - 1) X_{A1}} \qquad (J20\text{-}7)$$

Simplifying,

$$X_{A1} - X_{A2} = \frac{.5 X_{A1} - X_{A1}^2}{.5 + (\gamma - 1) X_{A1}} \qquad (J20\text{-}8)$$

(J21) To obtain the value of X_{A1} at which the molar flow rate of xylene will be at a maximum, we proceed as with a plug flow reactor. That is, set the derivative of F_X equal to zero w.r.t. X_{A1} and solve for X_{A1}:

$$F_X = F_{A0} \frac{(.5 - X_{A1}) X_{A1}}{.5 + (\gamma - 1) X_{A1}} \qquad (9\text{-}48)$$

Differentiating Equation (9-48) w.r.t. X_{A1},

$$\frac{dF_X}{dX_{A1}} = 0 = \frac{F_{A0} d(X_{A1} - X_{A2})}{dX_{A1}} = F_{A0} d\left[\frac{\frac{.5 X_{A1} - X_{A1}^2}{.5 + (\gamma - 1) X_{A1}}}{dX_{A1}}\right] \qquad (J21\text{-}1)$$

$$0 = F_{A0} \frac{(.5 - 2X_{A1})[.5 + (\gamma - 1) X_{A1}] - (\gamma - 1)(.5 X_{A1} - X_{A1}^2)}{[.5 + (\gamma - 1) X_{A1}]^2} \qquad (J21\text{-}2)$$

Rearranging,

$$(.5 - 2X_{A1})[.5 + (\gamma - 1) X_{A1}] = (\gamma - 1)(.5 X_{A1} - X_{A1}^2) \qquad (J21\text{-}3)$$

and simplifying, we obtain the quadratic equation,

$$(\gamma - 1) X_{A1}^2 + X_{A1} - .25 = 0 \qquad (J21\text{-}4)$$

whose roots are

$$X_{A1} = \frac{-1 \pm \sqrt{1 + (\gamma - 1)}}{2(\gamma - 1)} \qquad (J21\text{-}5)$$

Rejecting the negative root, we find the value of X_{A1} at which F_X is at a maximum:

$$X_{A1} = \frac{-1 + \sqrt{\gamma}}{2(\gamma - 1)} = \frac{-1 + \sqrt{.546}}{2(.546 - 1)} = .287 \qquad (J21\text{-}6)$$

The corresponding value of X_{A2} from Equation (J20-6) is

$$X_{A2} = \frac{(.546)(.287)^2}{(.5) - (.454)(.287)} = .121$$

Calculate the fluidized bed CSTR volume necessary to achieve the optimum conversion $X_{A1,\,opt}$. [Recall that $F_{A0} = 10$ lb-mole/hr, $k_1 = 55.20$ (cu ft/lb mole)$^{.5}$/hr), and $C_{A0}^{1.5} = (.021)^{1.5} = .00304$.]

(J22)

(J22) Recalling the mole balance on mesitylene in Equation (J20-2),

$$V = \frac{F_{A0}X_{A1}}{-r_M} = \frac{F_{A0}X_{A1}}{k_1 C_M C_H^{.5}} \tag{J22-1}$$

and substituting for concentration in terms of conversion, we find that

$$V = \frac{F_{A0}X_{A1}}{(k_1)(C_{A0}^{1.5})(.5 - X_{A1})(1 - X_{A1} - X_{A2})^{.5}} \tag{J22-2}$$

$$= \frac{(10)(.287)}{(55.2)(.00304)(.5 - .287)(1 - .287 - .121)^{.5}}$$

$$= \frac{2.87}{(55.2)(.00304)(.213)(.769)}$$

$$= 104 \text{ cu ft}$$

SUMMARY

For the competing reactions,

$$\text{Reaction 1:} \quad A + B \xrightarrow{k_D} D \tag{S9-1}$$

$$\text{Reaction 2:} \quad A + B \xrightarrow{k_U} U \tag{S9-2}$$

the rate expressions are

$$r_D = A_D e^{\frac{-E_D}{RT}} C_A^{\alpha_1} C_B^{\beta_1} \tag{S9-3}$$

$$r_U = A_U e^{\frac{-E_U}{RT}} C_A^{\alpha_2} C_B^{\beta_2} \tag{S9-4}$$

$$S = \frac{r_D}{r_U} = \frac{A_D}{A_U}\left(\exp -\frac{E_D - E_U}{RT}\right) C_A^{\alpha_1 - \alpha_2} C_B^{\beta_1 - \beta_2} \tag{S9-5}$$

1. If $E_D > E_U$, the selectivity parameter S will increase with increasing temperature.
2. If $\alpha_1 > \alpha_2$, and $\beta_1 > \beta_2$, the reaction should be carried out at high concentrations of both A and B to minimize the amount of unwanted product formed. Use a batch or tubular reactor.
3. If $\alpha_1 > \alpha_2$, and $\beta_2 > \beta_1$, the reaction should be carried out at high concentrations of A and low concentrations of B to maintain the selectivity parameter S at a high value. Use a semibatch reactor with pure A initially or a tubular reactor in which B is fed at different locations down the reactor. Other cases discussed in the text were $\alpha_2 > \alpha_1$, $\beta_2 > \beta_1$, and $\alpha_2 > \alpha_1$, $\beta_1 > \beta_2$.
4. The conversions for each reaction were

$$X_{A1} = \frac{\text{moles of A consumed in reaction 1}}{\text{moles of A initially}} \tag{S9-6}$$

$$X_{A2} = \frac{\text{moles of A consumed in reaction 2}}{\text{moles of A initially}} \tag{S9-7}$$

5. The number of moles of each species in reactions (S9-6) and (S9-7) remaining after a time t is given by the following equations:

$$N_A = N_{A0}(1 - X_{A1} - X_{A2}) \tag{S9-8}$$

$$N_B = N_{A0}(\theta_B - X_{A1} - X_{A2}) \tag{S9-9}$$

$$N_U = N_{A0}(\theta_U + X_{A2}) \tag{S9-10}$$

$$N_D = N_{A0}(\theta_D + X_{A1}) \tag{S9-11}$$

6. For gas phase reactions carried out under constant pressure, the volume (or volumetric flow rate) is

$$V = V_0(1 + \epsilon_1 X_{A1} + \epsilon_2 X_{A2}) \tag{S9-12}$$

$$\epsilon_1 = y_{A0}\delta_1 \tag{S9-13}$$

$$\epsilon_2 = y_{A0}\delta_2 \tag{S9-14}$$

where the subscripts 1 and 2 refer to reactions 1 and 2, respectively.

7. The yield at a point is defined as the ratio of the rate of formation of a specified product D to the rate of reaction of the key reactant A:

$$Y_C = \frac{r_D}{-r_A} = \frac{r_D}{-r_{A1} - r_{A2}} \tag{S9-15}$$

8. The exit or end yield is the ratio of the number of moles of a product at the end of the reaction to the number of moles of the key reactant that have been consumed:

$$Y_E = \frac{N_D}{N_{A0} - N_A} \tag{S9-16}$$

For a flow system this yield is

$$Y_E = \frac{F_D}{F_{A0} - F_A} \tag{S9-17}$$

9. The selectivity, based on the molar flow rates leaving the reactor, for the reactions given by Equations (S9-1) and (S9-2) is

$$S_E = \frac{F_D}{F_U} \tag{S9-18}$$

For a batch system, S_E is based on the ratio of the number of moles of D present at the end of the reaction to the number of moles of U present at the end of the reaction:

$$S_E = \frac{N_D}{N_U} \tag{S9-19}$$

QUESTIONS AND PROBLEMS

P9-1. A liquid feed to a well mixed reactor consists of .4 gmole/l of A and the same molar concentration of F. The product C is formed from A by two different reaction mechanisms; either by direct transformation or through intermediate B. The intermediate is also formed from F. Along with C, which remains in solution, an insoluble gas D is formed which separates in the reactor.

All reaction steps are irreversible and first order, except for the formation of B from F, which is second order in F. The liquid carrier for reactants and products is an inert solvent, and no volume change results from the reaction:

$$k_1 = .01 \text{ min}^{-1}, \quad k_2 = .02 \text{ min}^{-1}$$
$$k_3 = .07 \text{ min}^{-1}, \quad k_4 = .50 \ l/\text{gmole/min}$$
$$\text{reactor volume} = 120 \ l$$

    a. What is the maximum possible molar concentration of C in the product?

    b. If the feed rate is 2.0 l/min, what is the yield of C (expressed as a percentage of the maximum), and what is the mole fraction of C in the product on a solvent-free basis?
                                                                  Ans. b: .583   [Calif. PECEE (August, 1970)]

P9-2. What reaction schemes and conditions would you use to maximize the selectivity parameters S for the following parallel reactions:

$$A + C \longrightarrow D \quad r_D = 15e^{-(273/T)}C_A^{.5}C_C$$
$$A + C \longrightarrow U \quad r_U = 200e^{-(2000/T)}C_A C_C$$
$$D = \text{desired product}$$
$$U = \text{undesired product}$$

P9-3. Two gas phase reactions are occurring in a plug flow tubular reactor, which is operated isothermally at a temperature of 440°F and a pressure of 5 atm. The first reaction is first order:

$$A \longrightarrow B, \quad -r_A = k_1 C_A, \quad k_1 = 10 \text{ sec}^{-1}$$

and the second reaction is zero order:

$$C \longrightarrow D + E, \quad -r_C = k_2 = 0.03 \text{ lb-moles/cu ft/sec}$$

The feed, which is equimolar in A and C, enters at a flow rate of 10 lb-moles/sec. What reactor volume is required for a 50% conversion of A to B?
    Data: Feed: 50% A, 50% C at 10 lb-moles/sec of total molar flow rate.   [CU]

P9-4. Consider the system of gas phase reactions

$$A \longrightarrow X \quad r_X = k_1 = 0.0005 \text{ lb-mole/cu ft/min}$$
$$A \longrightarrow B \quad r_B = k_2 C_A, \quad k_2 = 1 \text{ min}^{-1}$$
$$A \longrightarrow Y \quad r_Y = k_3 C_A^2, \quad k_3 = 60 \text{ cu ft/lb-mole/min}$$

B is the desired product; X and Y are foul pollutants that are expensive to get rid of.

a. What kind of reactor system would you recommend for this reaction scheme?

b. Calculate the size of the reactors involved to achieve a 90% conversion of A for a feed rate of 1 lb-mole of pure A per minute. The reactor is operated at 4 atm and 440°F. Hint: Consider a series scheme of reactors and what should be done when $S_{BX} = S_{BY}$. [CU]

P9-5. You are designing a plug flow reactor for the gas phase reaction

$$A \longrightarrow B, \qquad r_B = k_1 C_A^2, \qquad k_1 = 15 \text{ cu ft/lb-mole/sec}$$

Unfortunately, there is also a side reaction:

$$A \longrightarrow C, \qquad r_C = k_2 C_A, \qquad k_2 = 0.015 \text{ sec}^{-1}$$

C is a pollutant and costs money to dispose of; B is the desired product. What size reactor will provide an effluent stream with the maximum dollar value? B has a value of $60/lb-mole; it costs $15/lb-mole to dispose of C. A has a value of $10/lb-mole.
Other data:

  Feed: 22.5 Standard Cubic Feet/sec of pure A
  Reaction conditions: 460°F, 3 atm pressure
  Volumetric flow rate: 15 cu ft/sec at reaction conditions
  Concentration of A in feed: 4.47×10^{-3} lb-mole/cu ft [CU]

P9-6. The series reaction

$$A \xrightarrow{k_1} B \xrightarrow{k_2} C \qquad (P9\text{-}6\text{-}1)$$

is carried out in a batch reactor in which there is pure A initially.

a. Derive an equation in which the concentration of A is a function of time. If $k_1 = .01$ sec^{-1}, what is the ratio of C_A/C_{A0} after 1.5 min?

b. Derive an equation that gives the concentration of B as a function of time. If $k_2 = .005$ sec^{-1}, and $C_{A0} = .2$ gmole/l, what is the concentration of B after 2 min?

c. What is the concentration of C after 1 min? After 2 min?

d. Sketch the concentrations of A, B and C as functions of time. At what time is the concentration of B at a maximum?

P9-7. If the series reaction given in Equation (P9-6-1) is carried out in a CSTR, determine the reactor volume that will maximize the production of B for a volumetric flow rate of 20 l/min. The rate constants and initial concentration of A are given in Problem 9-6.

P9-8. Reconsider the hydrodealkylation reactions that form xylene and toluene:

$$H + M \longrightarrow X + Me \qquad (P9\text{-}8\text{-}1)$$
$$H + X \longrightarrow T + Me \qquad (P9\text{-}8\text{-}2)$$

a. Determine how the maximum molar flow rate of xylene varies with the ratio of mesitylene to hydrogen in the feed θ_M, for
   1. a tubular flow reactor, and
   2. a CSTR.

b. What is the optimum value of this ratio for each type of reactor?

P9-9. The toluene formed in the reaction given in Equation (P9-8-2) can also undergo hydrodealkylation to yield benzene, B:

$$H + T \longrightarrow B + Me \qquad (P9\text{-}9\text{-}1)$$

The rate law is

$$r_B = k_3 C_T C_H^{1/2} \qquad (P9\text{-}9\text{-}2)$$

with $\qquad k_3 = 11.2 \text{ (cu ft/lb-mole)}^{.5}/\text{hr}$

The feed conditions given for Frame (J15) will apply to the case where all three reactions are taking place.

    a. Set up a stoichiometric table that will include a column giving the concentration of each species in terms of the conversions X_1, X_2 and X_3.

    b. What is the CSTR reactor volume for which the molar flow rate of xylene will be at a maximum?

P9-10. The reactions

$$A \longrightarrow B \begin{array}{c} \nearrow C \\ \searrow R \end{array}$$

are taking place in two CSTRs connected in series. The volume of each reactor is 10 cu ft. The rate laws are

$$-r_A = k_1 \qquad k_1 = .06 \text{ lb-moles/cu ft/sec}$$
$$r_C = k_2 C_B \qquad k_2 = .082 \text{ sec}^{-1}$$
$$r_R = k_3 C_B \qquad k_3 = .091 \text{ sec}^{-1}$$

Pure A is fed to the first reactor at a concentration of 2 lb-moles/cu ft and a flow rate of 1 cu ft/sec. Calculate the molar flow rate of R in the effluent. [CU]

P9-11. Consider the cyclic reaction model in which A is formed directly from product C. Each reaction step is elementary:

$$A \underset{k_3}{\overset{k_1}{\underset{\longleftarrow}{\overset{\nearrow B \searrow k_2}{}}}} C \overset{k_4}{\longrightarrow} D$$

Species C can also decompose to form D. The model for this system is a liquid phase reaction that is carried out in a batch reactor containing pure A initially.

    a. From mole balances on species A, B, C, and D, derive four differential equations, giving the time rate of change of the concentrations of A, B, C, and D.

    b. Without solving any equations, sketch the concentrations of A, B, C, and D as a function of time:

        1. when $k_4 = 0$,

        2. when k_4 is small,

        3. when k_4 is large.

    c. Using an analog computer, solve the differential equations to obtain the equation giving the concentration of A as a function of time.

P9-12. You are designing a CSTR to produce a substance B from a feed composed primarily of substance A. The liquid phase reaction is first order and irreversible:

$$A \longrightarrow B: \qquad r_B = k_1 C_A, \qquad k_1 = 1.2 \text{ sec}^{-1}$$

Unfortunately, there is also a substance X in the feed stream; it reacts slowly with B to form a foul pollutant S:

$$X + B \longrightarrow 2S: \qquad r_S = k_2 C_X C_B, \qquad k_2 = .02 \text{ cu ft/lb-mole/sec}$$

The amount of B used up in the second reaction is negligible when compared with the amount formed in the first; also the amount of S formed is negligible when compared with the amount of X in the feed. The substance B is worth $15/lb-mole; the substance S represents a loss of $500/lb-mole; A and X are of very small value. What reactor size will maximize the value of the effluent? Data:

        Concentration of A in feed: .5 lb-mole/cu ft.
        Concentration of X in feed: .5 lb-mole/cu ft.
        Volumetric flow rate of feed stream: .25 cu ft/sec. [CU]

The following problems require a knowledge of energy balances (Chapter 8).

P9-13. The liquid phase reactions

$$\text{Reaction 1:} \quad A + B \xrightarrow{k_1} D \quad \text{(Desired reaction)}$$

$$\text{Reaction 2:} \quad A + B \xrightarrow{k_2} U \quad \text{(Undesired reaction)}$$

are carried out in a perfectly insulated CSTR. The desired reaction is first order in A and zero order in B, while the undesired reaction is zero order in A and first order in B. The feed rate is equimolar in A and B. Species A enters the reactor at a temperature of 100°C and species B enters at a temperature of 50°C. The operating temperature of the reactor is 400°K. The molar flow rate of A entering the reactor is 60 gmoles/min:

$$C_{P_A} = 20 \text{ cal/gmole/°K}.$$
$$C_{P_B} = 30 \text{ cal/gmole/°K}.$$
$$C_{P_D} = 50 \text{ cal/gmole/°K}.$$
$$C_{P_U} = 40 \text{ cal/gmole/°K}.$$

For reaction 1, $\Delta H_R = -3000$ cal/gmole of A at 300°K
For reaction 2, $\Delta H_R = -5000$ cal/gmole of A at 300°K

$$\left.\begin{array}{l} k_1 = 1000 \exp\left[\dfrac{-2000}{T}\right] \text{min}^{-1} \\ k_2 = 2000 \exp\left[\dfrac{-3000}{T}\right] \text{min}^{-1} \\ C_{A0} = .01 \text{ gmole}/l \end{array}\right\} \quad \text{where T is in °K}$$

a. What will be the exit molar flow rates of U and D from the reactor?

b. What is the CSTR reactor volume for the conditions specified above?

c. Is there a more effective way to maximize S? Explain.

(*Hint*: Start with a mole balance on A. Outline your method of solution before beginning any calculations.) [Exam. U. of M.]

P9-14. You are the engineer who is to design a CSTR for the elementary consecutive gas phase reactions

$$A \xrightarrow{k_1} B \xrightarrow{k_2} C$$

Both reaction rate constants show the Arrhenius temperature dependence. The feed conditions and the desired product specifications are known, along with the temperature of the heating medium. It is your job to design the reactor, i.e., to specify the reactor volume and the area of the heating coil inside the reactor.

a. Calculate the desired operating temperature inside the reactor.

(*Ans*: 269°F)

b. Calculate the volume of the reactor.

(*Ans*: 6.58 cu ft)

c. Calculate the area of the heating coil

(*Ans*: 773 sq ft)

Data:

Product: The ratio C_B/C_C is equal to 10.
50% of A in the feed is converted.

Feed: The feed is gas phase and pure A.
The molar flow rate is 0.05 lb-moles/sec.
The volumetric flow rate is 7.85 cu ft/sec.
The temperature is 400°F.
The pressure in the reactor is 4 atm.

Heat capacities: C_ps of A, B, and C are all 25 Btu/lb-mole-°F

Reactions:
$A_1 = 2 \times 10^9$ sec^{-1}
$A_2 = 1 \times 10^{11}$ sec^{-1}
$E_1 = 31,000$ Btu/lb-mole
$E_2 = 42,000$ Btu/lb-mole
$\Delta H_{r1} = +15,000$ Btu/lb-mole
$\Delta H_{r2} = -20,000$ Btu/lb-mole
$k = Ae^{-E/RT}$

Heat transfer: The heating medium is saturated high pressure steam at 350°F.
The overall heat transfer coefficient between the heating medium and the reaction mixture is 25 Btu/hr-sq ft-°F. [CU]

P9-15. You wish to use an existing CSTR for the liquid phase, exothermic, first-order reaction

$$A \longrightarrow B, \quad r_B = k_1 C_A$$

Unfortunately, there is also an undesired side reaction,

$$A \longrightarrow C, \quad r_C = k_2 C_A$$

which is also liquid phase, exothermic, and first order.

You have available as a heating medium saturated steam at 212°F. The feed is available at 130°F.

What area of heating coil (if any) will you specify to maximize the production of B? Other data:

Feed: volumetric flow rate: 2 cu ft/sec
density: 45 lb/cu ft
heat capacity: 0.8 Btu/lb-°F
pure A fed at a concentration of 1.1 lb-moles/cu ft

Reaction 1:
$A_1 = 3 \times 10^8$ sec^{-1}
$E_1 = 27,000$ Btu/lb-mole
$\Delta H_{r1} = -15,000$ Btu/lb-mole
$k = Ae^{-E/RT}$

Reaction 2:
$A_2 = 2 \times 10^{14}$ sec^{-1}
$E_2 = 45,000$ Btu/lb-mole
$\Delta H_{r2} = +25,000$ Btu/lb-mole

Reactor: volume: 30 cu ft
overall heat transfer coefficient between condensing steam and reacting fluid: 400 Btu/hr-sq ft-°F, i.e., 0.111 Btu/sec-sq ft-°F

Reasonable assumptions: The density and specific heat of fluid are unchanged by reaction or temperature change. The effect of the coil on the effective reactor volume is negligible. [CU]

P9-16. You are the engineer in charge of a CSTR in which the following endothermic elementary liquid phase reactions are occurring:

You adjust the temperature in the reactor by the addition of an inert heating medium I. Orders come down from the Economic Analysis Division for you to run at 90% conversion of A, to produce X and Y in a ratio of 55/45. What should be the inlet temperature and flow rate (in cu ft/sec) of the inert medium I? Both reactions exhibit the Arrhenius temperature dependence $k = Ae^{-E/RT}$. Data:

> Volume of reaction: 10 cu ft
> Inlet flow rate of A: .5 lb-moles/sec
> Inlet concentration of A: .5 lb-moles/cu ft
> Inlet flow rate of reacting material: 1 cu ft/sec
> Temperature of incoming reacting material: 140°F
> C_p of all materials: .5 Btu/lb-°R (Note: this is in units of mass)
> ρ of all materials: .8 lb/cu ft
> $A_1 = 10^{10}$ sec^{-1}
> $E_1 = 30,000$ Btu/lb-mole
> $A_2 = 10^7$ sec^{-1}
> $E_2 = 20,000$ Btu/lb-mole
> $\Delta H_1 = +15,000$ Btu/lb-mole
> $\Delta H_2 = -10,000$ Btu/lb-mole, [CU]

SUPPLEMENTARY READING

1. Selectivity, reactor schemes, and staging for multiple reactions, along with evaluation of the corresponding design equations, are presented in Chapter 7 of

> LEVENSPIEL, O., *Chemical Reaction Engineering*, 2nd ed. New York: John Wiley and Sons, 1972.

and Chapter 6 of

> DENBIGH, K. G. and J. C. R. TURNER, *Chemical Reactor Theory*, 2nd ed. London: Cambridge University Press, 1971.

Some example problems on reactor design for multiple reactors are presented in Chapter XVIII of

> HOUGEN, O. A. and K. M. WATSON, *Chemical Process Principles: Part 3: Kinetics and Catalysis*, New York: John Wiley and Sons, 1947.

and Chapter 4 of

> SMITH, J. M., *Chemical Engineering Kinetics*, New York: McGraw-Hill Book Co., 1970.

2. The use of solvents to increase the selectivity of liquid phase reactions has been discussed by

> WONG, K. F. and C. A. ECKERT, *Transactions of the Faraday Society*, 66, No. 573, Part 9 (1970), 2313.

> WONG, K. F. and C. A. ECKERT, *I.E.C. Process Design and Development* 8, (1969), 568.

3. A brief discussion of a number of pertinent references on parallel and series reactions is given at the end of Chapter 5 of

> ARIS, R., *Elementary Chemical Reactor Analysis*, Englewood Cliffs, N. J.: Prentice-Hall, Inc., 1969.

10
Diffusion Limitations in Heterogeneous Reactions

In the rate laws and catalytic reaction steps (i.e., diffusion, adsorption, surface reaction, desorption, and diffusion) presented in Chapter 6, we neglected the effects of mass transfer on the overall rate of reaction. In this chapter we shall discuss the effects of diffusion (mass transfer) limitations or *resistance* on the overall rate of processes involving both chemical reaction and mass transfer. The two types of diffusion resistances on which we shall focus attention are (1) *external resistance:* diffusion of the reactants or products between the bulk fluid and the external surface of the catalyst pellet and (2) *internal resistance:* diffusion of the reactants or products between external pellet surface (pore mouth) and the interior of the pellet. After a brief presentation of the fundamentals of diffusion, including *Fick's first law*, models will be developed which describe the behavior of catalysts in which the phenomenon of internal resistance to mass transfer is limiting. Also, the parameters commonly used to describe internal diffusive effects, i.e., the *Thiele modulus* and the *effectiveness factor*, will be presented. Finally, we shall discuss representative correlations of mass transfer rates in terms of *mass transfer coefficients* for

catalyst beds in which the external resistance is limiting. Qualitative observations will be made about the effects of fluid flow rate, pellet size, and pressure drop on reactor performance.

As was the case in Chapter 6, the application of these principles is not limited to industrial catalytic reactors. They can also be applied directly or extended to such systems as immobilized and/or encapsulated enzymes. These systems have recently found application in bioengineering and environmental engineering.

10.1–Mass Transfer Fundamentals

Mass transfer usually refers to any process in which diffusion plays a role. Diffusion is the spontaneous intermingling or mixing of atoms or molecules by random thermal motion. It gives rise to motion of the species *relative* to the motion of the mixture. In the absence of other gradients (such as temperature, electric potential, or gravitational potential) molecules of a given species within a single phase will always diffuse from regions of higher concentrations to regions of lower concentrations. This results in a molar flux of the species (e.g., A), $\vec{W}_A$, in the direction of the concentration gradient. The flux $\vec{W}_A$ (moles/area/time) is a vector quantity whose typical units are gmoles/sq cm/sec. For example, in the system in Figure 10-1, ammonia is diffusing at steady state from point 1 to point 2. If 12.04×10^{23} molecules of ammonia pass through an area bounded by a stationary frame 1 cm by 2 cm in the direction of the lower concentration and over a time interval of 10 sec, determine the molar flux of ammonia, $\vec{W}_A$, and the mass flux of ammonia, $\vec{m}_A$, through the stationary frame at steady state.

(K1)

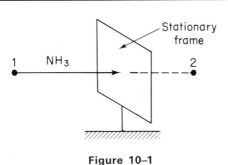

Figure 10–1

10.1A The Molar Flux

The molar flux of A, $\vec{W}_A$, is the result of two contributions: the molecular diffusion flux, $\vec{J}_A$, and the flux resulting from the bulk motion of the fluid, $\vec{B}_A$.

$$\vec{W}_A = \vec{J}_A + \vec{B}_A \qquad (10\text{-}1)$$

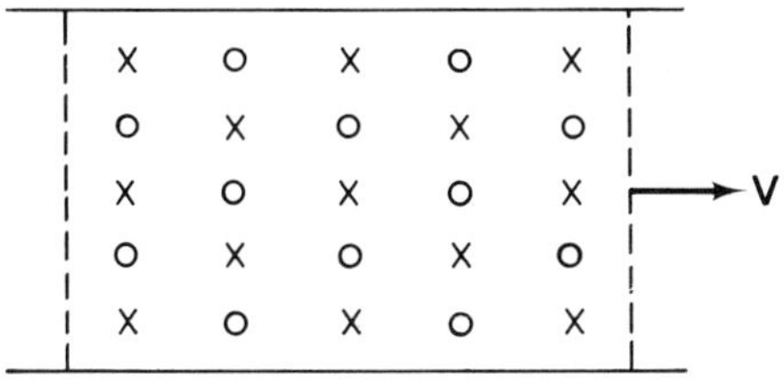

To illustrate the difference between these two contributions, consider the situations in Figure 10-2, in which a gaseous mixture of species A and B is well mixed and moving with velocity, $\vec{V}$, where × represents A and ○ represents B.

Since there are no concentration gradients, there will be no net diffusion of A or B, $\vec{J}_A = 0$; the flux of

Figure 10–2

411

(K1)

From the units of $\vec{W}_A$ given above Frame K1, one can deduce the molar and mass fluxes. The mass flux of A is the molar flux of A times the molecular weight of A, M_A:

$$\vec{m}_A = \vec{W}_A \times (\text{mol. wt.})$$

The number of gmoles of ammonia crossing the boundary is

$$\text{number of gmoles} = \frac{\text{number of molecules}}{\text{Avogadros number}} = \frac{12.04 \times 10^{23}}{6.02 \times 10^{23}} = 2$$

Two gmoles pass through an area of 2 sq cm in 10 sec. Therefore, the number of moles passing through one sq cm/sec is

$$\vec{W}_A = \frac{2 \text{ moles}}{(2 \text{ sq cm})(10 \text{ sec})} = \frac{.1 \text{ gmole}}{\text{sq cm sec}}$$

The mass flux is

$$\vec{m}_A = \left(\frac{17 \text{ g}}{\text{gmole}}\right)\left(\frac{.1 \text{ gmole}}{\text{sq cm sec}}\right) = \frac{1.7 \text{ g}}{\text{sq cm sec}}$$

A will be only that resulting from the bulk motion; and it is equal to the mole fraction of A, y_A, times the total flux, $\vec{W}$:

$$\vec{B}_A = y_A \vec{W} \qquad (10\text{-}2)$$

$$\vec{W}_A = 0 + \vec{B}_A = y_A \vec{W}$$

For a binary system the total flux $\vec{W}$ is the sum only of the fluxes of A and B:

$$\vec{W} = \vec{W}_A + \vec{W}_B$$

$$\vec{W}_A = \vec{B}_A = y_A(\vec{W}_A + \vec{W}_B)$$

$\vec{B}_A$ can also be expressed in terms of the concentration of A and the molar average velocity $\vec{V}$:

$$\vec{B}_A = C_A \vec{V}$$

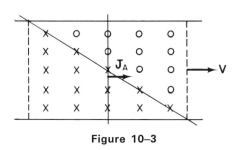

Figure 10–3

Now reconsider the situation: Instead of the mixture being spatially uniform, as in Figure 10-2, the concentration gradient in this mixture is moving with velocity $\vec{V}$ (see Figure 10-3).

In this case there will be diffusive flux of A, $\vec{J}_A$, relative to the motion of the mixture, in addition to the bulk flux of A. The total molar flux of A is given by

$$\boxed{\vec{W}_A = \vec{J}_A + \vec{B}_A} \qquad (10\text{-}1)$$

$\vec{B}_A$ can be expressed either in terms of the concentration of A, in which case

$$\vec{W}_A = \vec{J}_A + C_A \vec{V} \qquad (10\text{-}3)$$

or in terms of the mole fraction of A:

$$\boxed{\vec{W}_A = \vec{J}_A + y_A(\vec{W}_A + \vec{W}_B)} \qquad (10\text{-}4)$$

A rough but somewhat useful analogy to Equation (10-1) is the following: Consider a boy walking on a long flat car of a moving train. His velocity w.r.t. the ground, $\vec{V}_g$, (analogous to $\vec{W}_A$) is *equal to* his walking velocity on the flat car w.r.t. the flat car, $\vec{V}_W$ (analogous to $\vec{J}_A$), *plus* the velocity of the flat car w.r.t. the ground, $\vec{V}_F$ (analogous to $\vec{B}_A$):

$$\vec{V}_g = \vec{V}_W + \vec{V}_F$$

For discussion on this point (i.e. Equation (10-4)) see Problem P10-4.

10.1B Fick's First Law

Our discussion on diffusion will be restricted mainly to binary systems containing species A and B. We now wish to determine *how* the molar flux of a species is related to its concentration gradient. As an aid in the discussion of the rate law that is ordinarily used to describe diffusion, recall some similar rate laws from other transport processes. For example, in conductive heat transfer, the relation between the heat flux $\vec{q}$ and the temperature gradient is given by *Fourier's law*:

$$\vec{q} = -k_t \nabla T$$

where k_t is the thermal conductivity. The one-dimensional form of this equation is

$$q_x = -k_t \frac{\partial T}{\partial x}$$

In momentum transfer, the relationship between shear stress and shear rate for plainar simple shearing flow is given by *Newton's law of viscosity*:

$$\tau = -\mu \frac{\partial u}{\partial x}$$

The mass transfer flux law is analogous to the laws for heat and momentum transport. $\vec{J}_A$ is the diffusional flux of A resulting from a concentration difference and is related to the concentration gradient by Fick's first law:

$$\vec{J}_A = -cD_{AB}\nabla y_A \tag{10-5}$$

where c is the total concentration (gmoles/cu cm), and D_{AB} is the diffusivity of A in B (sq cm/sec). Combining Equations (10-1), (10-4), and (10-5) we obtain an expression for the molar flux of A:

$$\boxed{\vec{W}_A = -cD_{AB}\nabla y_A + y_A(\vec{W}_A + \vec{W}_B)} \tag{10-6}$$

We shall now consider four cases in which we can simplify Equation (10-6).

10.1C Evaluating the Bulk Flow Term

TABLE 10–1

1. Equimolar counter diffusion (EMCD).

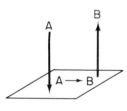

Figure 10–4

In EMCD, for every mole of A that diffuses in a given direction, 1 mole of B diffuses in the opposite direction. For example, consider a species A that is diffusing at steady state from the bulk fluid to a catalyst surface, where it isomerizes to form B. Species B then diffuses back into the bulk (see Figure 10-4). For every mole of A that diffuses to the surface, 1 mole of B diffuses away from the surface. The fluxes of A and B are equal in magnitude and flow counter to each other. Stated mathematically,

$$\vec{W}_A = -\vec{W}_B$$

Write an expression for $\vec{W}_A$ in terms of the concentration of A, C_A, for the case of EMCD. Also state what additional assumption(s) is (are) necessary to achieve this.

(K2)

2. Dilute concentrations.

a. When the mole fraction of the diffusing solute and the bulk motion in the direction of the diffusion are small, the second term on the right-hand side of Equation (10-6) can usually be neglected when compared with the first term. Under these conditions, along with the condition of constant total concentration, the flux of A is identical to Equation (K2-6), i.e.,

$$\vec{W}_A \doteq \vec{J}_A = -D_{AB}\nabla C_A \tag{K2-6}$$

(K2)

The following equations have been given:

$$\vec{W}_A = \vec{J}_A + \vec{B}_A \qquad (10\text{-}1)$$

$$\vec{B}_A = y_A \vec{W} \qquad (K2\text{-}1)$$

$$\vec{W} = \vec{W}_A + \vec{W}_B \qquad (10\text{-}4)$$

For EMCD
$$\vec{W}_A = -\vec{W}_B \qquad (K2\text{-}2)$$

then
$$\vec{W} = 0, \quad \therefore \quad \vec{B}_A = 0 \qquad (K2\text{-}3)$$

and
$$\vec{W}_A = \vec{J}_A \qquad (K2\text{-}4)$$

$$\vec{J}_A = -cD_{AB}\nabla y_A \qquad (10\text{-}2)$$

$$\vec{W}_A = -cD_{AB}\nabla y_A \qquad (K2\text{-}5)$$

For constant total concentration,

$$\vec{W}_A = -D_{AB}\nabla C_A \qquad (K2\text{-}6)$$

In rectangular coordinates, the gradient is in the form

$$\nabla = \left(i\frac{\partial}{\partial x} + j\frac{\partial}{\partial y} + k\frac{\partial}{\partial z} \right)$$

In most liquid systems the concentration of the diffusing solute is small, and Equation (K2-6) is used to relate $\vec{W}_A$ and the concentration gradient within the boundary layer.

b. Equation (10-6) also reduces to Equation (K2-6) for porous catalyst systems in which the total concentration is very low. Diffusion under these conditions is known as *Knudsen diffusion* and occurs when the mean free path of the molecule is greater than the diameter of the catalyst pore. Here, the reacting molecules collide more often with pore walls than with each other. The flux of species A for Knudsen diffusion (where bulk flow can usually be neglected) is

$$\vec{W}_A \doteq \vec{J}_A = -D_K \nabla C_A, \quad (10\text{-}7)$$

where D_K is the Knudsen diffusivity.

3. *Diffusion through a stagnant gas.*

The diffusion of a solute through a stagnant gas often occurs in systems in which there are two phases present. Evaporation and gas absorption are typical processes in which this type of diffusion can be found. Starting with Equation (10-6), show that the equation for the flux of A can be written

$$\vec{W}_A = cD_{AB} \nabla \ln(1 - y_A) = cD_{AB} \nabla \ln y_B \quad (10\text{-}8)$$

for the diffusion of species A through species B, which is stagnant.

(K3)

4. *Forced convection.*

In systems where the flux of A results primarily from forced convection, we assume that the diffusion contribution in the direction of the flow (e.g., axial or z direction), J_{Az}, is small in comparison to the bulk flow contribution, B_{Az},

$$J_{Az} \simeq 0$$

$$W_{Az} = B_{Az} = C_A U_z = \frac{v}{A_c} C_A \quad (10\text{-}9)$$

$$\vec{V} = U_x i + U_y j + U_z k$$

where A_c is the cross-sectional area and v is the volumetric flow rate. Although the component of the diffusional flux vector of A in the direction of flow, J_{Az}, is neglected, the component of the flux of A in the "x direction," J_{Ax}, which is normal to the direction of flow, may not be neglected.

In the material in which the molar flow rate, F_A, was discussed, any diffusional effects were neglected, and F_A was written as the product of the volumetric flow rate and the concentration:

$$F_A = vC_A$$

The molar flow rate of A, F_A, in a specified direction z, is the product of molar flux in that direction, W_{Az}, and the cross-sectional area normal to the direction of flow, A_c:

$$F_A = F_{Az} = W_{Az} A_c = \left[-cD_{AB} \frac{\partial y_A}{\partial z} + y_A (W_{Az}) \right] A_c$$

In developing mathematical models for chemically reacting systems in which diffusional effects are important, the first steps are to

1. perform a differential mass balance on a particular species A,

2. substitute for F_A in terms of W_{Az}, and then

3. replace W_{Az} by the appropriate expression for the concentration gradient.

(K3)

$$\vec{W}_A = -cD_{AB}\nabla y_A + y_A(\vec{W}_A + \vec{W}_B) \tag{10-6}$$

If the gas B is stagnant, there is no net flux of B w.r.t. a fixed coordinate, i.e.,

$$\vec{W}_B = 0$$

$$\vec{W}_A = -cD_{AB}\nabla y_A + y_A\vec{W}_A \tag{K3-1}$$

$$= \frac{-1}{1 - y_A} cD_{AB}\nabla y_A \tag{K3-2}$$

$$= cD_{AB}\nabla \ln(1 - y_A) = cD_{AB}\nabla \ln y_B \tag{K3-3}$$

10.1D Diffusion Through a Stagnant Film

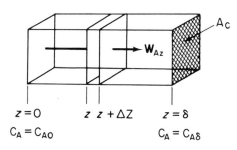

$z = 0$ $z\ z + \Delta Z$ $z = \delta$
$C_A = C_{A0}$ $C_A = C_{A\delta}$

Figure 10-5

As a first example, consider the steady-state diffusion of A through a film of thickness δ (Figure 10-5). The concentrations of A at the left and right boundaries are C_{A0} and $C_{A\delta}$, respectively with $C_{A0} > C_{A\delta}$. In Frame (K4) perform a differential mole balance on species A over a differential element of width Δz and cross-sectional area A_c and then arrive at a first-order differential equation in W_{Az}.

(K4)

Next, substitute the appropriate expression for W_{Az} into the mole balance, Equation (K4-3), for the case when the concentration of the diffusing solute is small and the total concentration is constant. Also, state mathematically the appropriate boundary conditions for the second-order differential equation you have just derived.

(K5)

Equation (K5-2) is an elementary differential equation which can be solved directly by integrating twice with respect to z. The first integration yields

$$\frac{dC_A}{dz} = K_1$$

and the second yields

$$C_A = K_1 z + K_2$$

where K_1 and K_2 are arbitrary constants of integration.

(K4) The general mole balance equation is

$$[\text{Rate in}] - [\text{Rate out}] + [\text{Rate gen}] = [\text{Rate accum.}] \tag{K4-1}$$

$$F_A|_z - F_A|_{z+\Delta z} + 0 = 0 \tag{K4-2}$$

Divide by Δz:

$$\frac{F_A|_{z+\Delta z} - F_A|_z}{\Delta z} = 0$$

Take the limit as $\Delta z \to 0$ to obtain

$$\frac{dF_A}{dz} = 0$$

Substitute for F_A in terms of W_{Az} and divide by A_c

$$F_A = W_{Az} A_c,$$

to get

$$\frac{dW_{Az}}{dz} = 0 \tag{K4-3}$$

(K5) For dilute concentration of the diffusing solute,

$$W_{Az} = J_{Az} = -cD_{AB}\frac{dy_A}{dz}$$

and for constant total concentrations,

$$W_{Az} = -D_{AB}\frac{dC_A}{dz} \tag{K5-1}$$

Substituting Equation (K5-1) into (K4-3),

$$\frac{dW_{Az}}{dz} = -D_{AB}\frac{d^2 C_A}{dz^2}$$

$$\frac{dW_{Az}}{dz} = 0, \quad \text{therefore} \quad \frac{d^2 C_A}{dz^2} = 0 \tag{K5-2}$$

Boundary conditions:

$$\text{when } z = 0, \quad C_A = C_{A0}$$
$$\text{when } z = \delta, \quad C_A = C_{A\delta}$$

In Frame (K6),

1. evaluate the constants K_1 and K_2, utilizing the boundary conditions, thereby obtaining an equation for the concentration profile, and

2. differentiate the profile w.r.t. z and obtain an equation for the flux of A across the film.

(K6)

10.1E Modeling Diffusion with Chemical Reaction

The method used in solving diffusion problems similar to the one above is shown in Table 10–2.

TABLE 10–2

Steps in modeling chemical systems with diffusion and reaction

1. State the problem and assumptions.
2. Define the system on which the balances are to be made.
3. Perform a differential mole balance on a particular species.
4. Obtain a differential equation in $\vec{W}_A$ by rearranging your balance equation properly and by taking the limit as the volume of the element goes to zero.
†5. Substitute the appropriate expression involving the concentration gradient for $\vec{W}_A$ from Table 10–1 to obtain a second-order differential equation for the concentration of A.
6. Express the reaction rate r_A (if any) in terms of concentration and substitute in the differential equation.
7. State the appropriate boundary and initial conditions.
8. Solve the resulting differential equation for the concentration profile.
9. Differentiate this concentration profile to obtain an expression for the molar flux of A.
10. Substitute numerical values for symbols.

†In some instances it may be easier to integrate the resulting differential equation in step 4 before substituting for $\vec{W}_A$.

The most common boundary conditions are presented in Table 10–3.

(K6)

$$C_A = K_1 z + K_2$$

At $z = 0$, $C_A = C_{A0}$; therefore,

$$C_{A0} = K_2$$

At $z = \delta$, $C_A = C_{A\delta}$; therefore,

$$C_{A\delta} = K_1 \delta + K_2 = K_1 \delta + C_{A0}$$

Eliminating K_1 and rearranging, we obtain

$$\frac{C_A - C_{A0}}{C_{A\delta} - C_{A0}} = \frac{z}{\delta} \tag{K6-1}$$

For dilute solute concentrations and constant total concentration,

$$W_{Az} = -D_{AB} \frac{dC_A}{dz} \tag{K6-2}$$

Rearranging (K6-1),

$$C_A = C_{A0} + (C_{A\delta} - C_{A0}) \frac{z}{\delta} \tag{K6-3}$$

To determine the flux, differentiate Equation (K6-3) w.r.t. z and then multiply by $-D_{AB}$:

$$W_{Az} = -D_{AB} \frac{(C_{A\delta} - C_{A0})}{\delta}$$

$$= \frac{D_{AB}}{\delta} (C_{A0} - C_{A\delta}) \tag{K6-4}$$

TABLE 10-3
TYPES OF BOUNDARY CONDITIONS

1. Specify a concentration at a boundary (e.g., $z = 0$, $C_A = C_{A0}$). For an instantaneous reaction at a boundary, the concentration of the reactants at the boundary is taken to be zero. ($C_{AS} = 0$)
2. Specify a flux at a boundary:
   a. No mass transfer to a boundary:
   $$\vec{W}_A = 0 \tag{10-10}$$
   e.g., at the wall of a nonreacting pipe,
   $$\frac{dC_A}{dr} = 0 \quad \text{at } r = R$$
   b. Set the molar flux to the surface equal to the rate of reaction on the surface:
   $$W_A|_{\text{surface}} = r''_A|_{\text{surface}} \tag{10-11}$$
   c. Set the molar flux to the boundary equal to convective transport away from the boundary:
   $$W_A|_{\text{boundary}} = k_c(C_{AS} - C_{AB}) \tag{10-12}$$
   where k_c is the mass transfer coefficient and C_{AS} and C_{AB} are the surface and bulk concentrations, respectively.
3. Planes of symmetry: When the concentration profile is symmetrical about a plane, the concentration gradient is zero in that plane of symmetry. For example, in the case of radial diffusion in a pipe, at the center of the pipe
   $$\frac{dC_A}{dr} = 0 \quad \text{at } r = 0$$

10.1F Temperature and Pressure Dependence of D_{AB}

Before closing this brief discussion on mass transfer fundamentals, further mention should be made of the diffusion coefficient. Equations for predicting gas diffusivities are given in the Appendix D.[†] The orders of magnitude of the diffusivities for gases, liquids, and solids, and the

TABLE 10-4
DIFFUSIVITY RELATIONSHIPS FOR GASES, LIQUIDS, AND SOLIDS

Phase	Order of Magnitude (sq cm/sec)	Temperature and Pressure Dependence
Gas: Bulk	10^{-1}	$D_{AB}(T_2, \Pi_2) = D_{AB}(T_1, \Pi_1)\frac{\Pi_1}{\Pi_2}\left(\frac{T_2}{T_1}\right)^{1.75}$
Knudsen	10^{-2}	$D_A(T_2) = D_A(T_1)\left(\frac{T_2}{T_1}\right)^{1/2}$
Liquid	10^{-5}	$D_{AB}(T_2) = D_{AB}(T_1)\frac{\mu_1}{\mu_2}\frac{T_2}{T_1}$
Solid	10^{-9}	$D_{AB}(T_2) = D_{AB}(T_1)\exp\left[\frac{E_D}{R}\left(\frac{T_2 - T_1}{T_2 T_1}\right)\right]$

where μ_1, μ_2 = liquid viscosities at temperatures T_1 and T_2 respectively, and E_D = diffusion activation energy.

[†]For further discussion of mass transfer fundamentals, see Chaps. 16 and 17 of *Transport Phenomena* by R. B. Bird, W. E. Stewart, and E. N. Lightfoot, (New York: John Wiley and Sons, 1960).
To estimate liquid diffusivities for binary systems, see K. A. Reddy and L. K. Doraiswamy, *I.E.C. Fundamentals* 6, (1967), 77.

10.2–Diffusion with First Order Homogeneous Reaction

Species A evaporates from a solid surface and undergoes isomerization to D as it diffuses away from the surface through a stagnant column of air (species B). The vapor pressure of A is very low; consequently the concentration of A in the gas phase is small. The reaction is irreversible and first order in A:

$$A \xrightarrow{k_1} D$$

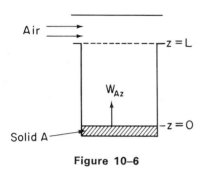

Figure 10-6

A gentle flow of air sweeps across the top of the container so that the concentration of A at this point is essentially zero (see Figure 10-6). Derive a differential equation in terms of the concentration of A, C_A, the diffusivity, D_{AB}, and the first order specific reaction rate constant k_1, and state the appropriate boundary conditions. Since A and D are present in dilute concentrations, you may neglect multicomponent effects. The temperature and pressure are spatially uniform and have values of 200°C and 3 atm, respectively.

(K7)

Equation (K7-5) can be solved directly (Appendix A4) to yield

$$C_A = A_1 \sinh(\alpha z) + B_1 \cosh(\alpha z) \qquad (10\text{-}13)$$

where $\alpha = \sqrt{k_1/D_{AB}}$

After evaluating the constants A_1 and B_1 with the aid of the boundary conditions, we find that

$$C_A = C_{A0}[\cosh(\alpha z) - \coth(\alpha L) \sinh(\alpha z)] \qquad (10\text{-}14)$$

Obtain an equation for the flux of A at any point z in the container.

(K8)

Finally, calculate the rate of evaporation of A from the surface in gmoles/sq cm. Additional information:

$$D_{AB} \text{ at 25°C and 1 atm} = .08 \text{ sq cm/sec}$$
$$\text{Vapor pressure of A at 200°C} = 76 \text{ mm Hg}$$
$$k_1 \text{ at 200°C} = .0005 \text{ sec}^{-1}$$
$$\text{Depth of container} = 10 \text{ cm}$$

(K7)

The steady-state differential mole balance of A is

$$F_A|_z - F_A|_{z+\Delta z} + r_A \Delta z A_c = 0$$

Divide by $-A_c \Delta z$ and take the limit as $\Delta z \to 0$ to get

$$\frac{1}{A_c}\frac{dF_A}{dz} - r_A = 0 \tag{K7-1}$$

Substituting for F_A in terms of W_{Az},

$$\frac{dW_{Az}}{dz} - r_A = 0 \tag{K7-2}$$

For dilute concentrations of A, $\vec{B}_A \simeq 0$ and

$$W_{Az} \doteq J_{Az} = -cD_{AB}\frac{dy_A}{dz}$$

The total concentration can be determined from the ideal gas law:

$$c = \frac{\Pi}{RT}$$

Since the temperature and pressure are constant, the total concentration is constant, and the flux can be written as

$$W_{Az} = -D_{AB}\frac{dC_A}{dz} \tag{K7-3}$$

The rate law is

$$-r_A = k_1 C_A \tag{K7-4}$$

By substituting Equations (K7-3) and (K7-4) into Equation (K7-2), we arrive at a second-order differential equation describing the diffusion with a first-order homogeneous chemical reaction:†

$$D_{AB}\frac{d^2 C_A}{dz^2} - k_1 C_A = 0 \tag{K7-5}$$

The boundary conditions are

$$\text{At } z = 0, \quad C_A = C_{A0}$$
$$\text{At } z = L, \quad C_A = C_{AL} = 0$$

(K8)

From Equation (K7-3),

$$W_{Az} = -D_{AB}\frac{dC_A}{dz}$$

Let $\alpha = \sqrt{k_1/D_{AB}}$:

$$\frac{dC_A}{dz} = C_{A0}\frac{d}{dz}(\cosh\alpha z - \coth\alpha L \sinh\alpha z) \tag{K8-1}$$

$$= C_{A0}\sqrt{\frac{k_1}{D_{AB}}}(\sinh\alpha z - \coth\alpha L \cosh\alpha z)$$

$$W_{Az} = C_{A0}\sqrt{D_{AB}k_1}(\coth\alpha L \cosh\alpha z - \sinh\alpha z) \tag{K8-2}$$

†Methods of solving differential equations of the type $d^2 y/dx^2 - \beta y = 0$ are given in Appendix A4.

Before proceeding to Section 10-3 one might want to work through some of Problems P10-1 through P10-8 at the end of the chapter.

10.3–INTERNAL DIFFUSION AND REACTION IN CATALYST PORES

In our discussion in Chapter 6 on surface reactions, we assumed that each point on the entire catalyst surface was *accessible* to the same concentration. However, where the reactants diffuse into the pores within the catalyst pellet, the concentration at the pore mouth will be higher than inside the pore, and we see that the entire catalytic surface is *not accessible* to the same concentration. To account for the fact that, because of diffusion limitations, the entire catalytic surface is not accessible to the maximum concentration present in the bulk, we introduced a parameter known as the effectiveness factor.

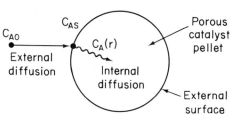

Figure 10–7

In the heterogeneous reaction sequence, mass transfer of reactants first takes place from the bulk fluid to the external surface of the pellet. The reactants then diffuse from the external surface into and through the pores within the pellet, with reaction taking place only on the catalytic surface of the pores. A schematic representation of this two step diffusion process is shown in Figures 6-3 (page 189) and 10-7.

We can express the overall rate of reaction (gmoles/sq cm/sec) in terms of either the overall or the internal effectiveness factor. The overall or actual rate of reaction in the catalyst bed or pellet is equal to the *overall effectiveness factor* Ω times the rate of reaction, r''_{A0}, that would result if the entire surface were accessible to the bulk concentration, C_{A0}

$$r''_A = \Omega \times r''_{A0} \tag{10-15}$$

[For example, for a second order reaction, $-r''_{A0} = kC^2_{A0}$.] The *internal effectiveness factor* η, based on the external surface concentration C_{AS}, is defined in an analogous manner.

10.3A First Order Reaction in Spherical Catalyst Pellets

As an example, we shall develop the internal effectiveness factor for spherical catalyst pellets. The development of models which treat pellets of different shapes and individual pores is undertaken in the problems at the end of this chapter. We shall first look at the internal mass transfer resistance to either the products or reactants occurring between the external pellet surface and the interior of the pellet. To illustrate the salient principles of this model, consider that the irreversible isomerization

$$A \longrightarrow B$$

occurs on the surface on the pore walls within the spherical pellet of radius R.

As illustrated in Figure 10-8, the pores in the pellet are not straight and cylindrical; rather, they are a series of tortuous, interconnecting paths of varying cross-sectional areas.

427

(K9) The rate of evaporation of A is equal to the flux from the surface at $z = 0$:

$$W_{Az} = C_{A0}\sqrt{D_{AB}k_1}\left(\coth\sqrt{\frac{kL^2}{D_{AB}}}\right) \tag{K9-1}$$

First calculate the diffusivity of A in B at 200°C and 3 atm.

$$D_{AB}(200) = D_{AB}(25)\frac{\Pi_1}{\Pi_2}\left(\frac{T_2}{T_1}\right)^{1.75}$$

$$= (.08)(\tfrac{1}{3})\left(\frac{473}{298}\right)^{1.75} = .06$$

$$\sqrt{\frac{k_1 L^2}{D_{AB}}} = \sqrt{\frac{(.0005)(10)^2}{.06}} = \sqrt{0.833} = .913$$

Coth $(0.913) = 1.38$

$$C_{A0} = y_{A0}c = \frac{P_{A0}}{\Pi}\frac{\Pi}{RT} = \frac{P_{A0}}{RT} = \frac{\frac{76}{760}}{(.083)(473)}$$

$$= (2.54 \times 10^{-3} \text{ gmole}/l) \times (1\ l/1000 \text{ cu cm}) = 2.54 \times 10^{-6} \text{ gmole/cu cm}$$

Substituting the above values into (K9-1),

$$W_{Az} = 2.54 \times 10^{-6} \text{ gmole/cu cm}\sqrt{(.06)(.0005) \text{ sq cm/sq sec}} \times 1.38$$

$$= 1.92 \times 10^{-8} \text{ gmole/sq cm-sec}$$

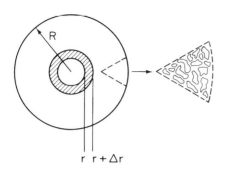

Figure 10-8

It would not be feasible to describe diffusion within each or any of the tortuous pathways; consequently, we shall define our diffusion coefficient so as to describe the *average* diffusion taking place at any position r in the pellet. We shall consider only radial variations in the concentration; the radial flux W_{Ar} will be based on the total area (voids and solid) *normal* to diffusion transport (i.e., $4\pi r^2$) rather than void area alone. This basis for W_{Ar} can be made possible by proper definition of an *effective diffusivity* D_e.

10.3A1 Derivation of the Differential Equation. In Frame (K10) perform a steady-state mole balance on species A in a spherical shell of inner radius r and outer radius $r + \Delta r$ to obtain an equation in terms of the flux W_{Ar}, the rate of reaction per unit surface area r_A'', and a, the surface area per unit volume (solids and voids) of the pellet. Note that even though A is diffusing inwards towards the center of the pellet, the convention of our shell balance dictates that the flux be in the direction of increasing r.

(K10)

| In at r | $-$ | Out at $r + \Delta r$ | $+$ | ? | $= 0$ |

After dividing by $-4\pi\Delta r$ and taking the limit as $\Delta r \to 0$, we obtain the differential equation

$$\frac{d(W_{Ar}r^2)}{dr} - r_A'' a r^2 = 0 \tag{10-16}$$

Select which of the following steps you would carry out next.

(K11)

1. Integrate the differential equation, (10-16).
2. Express r_A'' and W_{Ar} as functions of concentration.
3. State boundary conditions.
4. Not enough information known to proceed further.

For the present we shall take the surface reaction to be nth order in the gas phase concentration of A within the pellet

$$-r_A'' = k_n C_A^n \tag{10-17}$$

Since 1 mole of A reacts under conditions of constant temperature and pressure to form 1 mole of B, we have EMCD at constant total concentration (Table 10-1):

$$W_{Ar} = -cD_e \frac{dy_A}{dr} = -D_e \frac{dC_A}{dr} \tag{10-18}$$

In systems where one does not have EMCD in catalyst pores, the flux may still be given by Equation (10-18), because of the low concentrations of the reactant gases present.

(K10)
We choose the flux of A to be positive in the direction of increasing r, i.e., the outward direction. Since A is actually diffusing inward, the flux of A will have some negative number value, such as -10 gmoles/sq cm-sec, indicating that the flux is actually in the direction of decreasing r. We now proceed to perform our shell balance on A. The area which appears in the balance equation is the total area (voids and solids) normal to the direction of the mass flux:

$$(\text{Rate of A in at } r) = W_{Ar} \cdot \text{area} = W_{Ar} \times 4\pi r^2 \big|_r$$

$$(\text{Rate of A out at } r + \Delta r) = W_{Ar} \cdot \text{area} = W_{Ar} \times 4\pi r^2 \big|_{r+\Delta r}$$

$$\frac{\text{Rate of Generation of A with}}{\text{a shell thickness of } \Delta r} = \frac{\text{rate of Rxn}}{\text{unit surface area}} \cdot \frac{\text{surface area}}{\text{unit volume}} \cdot \text{volume of shell}$$

$$= r''_A \times a \times 4\pi r_m^2 \Delta r$$

where r_m is some mean radius between r and $r + \Delta r$.

$$\underline{\text{In at } r} \quad - \quad \underline{\text{Out at } r + \Delta r} \quad + \quad \underline{\text{Generation within } \Delta r} = 0$$

$$W_{Ar} \times 4\pi r^2 \big|_r - W_{Ar} \times 4\pi r^2 \big|_{r+\Delta r} + \quad r''_A \times a \times 4\pi r_m^2 \Delta r = 0 \qquad (\text{K10-1})$$

(K11)
It is debatable whether step 2 or step 3 should come next. Many times it is best to obtain the differential equation in terms of a single dependent variable (in this case the concentration of A) in order to determine which variables should be used to express the boundary conditions properly. A preferred order of procedure might be
1. Integrate the differential equation incorporating the appropriate boundary conditions.
2. Express W_{Ar} and r''_A as functions of the gas phase concentration.
3. State the boundary conditions.

10.3A2 The Effective Diffusivity.
The effective diffusivity D_e takes into account that:[†]

1. not all of the area normal to the direction of the flux is available (i.e., void) for the molecules to diffuse,
2. the paths are tortuous, and
3. the pores are of varying cross-sectional area.

An equation that relates D_e to the bulk or the Knudsen diffusivity is

$$D_e = \frac{D_A \epsilon \sigma}{\tau} \tag{10-19}$$

where

$\tau = \text{tortuosity} = \dfrac{\text{actual distance a molecule travels between two points}}{\text{shortest distance between those two points}}$

$\epsilon = \text{porosity} = \dfrac{\text{volume of void space}}{\text{total volume (voids \& solids)}}$

$\sigma = \text{constriction factor}$

The constriction factor accounts for the variation in the cross-sectional area that is normal to diffusion.[‡] It is a function of the ratio of maximum to minimum pore area (see Figure 10-9).

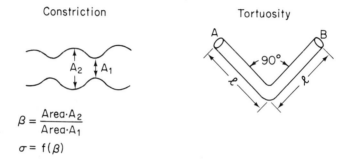

Figure 10–9

When the two areas, A_1 and A_2, are equal, the constriction factor is unity, and when $\beta = 10$, the constriction factor is approximately .5. In Frame (K12) calculate the tortuosity for the hypothetical pore of length, l, shown at the right in Figure 10-9.

(K12)

[†]See Chap. 3 of E. E. Peterson, *Chemical Reaction Analysis* (Englewood Cliffs, N.J.: Prentice-Hall, Inc., 1965), or Chap. 1 of C. N. Satterfield and T. K. Sherwood, *The Role of Diffusion in Catalysis* (Reading, Mass.: Addison-Wesley, 1963).

[‡]E. E. Peterson, *Chemical Reaction Analysis*, (Englewood Cliffs, N.J.: Prentice-Hall, Inc., 1965).

(K12)

$$\tau = \frac{\text{actual distance molecule travels from A to B}}{\text{shortest distance between A and B}}$$

The shortest distance between points A and B is $2l/\sqrt{2}$. The actual distance the molecule travels from A to B is $2l$:

$$\tau = \frac{2l}{\left(\frac{2l}{\sqrt{2}}\right)} = \sqrt{2} = 1.414$$

Typical values of the constriction factor, the tortuosity, and the porosity are, respectively, $\sigma = .8$, $\tau = 1.5$, and $\epsilon = .15$.

After substituting Equations (10-17) and (10-18) into Equation (10-16) and dividing through by D_e, we arrive at the differential equation describing diffusion with reaction in a catalyst pellet:

$$-\frac{d\left(r^2 \frac{dC_A}{dr}\right)}{dr} + \frac{r^2 k_n a C_A^n}{D_e} = 0 \qquad (10\text{-}20)$$

By differentiating the first term and dividing through by $-r^2$, Equation (10-20) becomes

$$\boxed{\frac{d^2 C_A}{dr^2} + \frac{2}{r}\frac{dC_A}{dr} - \frac{k_n a}{D_e} C_A^n = 0} \qquad (10\text{-}21)$$

The boundary conditions are

1. The concentration remains finite at the center of the pellet:

$$C_A \text{ is finite,} \quad \text{at } r = 0.$$

2. At the external surface of the catalyst pellet, the concentration is C_{AS}:

$$C_A = C_{AS}, \quad \text{at } r = R$$

10.3A3 Writing the Equation in Dimensionless Form. We now introduce dimensionless variables ψ and λ so that we may arrive at a parameter that is frequently discussed in catalytic reactions, the *Thiele modulus*.

Let $\psi = C_A/C_{AS}$ and $\lambda = r/R$; rewrite the flux $W_{Ar} = -D_e \frac{dC_A}{dr}$ and the boundary conditions in terms of the dimensionless variables ψ and λ. (*Hint:* The chain rule may be helpful.)

$$\frac{dC_A}{dr} = \frac{dC_A}{d\lambda}\frac{d\lambda}{dr} = \frac{d\psi}{d\lambda}\frac{dC_A}{d\psi}\frac{d\lambda}{dr}$$

(K13)

(K13)

$$\psi = \frac{C_A}{C_{AS}} \tag{K13-1}$$

$$\lambda = \frac{r}{R} \tag{K13-2}$$

With the transformation of variables, the boundary condition

$$C_A = C_{AS}, \quad \text{at } r = R$$

becomes

$$\psi = \frac{C_A}{C_{AS}} = 1, \quad \text{at } \lambda = 1 \tag{K13-3}$$

and the boundary condition,

$$C_A \text{ is finite}, \quad \text{at } r = 0$$

becomes

$$\psi \text{ is finite}, \quad \text{at } \lambda = 0. \tag{K13-4}$$

We now rewrite the differential equation for the molar flux in terms of our dimensionless variables:

$$W_{Ar} = -D_e \frac{dC_A}{dr}$$

$$\frac{dC_A}{dr} = \frac{d\psi}{d\lambda}\left(\frac{dC_A}{d\psi}\frac{d\lambda}{dr}\right) \tag{K13-5}$$

Now differentiate Equation (K13-1) w.r.t. ψ and Equation (K13-2) w.r.t. r, and substitute the resulting expressions,

$$\frac{dC_A}{d\psi} = C_{AS}, \quad \text{and} \quad \frac{d\lambda}{dr} = \frac{1}{R}$$

into Equation (K13-5) to obtain

$$\frac{dC_A}{dr} = \frac{d\psi}{d\lambda}\frac{C_{AS}}{R} \tag{K13-6}$$

The flux of A in terms of the dimensionless variables, ψ and λ, is

$$W_{Ar} = -D_e \frac{dC_A}{dr} = -\frac{D_e C_{AS}}{R}\frac{d\psi}{d\lambda} \tag{K13-7}$$

The overall rate of reaction is equal to the total molar flow of A, $\bar{M}_{Ar}$, into the catalyst pellet and can be obtained by multiplying the molar flux at the outer surface by the external surface area of the pellet, $4\pi R^2$:

$$\bar{M}_{Ar} = -4\pi R^2 W_{Ar}\Big|_{r=R} = +4\pi R^2 D_e \frac{dC_A}{dr}\Big|_{r=R} = 4\pi R D_e C_{AS} \frac{d\psi}{d\lambda}\Big|_{\lambda=1} \qquad (10\text{-}22)$$

Consequently, to determine the overall rate of reaction, which is given by Equation (10-22), first solve Equation (10-21) for C_A, differentiate C_A w.r.t. r, and then substitute the resulting expression into Equation (10-22). Before proceeding along these lines, note that a more common way of expressing the internal surface area of the catalyst (other than in terms of a, the surface area per volume of pellet) is in terms of S_a, the internal surface area per unit mass of catalyst:

$$S_a = \frac{a}{\rho_p} = \frac{\text{sq m}}{\text{g of cat}} \qquad (10\text{-}23)$$

where ρ_p is the density of the pellet. The rate of reaction per unit mass of catalyst, $-r'_A$, and the rate of reaction per unit surface area are related through the equation

$$-r'_A = -r''_A \cdot S_a$$

A typical value of S_a might be 500 sq m/g of catalyst.† In Frame (K14) rewrite Equation (10-21)

$$\frac{d^2C_A}{dr^2} + \frac{2}{r}\frac{dC_A}{dr} - \frac{k_n a C_A^n}{D_e} = 0 \qquad (10\text{-}21)$$

in terms of the dimensionless variables, $\lambda = r/R$ and $\psi = C_A/C_{AS}$, and the specific surface area, S_a.

(K14)
_____

_____

The square root of the coefficient of ψ^n, ϕ_n, is called the *Thiele modulus* and is a measure of the ratio of surface reaction rate to the rate of diffusion through the catalyst pellet:

$$\boxed{\phi_n^2 = \frac{k_n C_{AS}^{n-1} \rho_p S_a R^2}{D_e} = \frac{k_n C_{AS}^n \rho_p S_a R}{D_e \frac{C_{AS}}{R}} = \frac{\text{surface reaction rate}}{\text{diffusion rate}}} \qquad (10\text{-}24)$$

When the Thiele modulus is large, diffusion usually limits the overall rate of reaction; when it is small, the surface reaction is usually rate limiting.

If the surface reaction of A $\rightarrow$ B were rate limiting w.r.t. the adsorption of A and to the desorption of B, the reaction rate expression would be

$$-r''_A = k'_1 \frac{P_A}{1 + k'_2 P_A + k'_3 P_B} = \frac{k_1 C_A}{1 + k_2 C_A + k_3 C_B}$$

where

$$P_A = C_A RT$$
$$k_i = k'_i RT$$

†For methods of measuring S_a, see Chap. 8 of *Chemical Engineering Kinetics*, 2nd ed., J. M. Smith, (New York: McGraw-Hill Inc., 1970).

(K14) After recalling Equation (K13-6),

$$\frac{dC_A}{dr} = \frac{d\psi}{d\lambda} \frac{C_{AS}}{R}$$

differentiating the concentration gradient,

$$\frac{d^2C_A}{dr^2} = \frac{d}{dr}\left(\frac{dC_A}{dr}\right) = \frac{d}{d\lambda}\left(\frac{d\psi}{d\lambda} \frac{C_{AS}}{R}\right)\frac{d\lambda}{dr} = \frac{d^2\psi}{d\lambda^2} \frac{C_{AS}}{R^2} \quad \text{(K14-1)}$$

and rearranging Equation (10-23),

$$a = \rho_p S_a$$

the dimensionless form of Equation (10-21) is written as

$$\frac{C_{AS}}{R^2} \frac{d^2\psi}{d\lambda^2} + \frac{2}{\lambda} \frac{d\psi}{d\lambda} \frac{C_{AS}}{R^2} - \frac{k_n \rho_p S_a C_{AS}^n}{D_e} \psi^n = 0 \quad \text{(K14-2)}$$

Finally, divide by C_{AS}/R^2 to get

$$\frac{d^2\psi}{d\lambda^2} + \frac{2}{\lambda} \frac{d\psi}{d\lambda} - \frac{k_n R^2 S_a \rho_p C_{AS}^{n-1}}{D_e} \psi^n = 0 \quad \text{(K14-3)}$$

Then

$$\boxed{\frac{d^2\psi}{d\lambda^2} + \frac{2}{\lambda} \frac{d\psi}{d\lambda} - \phi_n^2 \psi^n = 0} \quad \text{(K14-4)}$$

where

$$\boxed{\phi_n^2 = \frac{k_n R^2 S_a \rho_p C_{AS}^{n-1}}{D_e}} \quad \text{(K14-5)}$$

If species A and B are weakly adsorbed and present in very dilute concentrations, we can assume that

$$(k_2 C_A + k_3 C_B) \ll 1$$

in which case we obtain the *apparent* first-order reaction,

$$-r''_A \simeq k_1 C_A$$

For a first-order reaction, Equation (K14-4) becomes

$$\frac{d^2\psi}{d\lambda^2} + \frac{2}{\lambda}\frac{d\psi}{d\lambda} - \phi_1^2 \psi = 0 \tag{10-25}$$

with boundary conditions

$$\text{B.C. 1:} \quad \psi = 1 \quad \text{at } \lambda = 1 \tag{10-26}$$

$$\text{B.C. 2:} \quad \psi \text{ is finite} \quad \text{at } \lambda = 0 \tag{10-27}$$

10.3A4 Solution to the differential equation. This differential equation is readily solved with the aid of the transformation $y = \psi \lambda$

$$\frac{d\psi}{d\lambda} = \frac{1}{\lambda}\frac{dy}{d\lambda} - \frac{y}{\lambda^2}$$

$$\frac{d^2\psi}{d\lambda^2} = \frac{1}{\lambda}\frac{d^2y}{d\lambda^2} - \frac{2}{\lambda^2}\frac{dy}{d\lambda} + \frac{2y}{\lambda^3}$$

With these transformations, Equation (10-25) reduces to

$$\frac{d^2y}{d\lambda^2} - \phi_1^2 y = 0$$

This form of differential equation is identical to that of Equation (K7-5) and has the same type of solution:

$$y = A_1 \cosh \phi_1 \lambda + B_1 \sinh \phi_1 \lambda.$$

In terms of ψ,

$$\psi = \frac{A_1}{\lambda} \cosh \phi_1 \lambda + \frac{B_1}{\lambda} \sinh \phi_1 \lambda$$

At $\lambda = 0$, $(1/\lambda) \to \infty$, $(\cosh \phi_1 \lambda) \to 1$, and $(\sinh \phi_1 \lambda) \to 0$. Since the second boundary condition requires ψ to be finite at the center, A_1 must be zero.

The constant B_1 is evaluated for B.C. 1, and the dimensionless concentration profile is

$$\psi = \frac{C_A}{C_{AS}} = \frac{1}{\lambda} \frac{\sinh \phi_1 \lambda}{\sinh \phi_1} \tag{10-28}$$

10.3B Evaluation of the Internal Effectiveness Factor and the Thiele Modulus

The internal effectiveness factor is defined as

$$\eta = \frac{[\text{Actual overall rate of reaction}]}{\begin{bmatrix}\text{Rate of reaction that would result if}\\ \text{entire interior surface was exposed to}\\ \text{the external surface concentration } C_{AS}\end{bmatrix}} \tag{10-29}$$

The actual overall rate of reaction would be the rate at which A diffuses into the pellet at the outer radius R of the pellet. The rate of diffusion into the pellet was given by Equation (10-22). Determine the effectiveness factor as a function of the Thiele modulus, using Equations (10-22) and (10-28) along with the following relationship which gives the rate of reaction that would exist if the entire internal surface were accessible to the concentration at the external surface, C_{AS}.

$$-r'_{AS} \times \text{(mass of catalyst)} = k_1 C_{AS} \times \tfrac{4}{3}\pi R^3 \rho_p S_a$$

(K15)

(K15) If the entire surface was exposed to the concentration at the outer external surface C_{As}, the rate would be

$$\text{Rate} = (\text{Rate per unit area})(\text{surface area})$$

$$= (\text{Rate per unit area})\left(\frac{\text{surface area}}{\text{mass of catalyst}}\right)(\text{mass of catalyst})$$

$$= (k_1 C_{As})(S_a)(\rho_p \tfrac{4}{3}\pi R^3) \tag{K15-1}$$

$$= -r'_{As}(\text{mass of catalyst}) \tag{K15-2}$$

The actual rate is the rate at which the reactant diffuses into the pellet at the outer surface. That is, we sum all the fluxes entering the pellet normal to the surface over the entire external pellet area A_p:

$$\bar{M}_{Ar} = \int_{A_p} \vec{W}_{A_{\text{surface}}} \cdot \vec{n}\, dA_p \tag{K15-3}$$

where $\vec{n}$ is the unit normal vector. For a spherical catalyst pellet, the flux to the external pellet surface is given by Equation (10-22):

$$\bar{M}_{Ar} = -4\pi R^2 W_{Ar}|_{r=R} = +4\pi R^2 D_e \frac{dC_A}{dr}\bigg|_{r=R} = 4\pi R D_e C_{As} \frac{d\psi}{d\lambda}\bigg|_{\lambda=1} \tag{10-22}$$

Differentiating Equation (10-28) and then evaluating the result at $\lambda = 1$,

$$\frac{d\psi}{d\lambda}\bigg|_{\lambda=1} = \left(\frac{\phi_1}{\lambda}\frac{\cosh \lambda \phi_1}{\sinh \phi_1} - \frac{1}{\lambda^2}\frac{\sinh \lambda \phi_1}{\sinh \phi_1}\right)\bigg|_{\lambda=1} \tag{K15-4}$$

$$= (\phi_1 \coth \phi_1 - 1) \tag{K15-5}$$

Substituting Equation (K15-5) into (10-20)

$$\bar{M}_{Ar} = 4\pi R D_e C_{As}(\phi_1 \coth \phi_1 - 1) \tag{K15-6}$$

We now substitute Equations (K15-1) and (K15-6) into Equation (10-29), to obtain an expression for the effectiveness factor:

$$\eta = \frac{\bar{M}_{Ar}}{-r'_{As} \times \text{mass cat}} = \frac{4\pi R D_e C_{As}}{k_1 C_{As} S_a \rho_p \tfrac{4}{3}\pi R^3}(\phi_1 \coth \phi_1 - 1) \tag{K15-7}$$

$$= 3\frac{1}{\dfrac{k_1 S_a \rho_p R^2}{D_e}}(\phi_1 \coth \phi_1 - 1) \tag{K15-8}$$

$$\boxed{\eta = \frac{3}{\phi_1^2}(\phi_1 \coth \phi_1 - 1)} \tag{K15-9}$$

For large values of the Thiele modulus,

$$\eta \simeq \frac{3}{\phi_1} = \frac{3}{R}\sqrt{\frac{D_e}{k_1 \rho_p S_a}} \tag{10-30}$$

To express the overall rate of reaction in terms of the Thiele modulus, we combine Equations (10-29) and (K15-1), and write the actual rate of reaction in the form

$$\bar{M}_{Ar} = \eta \cdot (k_1 C_{As})(S_a)(\text{mass of catalyst})$$

Assuming that the reaction conditions and catalyst with which you are working have a low effectiveness factor (e.g., $\eta = .2$) indicating significant diffusion limitation, which of the choices in Frame (K16) would significantly increase the overall rate of reaction for a given mass of catalyst?

(K16)

1. Increase the temperature of the catalyst bed by 10°C.

2. Increase the flow rate past the pellet, thereby decreasing the external resistance to mass transfer.

3. Make the catalyst pellets smaller.

4. If the reaction order is greater than 1, increase the bulk concentration of the reactant.

5. If the reaction order is greater than 1, decrease the bulk concentration of the reactant.

6. Increase the rate of surface reaction.

7. Increase the internal surface area.

If the reaction conditions and catalyst with which you are working have an effectiveness factor very close to unity, which of the following would significantly increase the overall rate of reaction for a given mass of catalyst?[†]

[†]For further discussion on internal effectiveness factors, see Chap. 4 of E. E. Peterson, *Chemical Reaction Analysis* (Englewood Cliffs, N.J.: Prentice-Hall, Inc., 1965) or Chap. 3 of C. N. Satterfield, and T. K. Sherwood, *The Role of Diffusion in Catalysis* (Reading, Mass.: Addison-Wesley Pub. Co., 1963).

(K16)

If the internal effectiveness factor η is small, i.e., on the order of .2, then the reaction is diffusion limited within the pellet. Consequently, factors influencing the rate of external mass transport [e.g., (K16-item 2)] will have negligible effect on the overall reaction rate. The overall rate of reaction for a first-order reaction is

$$\bar{M}_{Ar} \simeq \frac{3}{R}\sqrt{\frac{D_e S_a k_1}{\rho_p}} C_{AS} \tag{K16-1}$$

Therefore, to increase $\bar{M}_{Ar}$ for a specified mass of catalyst:
  3. Decrease R (make pellets smaller).
  1, 6, 7. Increase the temperature, internal surface area and surface reaction rate.

For reactions of order n, where n is other than first order, we have, from Equation (K14-5)

$$\phi_n^2 \simeq C_{AS}^{n-1} \tag{K16-2}$$

For large values of the effectiveness factor,

$$\eta \simeq \frac{1}{\phi_n} \simeq C_{AS}^{(1-n)/2} \tag{K16-3}$$

Consequently, for reactor orders greater than 1, the effectiveness factor decreases with increasing concentration at the external pellet surface. However, the overall rate is the product of η and the rate law, kC_{AS}^n, i.e.,

$$\bar{M}_{Ar} \sim \eta C_{AS}^n \sim C_{AS}^{(1+n)/2} = C_{AS}^{n'} \tag{K16-4}$$

where the apparent reaction order n' is related to the true reaction order n by

$$\boxed{n' = \frac{1+n}{2}}$$

Thus, even though η decreases with increasing C_{AS}, the overall rate increases with increasing concentration. Therefore, (4) is also correct.

In addition to an apparent reaction order, there is also an apparent activation energy, E_{APP}. This is the activation energy one would calculate from the slope of a plot $\ln(\bar{M}_{Ar})$ as a function of $(1/T)$ at a fixed concentration of A. Recall

$$\bar{M}_{Ar} \sim k_{APP} C_A^{n'} \sim A_{APP} \exp(-E_{APP}/RT) C_A^{n'} \tag{K16-5}$$

Taking the log

$$\ln \bar{M}_{Ar} \sim \frac{E_{APP}}{RT} + \cdots \tag{K16-6}$$

From Equation (K16-1)

$$\bar{M}_{Ar} \sim k_1^{1/2} \sim A_1^{1/2} [\exp(-E_T/RT)]^{1/2} \tag{K16-7}$$

where E_T = true activation energy, taking the ln of Equation (K16-7)

$$\ln \bar{M}_{Ar} \sim \frac{E_T}{2RT} + \cdots \tag{K16-8}$$

Comparing Equations (K16-6) and (K16-8) we see that the true activation energy is equal to twice the apparent activation energy.

$$\boxed{E_T = 2 E_{APP}.} \tag{K16-9}$$

This measurement of the apparent reactor order and activation energy results when diffusion limitations are present and is referred to as *disguised* or *falsified* kinetics (see Pb10-18).

(K17)
1. Eliminate resistance to mass transfer.
2. Increase the temperature of the catalyst bed.
3. Increase the velocity of the bulk fluid past the pellets.
4. Increase the internal surface area.

10.4–External Resistance to Mass Transfer

To begin our discussion on the diffusion of reactants from the bulk fluid to the external surface of a catalyst, we shall focus attention on the flow past a single catalyst pellet. Reaction takes place only within the catalyst and not in the fluid surrounding it. The fluid velocity in the vicinity of the spherical pellet will vary with position around the sphere. The hydrodynamic boundary layer is usually defined as the distance from a solid object where the fluid velocity is 99% of the bulk velocity V_∞. Likewise, the mass transfer boundary layer thickness δ is defined as the distance from a solid object where the concentration of the diffusing species reaches 99% of the bulk concentration.

A reasonable representation of the concentration profile for a reactant A diffusing to the external surface is shown in Figure 10-10.

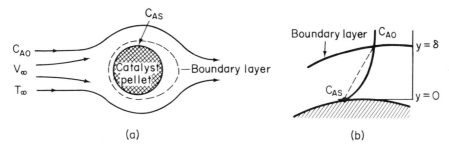

Figure 10–10

As illustrated, the change in concentration of A from C_{A0} to C_{AS}, at any position along the sphere, takes place in a very narrow fluid layer next to the surface of the sphere. Nearly all of the resistance to mass transfer is found in this layer.

10.4A The Mass Transfer Coefficient

A useful way of modeling diffusional transport is to treat the fluid layer next to a solid boundary as a stagnant film of thickness, δ. We say that all of the resistance to mass transfer is found within this hypothetical stagnant film, and the properties, (i.e. concentration, temperature) of the fluid at the outer edge of the film are identical to those of the bulk fluid. With this model, one can readily solve the differential equation for diffusion through a stagnant film. The dashed line in Figure 10-10(b) represents the concentration profile predicted by the hypothetical stagnant film model, while the solid line gives the actual profile. If the film thickness is much smaller than the radius of the pellet, curvature effects can be neglected. As a result only the one-dimensional diffusion equation need be solved, as was shown in Section 10.1 (also, see Figure 10-11).

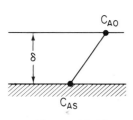

Figure 10–11

(K17) For effectiveness factors close to unity, the surface reaction rate is controlling and mass transfer resistances are negligible by comparison. Therefore only increases in (2) and (4), temperature and internal surface area respectively, will increase the overall rate of reaction. Of course, increases in C_A will also increase the reaction rate.

For either EMCD or dilute concentrations, the solution was shown to be in the form,

$$W_{A_r}(P) = \frac{D_{AB}}{\tilde{\delta}}(C_{A0} - C_{AS})$$

where $W_{A_r}(P)$ is the flux at a specified position or point on the sphere. The ratio of the diffusivity to the film thickness is the mass transfer coefficient, k_c. The tildas over $\tilde{k}_c$ and $\tilde{\delta}$ denote that they are, respectively, the *local* transfer coefficient and the boundary layer thickness at a particular point P on the sphere and not averaged over the entire sphere:

$$\tilde{k}_c = \frac{D_{AB}}{\tilde{\delta}} \tag{10-31}$$

The average mass transfer coefficient over the surface of area A is

$$k_c = \frac{\int_A \tilde{k}_c \, dA}{A}$$

The average molar flux from the bulk fluid to the surface is

$$W_{A_r} = k_c(C_{A0} - C_{AS}) \tag{10-32}$$

The mass transfer coefficient k_c is analogous to the heat transfer coefficient. The heat flux q from the bulk fluid at a temperature T_0 to a solid surface at T_s is

$$q_r = h(T_0 - T_s)$$

For forced convection, the heat transfer coefficient is normally correlated in terms of three dimensionless groups: the Nusselt number, Nu, the Reynolds number, Re, and the Prandtl number, Pr. For the single spherical pellets discussed here, these groups take the form

$$\text{Nu} = \frac{h d_p}{k_t} \tag{10-33}$$

$$\text{Pr} = \frac{\mu C_p}{k_t} = \frac{\mu}{\rho}\left(\frac{\rho C_p}{k_t}\right) = \frac{\nu}{\alpha_t} \tag{10-34}$$

$$\text{Re} = \frac{V_\infty \rho d_p}{\mu} \tag{10-35}$$

where
α_t = thermal diffusivity
ν = kinematic viscosity
d_p = diameter of pellet
V_∞ = free stream velocity
k_t = thermal conductivity,

and the other symbols are the same as previously defined.

The heat transfer correlation relating the Nusselt number to the Prandtl and Reynolds numbers for flow around a sphere is[†]

$$\text{Nu} = 2 + .236 \, \text{Re}^{.606} \, \text{Pr}^{1/3} \tag{10-36}$$

While this correlation can be used over a wide range of Reynolds numbers, it can be shown theoretically that if the sphere is immersed in a stagnant fluid (see Problem 10-7), then

$$\text{Nu} = 2$$

and that at high Reynolds numbers in which the boundary layer remains laminar, the Nusselt

[†]E. R. G. Eckert and R. M. Drake, *Analysis of Heat and Mass Transfer*, (New York: McGraw-Hill, Inc., 1972), p. 414.

number becomes

$$\text{Nu} = b\text{Re}^{1/2}\,\text{Pr}^{1/3}$$

$$b = f(\text{geometry and temperature distribution on sphere surface})$$

Consequently, we would expect some discrepancy in this correlation at very high and at very low Reynolds numbers.

Although further discussion of heat transfer correlations might be worthwhile, it will not help us to determine the mass transfer coefficient and the mass flux from the bulk fluid to the external pellet surface. However, the preceding discussion on heat transfer was not entirely wasted, since we would expect that *for similar geometries the heat and mass transfer correlations would be analogous*, and this is indeed the case. If a correlation for the Nusselt number exists, the mass transfer coefficient can be estimated by replacing the Nusselt and Prandtl numbers in this correlation by the Sherwood and Schmidt numbers, respectively:

$$\text{Sh} \longrightarrow \text{Nu}$$
$$\text{Sc} \longrightarrow \text{Pr}$$

In Frame (K18), suggest dimensionless mass transfer groups for the Schmidt and Sherwood numbers that are similar to their analogous heat transfer groups.

(K18)

For example,

$$\text{Nu} = \frac{h d_p}{k_t}, \qquad \text{Pr} = \frac{\nu}{\alpha_t}$$

$$\text{Sh} = \qquad\qquad \text{Sc} =$$

We now form the correlation for the mass transfer coefficient for flow around a spherical pellet:

$$\text{Sh} = 2 + .236\,\text{Re}^{.606}\,\text{Sc}^{1/3} \qquad (10\text{-}37)$$

10.4B Mass Transfer from a Single Particle

Calculate the mass flux of reactant A to a single catalyst pellet 1 cm in diameter suspended in a large body of fluid. The reactant is present in dilute concentrations, and the reaction is considered to take place instantaneously at the external pellet surface (i.e., $C_{AS} \simeq 0$). The bulk concentration of the reactant is 10^{-6} gmoles/cu cm, and the free stream velocity is 10 cm/sec. The kinematic viscosity is 0.5 centistokes (1 centistoke = 10^{-2} sq cm/sec), and the liquid diffusivity of A is 10^{-6} sq cm/sec.

(K19)

10.4C The Overall Effectiveness Factor

Imagine a situation where *external* and *internal resistance* to mass transfer to and within the pellet *are the same order of magnitude*. At steady state, the transport of the reactant(s) from the bulk fluid to the external surface of the catalyst, $W_{A_r}A_p$, is equal to the net rate of reaction of the reactant within and on the pellet, $-r''_A \times (A_s + A_p)$. For a single spherical pellet of radius R,

(K18)

The heat transfer and mass transfer coefficients are analogous. The corresponding fluxes are

$$q_z = h(T - T_S) \tag{K18-1}$$

$$W_{Az} = k_c(C - C_S) \tag{K18-2}$$

The one-dimensional differential forms of the mass flux for EMCD and the heat flux are, respectively,

$$W_{Az} = -D_{AB}\frac{\partial C_A}{\partial z}$$

$$q_z = -k_t \frac{\partial T}{\partial z}$$

If in Equation (10-33) we replace h by k_c and k_t by D_{AB},

$$\left.\begin{array}{c} h \longrightarrow k_c \\ k_t \longrightarrow D_{AB} \end{array}\right\} \text{Nu} \longrightarrow \text{Sh}$$

we obtain the mass transfer Nusselt number (i.e., the Sherwood number):

$$\text{Sh} = \frac{k_c d_p}{D_{AB}} = \frac{(\text{cm/sec})(\text{cm})}{(\text{sq cm/sec})}, \quad \text{dimensionless}$$

The Prandtl number is the ratio of the kinematic viscosity (i.e., the momentum diffusivity) to the thermal diffusivity. Since the Schmidt number is analogous to the Prandtl number, one would expect that Sc is the ratio of the momentum diffusivity, ν, to the mass diffusivity; this is indeed the case.

$$\alpha_t \longrightarrow D_{AB}$$

The Schmidt number is

$$\text{Sc} = \frac{\nu}{D_{AB}} = \frac{(\text{sq cm/sec})}{(\text{sq cm/sec})}, \quad \text{dimensionless}$$

(K19)

For dilute concentrations of the solute

$$W_{Ar} = k_c(C_{A0} - C_{AS}) \tag{10-32}$$

Since reaction is assumed to occur instantaneously on the external surface of the pellet, $C_{AS} = 0$:

$$\text{Sh} = \frac{k_c d_p}{D_{AB}} = 2 + .236\, \text{Re}^{.606}\text{Pr}^{1/3} \tag{K19-1}$$

$$k_c = \frac{2D_{AB}}{d_p} + .236\frac{D_{AB}}{d_p}\text{Re}^{.606}\text{Sc}^{1/3} \tag{K19-2}$$

$$\text{Re} = \frac{\rho d_p V_\infty}{\mu} = \frac{d_p V_\infty}{\nu} = \frac{(1)(10)}{.005} = 2000$$

$$\text{Sc} = \frac{\nu}{D_{AB}} = \frac{5 \times 10^{-3}}{10^{-6}} = 5000$$

Substituting the above values into Equation (K19-1),

$$\text{Sh} = 2 + .236(2000)^{.606}(5000)^{.33} = 405$$

$$k_c = \frac{D_{AB}}{d_p}\text{Sh} = \left(\frac{10^{-6}}{1}\right) \times 405 = 4.05 \times 10^{-4}\ \text{cm/sec}$$

$$C_{A0} = 10^{-6}\ \text{gmoles/cu cm}$$

Substituting for k_c and C_{A0} in Equation (10-32), the molar flux to the surface is

$$W_{Ar} = (4.05 \times 10^{-4})(10^{-6} - 0) = 4.05 \times 10^{-10}\ \text{gmole/sq cm-sec}$$

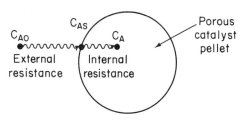

Figure 10-12

$A_p = 4\pi R^2$. For most catalysts, the internal surface area A_s is much greater than the external area A_p, i.e., $A_s \gg A_p$, in which case

$$\bar{M}_{Ar} = W_{Ar}A_p = -r''_A A_s \qquad (10\text{-}38)$$

where r''_A is the average overall rate of reaction within and on the pellet per unit surface area. The relationship for the rate of mass transport is (see Figure 10-12)

$$\bar{M}_{Ar} = W_{Ar}A_p = k_c(C_{A0} - C_{AS})A_p \qquad (10\text{-}39)$$

At this point, the mass flux to the surface *could not be* calculated using Equation 10-39, since the surface concentration C_{AS} is not known and cannot be easily determined experimentally. Consequently, to evaluate W_{Ar}, we must either determine C_{AS} from another relation or eliminate it from this equation. Since internal diffusion resistance is significant, not all of the interior surface of the pellet is accessible to the concentration at the external surface of the pellet, C_{AS}. We have already learned that the effectiveness factor is a measure of this surface accessibility, i.e., $r'_A = \eta r'_{AS}$. Assuming that the surface reaction is first order w.r.t. A, we can utilize the internal effectiveness factor to write

$$-r''_A = \eta k C_{AS} \qquad (10\text{-}40)$$

Use Equations (10-38)–(10-40) to eliminate C_{AS} from Equation (10-39), so that the total molar transport of A from the bulk fluid to the external pellet surface can be expressed solely in terms of the bulk concentration and other parameters (e.g., the mass transfer coefficient k_c and the specific reaction rate k) of the system.

(K20)

In discussing the surface accessibility, we defined the internal effectiveness factor η w.r.t. the concentration *at* the external surface of the pellet, C_{AS}:

$$\eta = \frac{\text{actual overall rate}}{\text{rate which would result if the entire surface were exposed to the external surface concentration } C_{AS}} \qquad (10\text{-}29)$$

We also defined an overall effectiveness factor that is based on the bulk concentration

$$\Omega = \frac{\text{actual overall rate}}{\text{rate that would result if the entire surface were exposed to the bulk concentration } C_{A0}} \qquad (10\text{-}41)$$

$$r'_A = \Omega r'_{A0} \qquad (10\text{-}15)$$

(K20)

The net rate of reaction given by Equation (10-38) is equal to the rate of mass transfer of A from the bulk fluid to the external pellet surface (Equation 10-39).

$$W_{Ar}A_P = -r''_A A_S = -r'_A \cdot (\text{mass of pellet}) \tag{10-38}$$

where
$$A_S = S_a \cdot (\text{mass of pellet})$$

$$W_{Ar}A_P = k_c(C_{A0} - C_{AS})A_P \tag{10-39}$$

We need to eliminate the surface concentration from any equation giving the rate of reaction or rate of mass transfer, since C_{AS} cannot be measured by standard techniques. To accomplish this, first substitute Equation (10-40) into Equation (10-38),

$$W_{Ar}A_P = \eta k A_S C_{AS} \tag{K20-1}$$

and then equate Equations (K20-1) and (10-39):

$$\eta k A_S C_{AS} = k_c A_P (C_{A0} - C_{AS}) \tag{K20-2}$$

Solve Equation (K20-2) for C_{AS},

$$C_{AS} = \frac{k_c A_P C_{A0}}{k_c A_P + \eta k A_S}, \tag{K20-3}$$

and substitute (Equation K20-3) into Equation (K20-1):

$$W_{Ar}A_P = \frac{k_c A_P C_{A0}}{k_c A_P + \eta k A_S} \eta k A_S \tag{K20-4}$$

Dividing by kA_S, we obtain the net rate of reaction (total molar flow of A to surface) in terms of the bulk fluid concentration, which is a measurable quantity:

$$\bar{M}_{Ar} = W_{Ar}A_P = \left(\frac{\eta k_c}{\eta + \frac{k_c A_P}{k A_S}}\right) A_P C_{A0} \tag{K20-5}$$

For example, for a second order reaction in a spherical catalyst pellet of radius R the two effectiveness factors take the form

$$\eta = \frac{-4\pi R^2 D_e \left.\frac{\partial C_A}{\partial r}\right|_{r=R}}{\frac{4}{3}\pi R^3 \rho_p S_a k C_{AS}^2} = \frac{-3 D_e \left.\frac{\partial C_A}{\partial r}\right|_{r=R}}{R \rho_p S_a k C_{AS}^2}$$

$$\Omega = \frac{4\pi R^2 k_c (C_{A0} - C_{AS})}{\frac{4}{3}\pi R^3 \rho_p S_a k C_{A0}^2} = \frac{3 k_c (C_{A0} - C_{AS})}{R \rho_p S_a k C_{A0}^2}$$

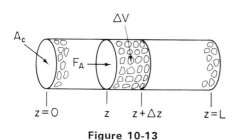

Figure 10-13

After reviewing Frame (K20), write an expression relating η and Ω for a packed bed of spherical catalyst pellets in which a first order reaction is taking place. (Hint: *Equate the total mass transfer rate of* A *from the bulk fluid to the external surface within a system volume* ΔV, *to the net rate of reaction of* A *inside the pellets.* Let a_x denote the external surface area of catalyst per unit volume of bed. See Figure 10-13).

(K21)

10.5–MASS TRANSFER AND REACTION IN A PACKED BED

We now consider the same isomerization reaction taking place in a *packed bed of catalyst pellets* rather than on one single pellet (see Figure 10-13).

We shall perform a balance on A over the volume element ΔV, neglecting any radial variations in concentration and assuming that the bed is operated at steady state. The following symbols will be used in developing our model:

A_c = cross-sectional area of the tube
a_x = the external surface area of catalyst per unit volume of bed (packing and voids) in packed beds
C_{A0} = bulk concentration of A
ρ = bulk density of the catalyst
v_0 = volumetric flow rate
U = superficial velocity = v_0/A_c

(Rate in) − (Rate out) + (Rate of formation of A) = 0
$A_c W_{Az}|_z - A_c W_{Az}|_{z+\Delta z} + \quad r'_A \rho \Delta V \quad = 0$

Dividing by Δz and taking the limit $\Delta z \to 0$,

$$-\frac{dW_{Az}}{dz} + r'_A \rho = 0 \tag{10-42}$$

Assuming that the total concentration c is constant,

$$W_{Az} = -D_{AB}\frac{dC_{A0}}{dz} + y_A(W_{Az} + W_{Bz})$$

(K21)

The total molar flow $\bar{M}_{Ar}$ from the gas phase to the external catalyst surface is equal to the overall rate of reaction on the internal catalyst surface. The total molar flow to the catalytic surface in a volume ΔV is

$$\bar{M}_{Ar} = (\text{molar flux})\left(\frac{\text{external surface area}}{\text{unit volume}}\right)(\text{volume})$$

$$= W_{Ar}a_x \Delta V \qquad \text{(K21-1)}$$
$$= k_c(C_{A0} - C_{AS})a_x \Delta V$$

where k_c is the mass transfer coefficient for the packed bed. The overall rate of reaction in a volume ΔV is

$$\bar{M}_{Ar} = \left(\frac{\text{rate}}{\text{internal surface area}}\right)\left(\frac{\text{surface area}}{\text{mass catalyst}}\right)\left(\frac{\text{mass catalyst}}{\text{volume}}\right)(\text{volume})$$

$$= -r_A'' S_a \rho_B \Delta V \qquad \text{(K21-2)}$$

where ρ_B is the *bulk* density. Equating Equations (K21-1) and (K21-2),

$$k_c(C_{A0} - C_{AS})a_x = -r_A'' S_a \rho_B \qquad \text{(K21-3)}$$

the reaction is first order and can be expressed in terms of either the internal effectiveness factor,

$$-r_A'' = \eta k C_{AS} \qquad \text{(K21-4)}$$

or the overall effectiveness factor,

$$-r_A'' = \Omega k C_{A0} \qquad \text{(K21-5)}$$

Substituting Equation (K21-4) into Equation (K21-3) and rearranging,

$$C_{AS} = \frac{k_c C_{A0} a_x}{k_c a_x + \eta k \rho_B S_a} \qquad \text{(K21-6)}$$

Substituting Equation (K21-6) into Equation (K21-4),

$$-r_A'' = \frac{\eta k_c a_x}{k_c a_x + \eta k \rho_B S_a} k C_{A0} \qquad \text{(K21-7)}$$

Comparing Equations (K21-7) and (K21-5), we arrive at the following relationship between Ω and η for a first-order reaction.

$$\Omega = \left(\frac{\eta}{1 + \eta \frac{k \rho_B S_a}{k_c a_x}}\right) \qquad \text{(K21-8)}$$

and writing the bulk flow term in the form
$$B_{Az} = y_A(W_{Az} + W_{Bz}) = V^* y_A C = UC_{A0}$$
Equation (10-42) can be written in the form,
$$D_{AB}\frac{d^2C_{A0}}{dz^2} - U\frac{dC_{A0}}{dz} + r'_A \rho = 0 \tag{10-43}$$

Suggest expressions or equations in which the overall reaction rate within the pellet, r'_A, can be related to the bulk concentration of A, C_{A0}.

(K22) _____

_____

Substituting Equation (K22-3) into Equation (10-43), we form the differential equation describing diffusion with a first-order reaction in a catalyst bed:

$$\boxed{D_{AB}\frac{d^2C_{A0}}{dz^2} - U\frac{dC_{A0}}{dz} - \Omega\rho k S_a C_{A0} = 0} \tag{10-44}$$

As an example, we shall solve this equation for the case in which the flow rate through the bed is very large and the axial diffusion can be neglected. Rewrite Equation (10-44), taking this simplification into account.

(K23) _____

_____

With the aid of the boundary condition at the entrance of the reactor,
$$C_{A0} = C_{AE} \quad \text{at } z = 0$$
Equation (K23-1) can be integrated to give
$$C_{A0} = C_{AE} e^{-(\rho k S_a \Omega z)/U} \tag{10-45}$$
The net rate of reaction within the bed is obtained by taking the difference between the inlet and outlet molar flow rates of A:
$$R_A = F_A|_{z=0} - F_A|_{z=L} \tag{10-46}$$
Since axial diffusion was neglected, the molar flow rate in the axial direction can be expressed by the product $C_{A0}v_0$, then:
$$R_A = v_0 C_{AE} - v_0 C_{AL}$$
$$C_{AL} = C_{AE} e^{-(kS_a L\Omega \rho)/U} \tag{10-47}$$
$$R_A = C_{AE} v_0 (1 - e^{-(kS_a L\Omega \rho)/U}) \tag{10-48}$$
The net rate of reaction could also have been found by integrating $-r'_A$ over the catalyst bed:
$$R_A = \int -r'_A \rho \, dV = \int -r'_A \rho A_c \, dz \tag{10-49}$$
$$-r'_A = \Omega k S_a C_{A0} = \Omega k S_a C_{AE} \exp\left(\frac{-\Omega \rho S_a k z}{U}\right) \tag{10-50}$$

(K22)

$-r'_A$ is the net rate of reaction within and on the catalyst per unit mass of catalyst. It is a function of the reaction concentration within the catalyst. The overall rate, $-r'_A$, can be related to the ratio of reaction of A that would exist if the entire surface were exposed to the bulk concentration C_{A0}, through the overall effectiveness factor Ω:

$$-r'_A = -r'_{A0} \times \Omega \tag{K22-1}$$

For the first-order reaction considered here,

$$-r'_{A0} = r''_{A0} \cdot S_a = k S_a C_{A0} \tag{K22-2}$$

Substituting Equation (K22-2) into Equation (K22-1), we obtain the overall rate of reaction per mass of catalyst in terms of the bulk concentration C_{A0}:

$$-r'_A = \Omega k S_a C_{A0} \tag{K22-3}$$

(K23)

Neglecting axial diffusion w.r.t. forced axial convection, i.e.,

$$U \frac{dC_{A0}}{dz} \gg D_{AB} \frac{d^2 C_{A0}}{dz^2}$$

Equation (10-44) can be arranged in the form

$$\frac{dC_{A0}}{dz} = -\left(\frac{\Omega \rho k S_a}{U}\right) C_{A0} \tag{K23-1}$$

$$R_A = \rho \Omega k A_c S_a C_{AE} \int_0^L \exp\left(\frac{-\Omega \rho S_a k z}{U}\right) dz \qquad (10\text{-}51)$$
$$= v_0 C_{AE}(1 - e^{-(k S_a \Omega L \rho)/U})$$

To evaluate the overall effectiveness factor, we need to determine the mass transfer coefficient. The mass transfer coefficient for a packed bed of pellets can be determined from the relationship,

$$\text{Sh}' = (\text{Re}')^{1/2} \text{Sc}^{1/3} \qquad (10\text{-}52)$$

where
$$\text{Sh}' = \frac{k_c d}{D_{AB}} \frac{\epsilon_v}{(1 - \epsilon_v)\gamma}$$

ϵ_v = void fraction

d = pellet diameter, or $d = \left(\dfrac{6V_p}{\pi}\right)^{1/3}$

V_p = volume of pellet

γ = shape factor = $\dfrac{A_p}{\pi d^2}$

$\text{Re}' = \dfrac{Ud}{\nu} \dfrac{1}{(1 - \epsilon_v)\gamma}$

The remaining symbols in Equation (10-52) are defined as before.

10.6–Other Forms of the Mass Transfer Coefficient

Mass transfer coefficients can be and commonly are expressed in driving forces other than concentration. Various types of mass transfer coefficients along with respective driving forces are given in Table 10–5.

TABLE 10–5

$$\text{COEFFICIENT} = \frac{\text{MASS TRANSFER RATE}}{\text{DRIVING FORCE}}$$

Driving Force	Coefficient	Units	Flux (gmoles/sq cm-sec)
Gas phase mole fraction, y_A	k_y	gmoles/sq cm-sec	$W_A = k_y(y_{A2} - y_{A1})$
Liquid phase mole fraction, x_A	k_x	gmoles/sq cm-sec	$W_A = k_x(x_{A2} - x_{A1})$
Partial pressure, P_A	k_g	gmoles/sq cm-sec-atm	$W_A = k_g(P_{A2} - P_{A1})$
Concentration, C_A	k_c	cm/sec	$W_A = k_c(C_{A2} - C_{A1})$

The gas phase mass transfer coefficient determined from relationships similar to Equations (10-37) and (10-52) can be directly applied only for dilute concentrations of solutes.[†] At high solute concentrations, it is necessary to correct for the *bulk flow* contribution to the mass flux. The solutions to the differential equations for mass transfer for diffusion through a stagnant

---

[†]To obtain correlations for mass transfer coefficients for a variety of systems and geometries see either R. E. Treybal, *Mass Transfer Operations*, 2nd ed. Chap. 3 (New York: McGraw-Hill, Inc., 1968), or W. L. McCabe and J. C. Smith, *Unit Operations in Chemical Engineering*, 2nd ed., (New York: McGraw-Hill, Inc., 1967).

film at high solute concentrations give the following equation for the molar flux of A:

$$W_{Az} = \frac{D_{AB}}{\delta y_{B_{ln}}}(C_{A0} - C_{AS}) = \frac{k_c}{y_{B_{ln}}}(C_{A0} - C_{AS}) \tag{10-53}$$

$$W_A = \dot{k}_c(C_{A0} - C_{AS}) \tag{10-54}$$

where
$$\dot{k}_c = \frac{k_c}{y_{B_{ln}}} \tag{10-55}$$

and
$$y_{B_{ln}} = \frac{y_{B1} - y_{B2}}{\ln\left(\frac{y_{B1}}{y_{B2}}\right)} = \frac{y_{A2} - y_{A1}}{\ln\left(\frac{1 - y_{A1}}{1 - y_{A2}}\right)} \tag{10-56}$$

It can be seen that for high solute concentrations the mass transfer coefficient k_c is a function of concentration, and solutions to the previous examples becomes slightly more difficult.[†]

SUMMARY

1. The molar flux of A in a binary mixture of A and B is

$$\vec{W}_A = -cD_{AB}\nabla y_A + y_A(\vec{W}_A + \vec{W}_B) \tag{S10-1}$$

   a. For equimolar counter diffusion or for dilute concentration of the solute,

$$\vec{W}_A = -cD_{AB}\nabla y_A \tag{S10-2}$$

   b. For diffusion through a stagnant gas,

$$\vec{W}_A = cD_{AB}\nabla \ln(1 - y_A) \tag{S10-3}$$

   c. For negligible diffusion,

$$\vec{W}_A = y_A \vec{W} = C_A \vec{V} \tag{S10-4}$$

2. The internal effectiveness factor is

$$\eta = \left[\frac{\text{actual overall rate of reaction}}{\text{rate of reaction that would result if the entire surface were accessible to the concentration at the external surface, } C_{AS}}\right] \tag{S10-5}$$

$$-r'_A = -r'_{AS} \times \eta$$

The internal effectiveness factor is usually expressed as a function of the Thiele modulus. The overall effectiveness factor is based on the bulk concentration.

$$-r'_A = -r'_{A0} \times \Omega$$

3. The rate of mass transfer from the bulk fluid to a boundary at concentration C_{AS} is

$$W_A = k_c(C_{A0} - C_{AS}) \tag{S10-6}$$

where k_c is the mass transfer coefficient.

4. The Sherwood and Schmidt numbers are, respectively,

$$\text{Sh} = \frac{k_c L}{D_{AB}} \tag{S10-7}$$

$$\text{Sc} = \frac{\nu}{D_{AB}} \tag{S10-8}$$

[†]For further discussion see Chapters 16–21 of R. B. Bird, W. E. Stewart and E. N. Lightfoot, *Transport Phenomena* (New York: John Wiley and Sons, 1960).

5. If a heat transfer correlation exists for a given system and geometry, the mass transfer correlation may be found by replacing the Nusselt number by the Sherwood number and the Prandtl number by the Schmidt number in the existing heat transfer correlation.

6. The equation for axial diffusion and flow for a first-order reaction in a packed bed is

$$D_{AB}\frac{d^2C_{A0}}{dz^2} - U\frac{dC_{A0}}{dz} - \Omega k \rho S_a C_{A0} = 0 \tag{S10-9}$$

The symbol Ω represents the overall effectiveness factor.

QUESTIONS AND PROBLEMS

P10-1. What is diffusion? What factors may cause diffusion to occur? Give the special forms of Fick's law for the flux $\vec{W}_A$, that apply to
   a. equimolar counterdiffusion
   b. diffusion of A through stagnant B.

P10-2. Diffusivities:
   a. Calculate the diffusion coefficient of CO_2 in air at 0°C and 1 atm, using the Fuller equation. Compare the calculated value given in *Perry's Chemical Engineering Handbook*. (See Supplementary Readings at the end of the chapter.)
   b. What is the diffusivity of CO_2 in air at $-25°C$ and 1.25 atm?
   c. What is the diffusivity of ethyl acetate in water at 20°C? (This information can be obtained from *Perry's Handbook*, *Handbook of Chemistry in Physics* (48th ed.), ed. R.C. Weast, Cleveland, Ohio: Chemical Rubber Co., 1967, or K. A. Reddy and L. K. Doraiswamy *IEC Fund* 6, 77 (1967).
   d. What is the diffusivity of ethyl acetate in water at 50°C?

P10-3. For a binary system (A and B) show that

$$D_{AB} = D_{BA}$$

[*Hint*: The flux for species B is $\vec{W}_B = -cD_{BA}\nabla y_B + y_B\vec{W}$. The flux for species A was given by Equation (10-6).]

P10-4. Consider a binary mixture of species A and B. Let $\vec{V}_A$ and $\vec{V}_B$ be the particle velocities of species A and B, respectively. The flux of species A, $\vec{W}_A$, is the product of the concentration of A and $\vec{V}_A$, i.e.,

$$\vec{W}_A = C_A \vec{V}_A \tag{P10-4-1}$$

The diffusive flux of A, $\vec{J}_A$, is the flux of A relative to the molar average velocity $\vec{V}^*$:

$$\vec{J}_A = C_A(\vec{V}_A - \vec{V}^*) \tag{P10-4-2}$$

$$\vec{V}^* = \sum y_i \vec{V}_i$$

where $\vec{V}_i$ is the particle velocity of species i and y_i is the mole fraction of species i.

Using Equation (P10-4-2) along with (P10-4-1) and any similar equation, show that

$$\vec{W}_A = \vec{J}_A + y_A(\vec{W}_A + \vec{W}_B) \tag{10-6}$$

P10-5. The dimer, A_2, diffuses at steady state from the bulk solution to a catalytic surface, where it dissociates instantaneously to form 2A. Species A then diffuses back into the bulk solution which contains only A and A_2 (see Figure P10-5).

   a. From a differential mole balance on A_2, derive a differential equation in terms of $\vec{W}_{A_2}$. State the appropriate boundary conditions.
   b. After determining the proper relationship between $\vec{W}_{A_2}$ and $\vec{W}_A$, use Equation (10-6) to substitute for $\vec{W}_A$ in the differential equation derived in part a.
   c. Next, obtain the concentration profile for A_2.

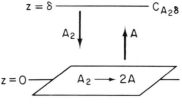

Figure P10–5

    d. Determine the flux of A_2 to the surface.

    e. If the boundary layer thickness is 10^{-2} cm and the bulk concentration of A_2 is 10^{-5} gmole/l, estimate the rate of reaction of A_2 on the solid surface.

P10-6. Species A is evaporating from a liquid surface into a stagnant film of B. The system is similar to that shown in the schematic diagram in Figure 10-11. The concentration of the diffusing solute is high; consequently, the bulk flow of A cannot be neglected.

    a. Obtain an equation for the concentration profile of A.

    b. Show that the flux of A from the solid surface can be given by Equation (10-53).

    c. Calculate the rate of evaporation of A, which is water, into B, which is air. The concentration of A at the top of the film is zero; the film thickness is 1 mm, and the temperature of the air-water system is 50°C.

P10-7. A spherical ball of napthalene of radius R is evaporating into a large body of stagnant air. The concentration of napthalene in air is to be considered dilute, and only radial variations in concentration will be considered.

    a. From a shell balance arrive at the differential equation,

$$\frac{1}{r^2}\frac{\partial}{\partial r}\left[D_{AB}\frac{\partial(r^2 C_A)}{\partial r}\right] = \frac{\partial C_A}{\partial t}$$

Next, assume steady state; also assume that the boundary conditions are

$$C_A = C_{AS} \quad \text{at } r = R$$
$$C_A = C_{A\infty} \quad \text{at } r \longrightarrow \infty$$

    b. Obtain the concentration profile for A.

    c. Obtain an equation for the flux of napthalene from the surface of the sphere.

    d. Show that, for diffusion from a sphere into stagnant fluid,

$$\text{Sh} = 2$$

P10-8. Repeat Problem 7, assuming a case where the evaporation takes place at a higher temperature, resulting in high concentrations of napthalene in the gas phase.

P10-9. Derive equations for the internal and overall effectiveness factors for the case of a zero-order reaction taking place in a spherical catalyst pellet.

P10-10. In Frame (K20) we derived an equation for the molar flux of A in terms of the bulk concentration C_{A0} for a first-order reaction. This was accomplished by the appropriate combination of equations needed to eliminate the surface concentration C_{AS}.

    a. Use a similar procedure to express the molar flux of A in terms of C_{A0} for the second-order reaction,

$$-r'_A = \eta k C_{AS}^2 \qquad \text{(P10-10-1)}$$

    b. Obtain an equation relating the overall and internal effectiveness factors for the second-order reaction whose rate equation is given in Equation (P10-10-1).

P10-11. Derive an equation for the effectiveness factor for the long straight cylindrical pore shown in Figure P10-11. The reaction takes place only on the interior walls of the pore;

the rate law is

$$-r'_A = kC_A \qquad \text{(P10-11-1)}$$

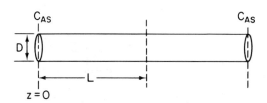

Figure P10–11

where

C_A is the gas phase concentration
D is the pore diameter
L is one-half the pore length
C_{AS} is the concentration at the pore mouth.

Mass transfer in the pore results only from molecular diffusion. Neglect any radial variations in concentration.

P10-12. Derive an equation for both the internal and the overall effectiveness factors for the rectangular porous slab shown in Figure P10-12. The reaction taking place inside the slab is first order and is given by Equation (P10-11-1). Assume dilute concentrations and neglect any variations in this axial (x) direction.

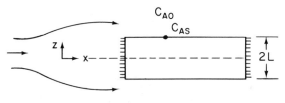

Figure P10–12

P10-13. Pure nitrogen at 160°C enters a cylinder whose inner wall is coated with napthalene. The nitrogen flow is turbulent, and the Dittus-Boelter relation for turbulent flow heat transfer in a pipe is

$$\text{Nu} = .023(\text{Re})^{.8}(\text{Pr})^{.4}$$

Derive the complete design equation in terms of the transfer coefficient at the tube entrance, k_x, the tube diameter, and the inlet velocity that would give the tube length necessary to obtain a mole fraction of 0.25 of napthalene in the exit stream.

You may assume constant total pressure, temperature, shear viscosity, and diameter of the tube; however, you may *not* assume a dilute mixture or constant kinematic viscosity.

Additional information:

Molecular weight of napthalene = 128
Vapor pressure of napthalene at 160°C = 200 mm Hg

List all other assumptions used in your analysis. Insert all known numerical values, and write the design equation in the simplest possible form.

P10-14. A single solid napthalene sphere is falling through dry air at 100°F and 1 atm. At this temperature, the vapor pressure of napthalene is 117 mm Hg. The sphere diameter is 5 mm. The diffusivity of napthalene in air is .235 sq ft/hr. What is the rate of mass transfer from the napthalene sphere in lbs/hr? (*Hint:* First determine the terminal velocity of the sphere).

P10-15. Pure oxygen is being absorbed by xylene in a catalyzed reaction in the experimental apparatus sketched in Figure P10-15. Under constant conditions of temperature and liquid composition the following data were obtained:

Stirrer Speed	Rate of uptake of O_2 (ml/hr)			
	System pressure (absolute)			
	1.2 atm	1.6 atm	2.0 atm	3.0 atm
400 rpm	15	31	75	152
800 rpm	20	39	102	205
1200 rpm	21	62	105	208
1600 rpm	21	61	106	207

No gaseous products are formed by the chemical reaction. What would you conclude about the relative importance of liquid phase diffusion and about the order of the kinetics of this reaction? (Calif. PECEE. November 1968)

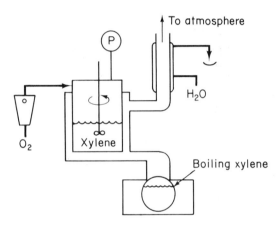

Figure P10–15

P10-16. The elementary isomerization reaction

$$A \longrightarrow B$$

is taking place on the walls of a cylindrical catalyst pore. In one run, a catalyst poison P entered the reactor along with the reactant A. To estimate the effect of poisoning, we assume that the poison renders the catalyst pore walls near the pore mouth ineffective up to a distance z_1, so that no reaction takes place on the walls in this entry region (see Figure P10-16).

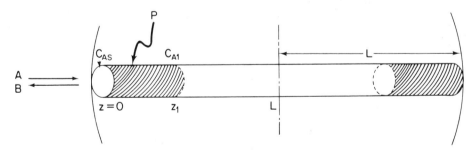

Figure P10–16

a. Derive an expression for the concentration profile and also for the molar flux of A in the *ineffective* region $0 < z < z_1$, in terms of z_1, D_{AB}, C_{A1}, and C_{AS}.

b. Before poisoning of the pore occurred, the effectiveness factor was given by

$$\eta = \frac{1}{\phi} \tanh(\phi)$$

where

$$\phi = L\sqrt{\frac{2k''}{rD_{AB}}}$$

with
k'' = reaction rate constant (length/time)
r = pore radius (length)
D_{AB} = effective molecular diffusivity (area/time)

Without solving any further differential equations, obtain the new effectiveness factor η' for the poisoned pore. [Exam, U. of M.]

P10-17. Ammonia is being absorbed in a wetted wall column similar to the one shown in Figure P10-17.

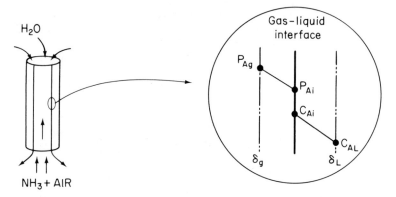

Figure P10–17

Ammonia is present in dilute concentrations.

a. Show that at any point in the column,

$$W_{Az} = k_g(P_{Ag} - P_{Ai}) = k_L(C_{Ai} - C_{AL})$$

where

$$k_g = \frac{D_{NH_3\text{-AIR}}}{\delta_g RT}$$

$$k_L = \frac{D_{NH_3\text{-}H_2O}}{\delta_L}$$

b. The equilibrium relationship at the gas-liquid interface is given by Henry's law:

$$P_{Ai} = HC_{Ai}$$

c. Show that the rate of absorption of ammonia at any point in the column can be written in the form

$$W_{Az} = K_{0G}(P_{Ag} - P^*_{AL}) \qquad (P8)$$

where

$$\frac{1}{K_{0G}} = \frac{1}{k_g} + \frac{H}{k_L}$$

$$P^*_{AL} = HC_{AL}$$

NOTE:

$$\begin{pmatrix}\text{The \% resistance}\\ \text{to mass transfer}\\ \text{in the gas phase}\end{pmatrix} = 100\,\frac{\text{gas phase resistance}}{\text{overall resistance}} = 100 \times \frac{\frac{1}{k_g}}{\frac{1}{K_{0G}}}$$

d. Derive the design equation from which the column height necessary to absorb 90% of the entering ammonia can be calculated.

P10-18. FALSIFIED KINETICS

The irreversible gas phase dimerization
$$2A \longrightarrow A_2$$
is carried out at 8.2 atm in a Carberry reactor to which only Pure A is fed. There were 40 grams of catalyst in each of the 4 spinning baskets.

The following runs were carried out at 227°C.

Total Molar Feed Rate F_{t0} (gmole/min)	Mole Fraction A in Exit y_A
1	.21
2	.33
4	.40
6	.57
11	.70
20	.81

The following run was carried out at 327°C.

9 (gmole/min)	.097

1. What is the apparent reaction order and the apparent activation of energy?
2. Determine the true reaction order, specific reaction rate, and activation energy.
3. Calculate the Thiele modulus and the effectiveness factor.
4. What diameter of pellets should be used to make the catalyst more effective?
5. Calculate the rate of reaction on a rotating disk made of the catalytic material when the gas phase reactant concentration is .01 gmole/liter and the temperature is 527°C. The disk is flat, non-porous, and 5 cm in diameter.

Additional information:

$$\text{Effective diffusivity} = 0.23 \text{ cm}^2/\text{sec}$$
$$\text{Surface area of porous catalyst} = 49 \text{ m}^2/\text{gm of cat}$$
$$\text{Density of catalyst pellets} = 2.3 \text{ gm/cc}$$
$$\text{Radius of catalyst pellets} = 1 \text{ cm}$$

SUPPLEMENTARY READING

1. The fundamentals of diffusional mass transfer can be found in Chapters 16 and 17 of

BIRD, R. B., W. E. STEWART, and E. N. LIGHTFOOT, *Transport Phenomena*. New York: John Wiley and Sons, 1960

or Chapters 1 and 2 of

GEANKOPLIS, C. J., *Mass Transport Phenomena*. New York: Holt Rinehart, and Winston, 1972

or Chapter 11 of

GREENKORN, R. A. and D. P. KESSLER, *Transfer Operations*. New York: McGraw-Hill Book Co., 1972

or Chapter 7 of

SISSOM, L. E. and D. R. PITTS, *Elements of Transport Phenomena*. New York: McGraw-Hill Book Co., 1972

or Chapter 2 of

LEVICH, V. G., *Physiochemical Hydrodynamics*. Englewood Cliffs, N.J.: Prentice-Hall, Inc., 1962.

2. Equations for predicting gas diffusivities are given in Appendix D. Experimental values of the diffusivity can be found in a number of sources, two of which are

> *Perry's Chemical Engineering Handbook*, 4th ed. New York: McGraw-Hill Book Co., 1963

and

> SHERWOOD, T. K. and R. L. Pigford, *Absorption and Extraction*. New York: McGraw-Hill Book Co., 1952.

3. A number of correlations for the mass transfer coefficient can be found in

> TREYBAL, R. E., *Mass Transfer Operations*, 2nd ed. New York: McGraw-Hill Book Co., 1968

and pp. 625ff. of

> MCCABE, W. L. and J. C. SMITH, *Unit Operations in Chemical Engineering*, 2nd ed. New York: McGraw-Hill Book Co., 1967.

Also Chapter 6 of

> GEANKOPLIS, C. J. *Op. Cit.*

and Chapter 30 of

> WELTY, J. R., C. E. WICKS and R. E. WILSON, *Fundamentals of Momentum, Heat, and Mass Transfer*. New York: John Wiley and Sons, 1969.

4. There are a number of books which discuss internal diffusion in catalyst pellets; however, one of the first books that should be consulted on this and other topics on heterogeneous catalysis is

> PETERSEN, E. E., *Chemical Reaction Analysis*. Englewood Cliffs, N.J.: Prentice-Hall, Inc., 1965.

In addition, see

> Satterfield, C. N. and T. K. SHERWOOD, *The Role of Diffusion in Catalysis*. Reading, Mass.: Addison-Wesley Publishing Co., 1963

and Chapter 6 of

> ARIS, R., *Elementary Chemical Reactor Analysis*, Englewood Cliffs, New Jersey: Prentice Hall, Inc., 1969. One should find references listed at the end of this chapter particularly useful.

The effects of mass transfer on reactor performance are also discussed in Chapter 7 of

> DENBIGH, K. and J. C. R. TURNER, *Chemical Reactor Theory*, 2nd ed. London: Cambridge University Press, 1971.

An extensive discussion of industrial catalytic reactors is presented in

> SHANKLAND, R. V. in *Advances in Catalysis*, Vol. VI. New York: Academic Press, 1954, p. 271.

5. Diffusion with homogeneous reaction is discussed in

> ASTRARITA, G., *Mass Transfer with Chemical Reaction*. New York: Elsevier Publishing Co., 1967

and

> DANCKWERTS, P. V., *Gas-Liquid Reactions*, New York: McGraw-Hill Book Co., 1970.

Gas-liquid reactor design is also discussed in

> SCHAFTLEIN R. W. and T. W. FRASER RUSSELL, *I.E.C. 60*, No. 5, 1968, 12

and

> MILLER, D. N., *I.E.C.* 56, No. 10, 1964, 18.

Appendices

A

GRAPHICAL AND NUMERICAL TECHNIQUES

A1.–Integration—The Graphical Form of Simpson's Rule

Given a set of values of a function $f_i(x_i)$ for a set of values x_i between x_1 and x_2, we wish to evaluate the integral

$$I = \int_{x_1}^{x_2} f(x)\, dx \tag{A1-1}$$

between points x_1 and x_2. Below we shall discuss how the graphical form of Simpson's rule may be used to determine the value of this integral.

1. Plot $f(x)$ vs. x and draw a smooth curve through all the points.

2. Connect the points at which the integration is to begin $[f(x_1), x_1]$ and end $[f(x_2), x_2]$ by a straight line AB.

3. Locate the midpoint $\bar{x}$ of the interval between x_1 and x_2,

$$\bar{x} = \frac{x_1 + x_2}{2} \tag{A1-2}$$

and sketch a vertical line CD from line AB to the curve.

4. Locate a point on the line CD that lies two thirds of the distance *from* line AB *to* the curve. This will be the mean value $\bar{f}$ of $f(x)$ over the interval (x_1, x_2). See Figure A-1.

Therefore, the value of the integral will be

$$I = \int_{x_1}^{x_2} f(x)\, dx = (x_2 - x_1)\bar{f} \tag{A1-3}$$

If the curve is convex upward, the rule is still two thirds of the distance *from* the straight line *to* the curve.

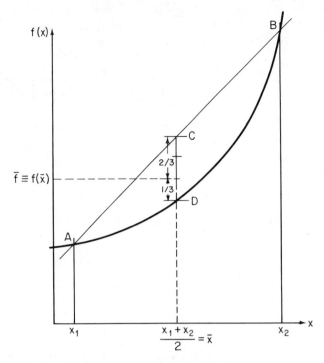

Figure A–1

Consider the following data,

x:	.0	.4	.8	1.2	1.6	2.0	2.5
$f(x)$:	100	149	223	332	495	740	1220

for which we want to evaluate the integral over the interval $x = 0$ and $x = 2.5$:

$$I = \int_0^{2.5} f(x)\,dx \tag{A1-4}$$

$f(x)$ is plotted as a function of x in Figure A-2. After connecting points (0, 100) and (2.5, 1220)

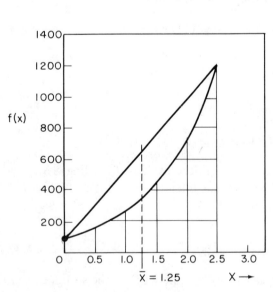

Figure A–2

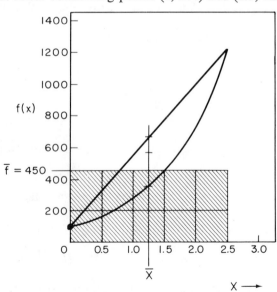

Figure A–3

by a straight line, we locate the mid point, $\bar{x}$, of the interval $0 \to 2.5$, which is 1.25. The mean value of the function, $\tilde{f}$, is found to be 450; therefore the value of the integral is

$$I = \int_0^{2.5} f(x)\,dx = \tilde{f}(2.5 - 0) \tag{A1-5}$$

The shaded areas in Figures A-3 and A-4 are equal.

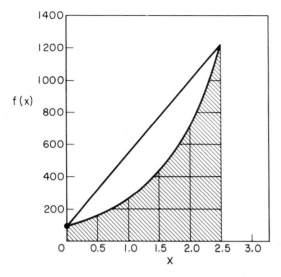

Figure A–4   Figure A–5

$$I = (450)(2.5) = 1125 \tag{A1-6}$$

The function used in this example was

$$f(x) = 100e^x \tag{A1-7}$$

$$I = \int_0^{2.5} 100e^x\,dx = 100[e^{2.5} - 1] \tag{A1-8}$$

$$I = 1120$$

The Simpson graphical technique is exact for curves which can be fit with a zero-, first-, second-, or third-order polynomial function. The proof of this exactness along with the percentage accuracy for higher-order polynomials has been developed by Curl.† When the curve does follow a higher-order polynomial, the integration over the entire region should be divided into smaller regions as shown in Figure A-5.

In each of these smaller regions, the curvature should follow a third-order or smaller polynomial, and the Simpson technique is used for each of the smaller regions. The total integral will be the sum of the integrals of each of the smaller regions.

A2.–Useful Integrals in Reactor Design

$$\int_0^x \frac{dx}{1-x} = \ln\left(\frac{1}{1-x}\right) \tag{A2-1}$$

$$\int_0^x \frac{dx}{(1-x)^2} = \frac{x}{1-x} \tag{A2-2}$$

†R. L. Curl (Unpublished dittoed notes, University of Michigan, Ann Arbor, 1967.)

$$\int_0^x \frac{dx}{1+\epsilon x} = \frac{1}{\epsilon} \ln(1+\epsilon x) \tag{A2-3}$$

$$\int_0^x \frac{1+\epsilon x}{1-x} dx = (1+\epsilon) \ln\left(\frac{1}{1-x}\right) - \epsilon x \tag{A2-4}$$

$$\int_0^x \frac{1+\epsilon x}{(1-x)^2} dx = \frac{(1+\epsilon)x}{1-x} - \epsilon \ln\left(\frac{1}{1-x}\right) \tag{A2-5}$$

$$\int_0^x \frac{(1+\epsilon x)^2}{(1-x)^2} dx = 2\epsilon(1+\epsilon)\ln(1-x) + \epsilon^2 x + \frac{(1+\epsilon)^2 x}{1-x} \tag{A2-6}$$

$$\int_0^x \frac{dx}{(1-x)(\theta_B - x)} = \frac{1}{\theta_B - 1} \ln\left(\frac{\theta_B - x}{\theta_B(1-x)}\right) \tag{A2-7}$$

$$\int_0^x \frac{dx}{ax^2 + bx + c} = \frac{-2}{2ax + b} + \frac{2}{b} \quad \text{for } b^2 = 4ac \tag{A2-8}$$

$$\int_0^x \frac{dx}{ax^2 + bx + c} = \frac{1}{a(p-q)} \ln\left(\frac{q}{p} \cdot \frac{x-p}{x-q}\right) \quad \text{for } b^2 > 4ac \tag{A2-9}$$

where p and q are the roots of the equation
$$ax^2 + bx + c = 0, \quad \text{i.e, } p, q = \frac{-b \mp \sqrt{b^2 - 4ac}}{2a}$$

$$\int_0^x \frac{a+bx}{c+gx} dx = \frac{bx}{g} + \frac{ag - bc}{g^2} \ln|c + gx| \tag{A2-10}$$

A3.–Equal Area Graphical Differentiation

There are many ways of differentiating numerical and graphical data. We shall confine our discussions to the technique of equal area differentiation. In the procedure delineated below we want to find the derivative of (y) w.r.t. (x).

1. Tabulate the (y_i, x_i) observations as shown in Table A2–1.

2. For each *interval*, calculate $\Delta x_n = x_n - x_{n-1}$ and $\Delta y_n = y_n - y_{n-1}$.

3. Calculate $\Delta y_n / \Delta x_n$ as an estimate of the *average* slope in an interval x_{n-1} to x_n.

4. Plot these values as a stepped curve vs. x_i. The value between x_2 and x_3, for example, is $(y_3 - y_2)/(x_3 - x_2)$. Refer to Figure A-6.

5. Next, draw in the *smooth curve* that best approximates the *area* under the stepped curve. That is, attempt in each interval to balance areas such as those labeled A and B, but when this is not possible, balance out over several intervals (as for the areas labeled C and D). From our definitions of Δx and Δy we know that

$$y_n - y_1 = \sum_{i=2}^{n} \frac{\Delta y}{\Delta x_i} \Delta x_i \tag{A3-1}$$

The equal area method attempts to estimate dy/dx so that

$$y_n - y_i = \int_{x_1}^{x_n} \frac{dy}{dx} dx \tag{A3-2}$$

that is, so that the area under $\Delta y/\Delta x$ is the same as that under dy/dx, everywhere possible.

6. Read estimates of dy/dx from this curve at the data points $x_1, x_2, \ldots$ and complete the table.

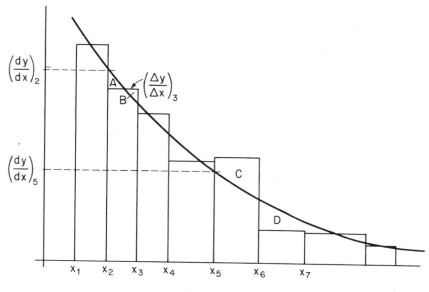

Figure A-6

TABLE A-1

x_i	y_i	Δx	Δy	$\dfrac{\Delta y}{\Delta x}$	$\dfrac{dy}{dx}$
x_1	y_1				$\left(\dfrac{dy}{dx}\right)_1$
		$x_2 - x_1$	$y_2 - y_1$	$\left(\dfrac{\Delta y}{\Delta x}\right)_2$	
x_2	y_2				$\left(\dfrac{dy}{dx}\right)_2$
		$x_3 - x_2$	$y_3 - y_2$	$\left(\dfrac{\Delta y}{\Delta x}\right)_3$	
x_3	y_3				$\left(\dfrac{dy}{dx}\right)_3$
		$x_4 - x_3$	$y_4 - y_3$	$\left(\dfrac{\Delta y}{\Delta x}\right)_4$	
x_4	y_4				$\left(\dfrac{dy}{dx}\right)_4$
		$x_5 - x_4$	$y_5 - y_4$	$\left(\dfrac{\Delta y}{\Delta x}\right)_5$	
x_5	y_5		Etc.		

To illustrate this technique, consider the following data from which we wish to determine df/dx as a function of x:

x:	0	.2	.4	.6	.8	1.0
$f(x)$:	0	182	330	451	551	631

First we calculate $\Delta f/\Delta x$ (Table A-2) and then plot it in the manner shown in Figure A-7.

TABLE A-2

x	$f(x)$	$\dfrac{\Delta f}{\Delta x}$
0	0	
		$\dfrac{182 - 0}{.2 - 0} = 910$
.2	182	
		$\dfrac{330 - 182}{.4 - .2} = 740$
.4	330	
		$\dfrac{451 - 330}{.6 - .4} = 605$
.6	451	
		$\dfrac{551 - 451}{.8 - .6} = 500$
.8	551	
		$\dfrac{631 - 551}{1.0 - .8} = 400$
1.0	631	

After drawing the smooth equal area curve, we can complete our table to find df/dx as a function of x (Table A–3).

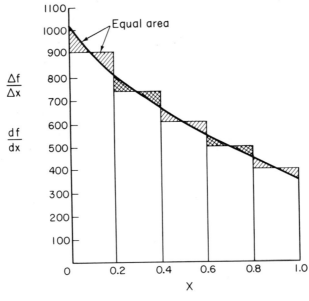

Figure A–7

The function used in this example was

$$f(x) = 1000(1 - e^{-x}) \tag{A3-3}$$

Differentiating Equation (A3-3) w.r.t. x,

$$\frac{df}{dx} = 1000 e^{-x}$$

The actual numerical values of the differential are given in column 5 of Table A–3.

TABLE A–3

x	$f(x)$	$\dfrac{\Delta f}{\Delta x}$	$\dfrac{df}{dx}$	Actual $\dfrac{df}{dx}$
0	0	—	1010	1000
.2	182	910	805	818
.4	332	740	670	670
.6	451	605	550	548
.8	551	500	450	449
1.0	631	400	360	368

Differentiation is, at best, less accurate than integration. This method also *clearly indicates bad data* and allows for compensation of such data. Differentiation is only valid, however, when the data is presumed to differentiate *smoothly*—for example, in rate-data analysis, interpretation of transient diffusion data, etc.

A4.–Solutions to Differential Equations

Methods of solving differential equations of the type

$$\frac{d^2 y}{dx^2} - \beta y = 0 \qquad (A4\text{-}1)$$

can be found in such texts as *Applied Differential Equations* by M. R. Spiegel, Englewood Cliffs, N. J.: Prentice-Hall, Inc., 1958 (see Chapter 4) or in *Differential Equations* by F. Ayres, Schaum Outline Series, New York: McGraw-Hill, Book Co., 1952. One method of solution is to determine the characteristic roots of

$$\left(\frac{d^2}{dx^2} - \beta\right) y = (m^2 - \beta) y \qquad (A4\text{-}2)$$

which are

$$m = \pm\sqrt{\beta}$$

The solution to the differential equation is

$$y = A_1 e^{-\sqrt{\beta}\,x} + B_1 e^{+\sqrt{\beta}\,x} \qquad (A4\text{-}3)$$

where A_1 and B_1 are arbitrary constants of integration. It can be verified that Equation (A4-3) can be arranged in the form

$$y = A \sinh \sqrt{\beta}\, x + B \cosh \sqrt{\beta}\, x \qquad (A4\text{-}4)$$

As an exercise you may want to verify that Equation (A4-4) is indeed a solution to Equation (A4-1).

B

IDEAL GAS CONSTANT AND CONVERSION FACTORS

$$R = \frac{.73 \text{ cu ft-atm}}{\text{lb-mole-°R}}$$

$$R = 10.73 \frac{\text{psi-cu ft}}{\text{lb-mole-°R}}$$

$$R = .082 \frac{l\text{-atm}}{\text{gmole-°K}}$$

$$R = \frac{1.98 \text{ cal}}{\text{gmole-°K}}$$

$$R = \frac{1.98 \text{ Btu}}{\text{lb-mole-°R}}$$

> 1 lb-mole of an ideal gas at 32°F and 1 atm occupies 359 cu ft
> 1 gmole of an ideal gas at 0°C and 1 atm occupies 22.4 l

Length

1 micron = 10^{-4} cm
1 inch = 2.54 cm
1 ft. = 30.48 cm
1 m = 100 cm
(1 ft. = 30.48 cm $\times \frac{1 \text{ inch}}{2.54 \text{ cm}}$ = 12 inches)

Volume

1 cu cm = .001 liters
1 cu inch = .0164 liters
1 fluid oz. = .0296 liters
1 U.S. gallon = 3.785 liters
1 cu ft. = 28.32 liters
1 cu m = 1000 liters
(1 cu ft = 28.32 liters $\times \frac{1 \text{ gallon}}{3.785 \text{ liters}}$ = 7.482 gal.)

Mass

1 lb. = 454 gram
1 kg = 1000 gram
1 grain = .0648 gram
1 oz. (Av) = 28.37 gram
1 ton = 908,000 gram

Pressure

1 torr (1 mm Hg) = 1.3158×10^{-3} atm
1 inch H_2O = 2.456×10^{-3} atm
1 in. Hg = .0334 atm
1 Psi = .06804 atm
1 megadyne/sq cm = .9869 atm, (1 megadyne/sq cm = 1 megabar)

Energy (work)

1 BTU = 252.16 calories
1 Joule = 0.239 calories, (1 Joule = 1 Kg·sq m/sq sec)
1 liter-atm = 24.22 calories
1 hp-hr = 641,617 calories
1 kw-hr = 860,421 calories

(1 liter atm. = 24.22 calories $\times \dfrac{1 \text{ Joule}}{.239 \text{ calories}}$ = 101.34 Joule

Temperature

°F = 1.8 × °C + 32
°R = °F + 459.69
°K = °C + 273.16
°R = 1.8 × °K

C

THERMODYNAMIC RELATIONSHIP INVOLVING THE EQUILIBRIUM CONSTANT[†]

For the gas phase reaction

$$A + \frac{b}{a}B \rightleftharpoons \frac{c}{a}C + \frac{d}{a}D \tag{2-2}$$

1. The pressure equilibrium constant K_P is

$$K_P = \frac{P_C^{c/a} P_D^{d/a}}{P_A P_B^{b/a}} \tag{AC-1}$$

P_i = partial pressure of species i in *atm*

2. The concentration equilibrium constant is

$$K_C = \frac{C_C^{c/a} C_D^{d/a}}{C_A C_B^{b/a}} \tag{AC-2}$$

3. For ideal gases, K_C and K_P are related by

$$K_P = K_C(RT)^\delta \tag{AC-3}$$

$$\delta = \frac{c}{a} + \frac{d}{a} - \frac{b}{a} - 1 \tag{AC-4}$$

4. K_P is a function of temperature only, and the temperature dependence of K_P is given by van't Hoff's equation

$$\frac{d \ln K_P}{dT} = \frac{\Delta H_{\text{rxn}}(T)}{RT^2} \tag{AC-5}$$

$$\frac{d \ln K_P}{dT} = \frac{\Delta H_{\text{rxn}}^\circ(T_R) + \Delta C_P(T - T_R)}{RT^2} \tag{AC-6}$$

5. If $\Delta C_P = 0$, we can integrate Equation (AC-6) to obtain

$$\ln \frac{K_{P2}}{K_{P1}} = \frac{\Delta H_{\text{rxn}}(T_R)}{R}\left[\frac{1}{T_1} - \frac{1}{T_2}\right] \tag{AC-7}$$

6. The equilibrium constant at temperature T can be calculated from the change in the Gibbs free energy using

$$-RT \ln [K_P(T)] = \Delta G^\circ(T) \tag{AC-8}$$

$$\Delta G^\circ = \left[\frac{c}{a}G_C^\circ + \frac{d}{a}G_D^\circ - \frac{b}{a}G_B^\circ - G_A^\circ\right] \tag{AC-9}$$

7. Tables that list the standard Gibbs free energy of formation of a given species G_i° are available in the literature.

[†]For the limitations and for further explanation of these relationships, see, for example, p. 138 of K. Denbigh, *Chemical Equilibrium*, Cambridge, England: Cambridge University Press, 1961.

D

PREDICTION OF BINARY GAS DIFFUSIVITIES

One of the most common equations used in predicting binary gas diffusivities, owing to its theoretical foundations, is the Hirschfelder-Bird-Spotz equation.[†]

A more recent empirical correlation has been developed by E. N. Fuller.[‡] Fuller used 308 experimental values of the diffusivities of various gases to determine the coefficients a, b, c, d, g, and f in the equation

$$D_{AB} = \frac{cT^b[(1/M_A) + (1/M_B)]^{1/2}}{P[(\Sigma V_A)^a + (\Sigma V_B)^g]^f} \quad \text{(AD-1)}$$

Using a nonlinear least-squares analysis, the empirical equation which gives the smallest standard deviation is

$$D_{AB} = \frac{10^{-3}T^{1.75}[(1/M_A) + (1/M_B)]^{1/2}}{P[(\Sigma V_A)^{1/3} + (\Sigma V_B)^{1/3}]^2} \quad \text{(AD-2)}$$

where P = total pressure (atm)
M_i = molecular weight
D_{AB} = diffusivity (sq cm/sec)
T = temperature (°K)
ΣV_i = sum of the diffusion volumes for component i, as given in Table D-1.

TABLE D-1
Special Atomic Diffusion Volumes[‡]

Atomic and Structural Diffusion Volume Increments			
C	16.5	(Cl)	19.5
H	1.98	(S)	17.0
O	5.48	Aromatic or heterocyclic rings	−20.2
(N)[b†]	5.69		
Diffusion Volumes of Simple Molecules			
H_2	7.07	CO_2	26.9
D_2	6.70	N_2O	35.9
He	2.88	NH_3	14.9
N_2	17.9	H_2O	12.7
O_2	16.6	(CCl_2F_2)	114.8
Air	20.1	(SF_6)	69.7
Ne	5.59	(Cl_2)	37.7
Ar	16.1	(Br_2)	67.2
Kr	22.8	(SO_2)	41.1
(Xe)	37.9		
CO	18.9		

[†]Parentheses indicate that the listed value is based on only a few data points.

[†]See R. B. Bird, W. E. Stewart, and E. N. Lightfoot, *Transport Phenomena*, (New York: John Wiley & Sons, 1960), p. 511.

[‡]E. N. Fuller, P. D. Schetter, and J. C. Giddings, *I.E.C. 58* (1966), 19. Several other equations for predicting diffusion coefficients can be found in R. C. Reid and T. K. Sherwood, *The Properties of Gases and Liquids*, 2nd ed., (New York: McGraw-Hill, Inc., 1966), Chap. 11.

Example. Calculate the diffusivity of CS_2 in air at 35°C:

$$D_{CS_2-\text{air}} = \frac{10^{-3}T^{1.75}\left(\frac{1}{M_A} + \frac{1}{M_{CS_2}}\right)^{1/2}}{P(\Sigma V_{CS_2}^{1/3} + \Sigma V_A^{1/3})^2} \tag{AD-3}$$

For air

1. $V_A = 20.1$.
2. Molecular weight $= 29.0$.

For CS_2, from Table D-1,

$$\begin{array}{ll} C = 16.5, & C = 16.5 \\ S = 17.0, & S_2 = 34.0 \\ & \overline{50.5} \end{array}$$

1. $V_{CS_2} = 50.5$.
2. Molecular weight $= 76$.

$$D_{CS_2-\text{air}} = \frac{10^{-3}(308)^{1.75}\left(\frac{1}{29} + \frac{1}{76}\right)^{1/2}}{1[(50.5)^{1/3} + (20.1)^{1/3}]^2}$$

$$= .12 \text{ sq cm/sec}$$

E

MEASUREMENT OF SLOPES

By plotting data directly on the appropriate log-log or semilog graph paper, a great deal of time may be saved over computing the logs of the data and then plotting them on linear graph paper. In this appendix we shall review the various techniques for plotting data and measuring slopes on log-log and semilog graphs.

We shall now consider the analysis and plotting of concentration time data from constant-volume batch systems for various reaction orders.

E1.–LINEAR PLOTS (ZERO ORDER REACTIONS)

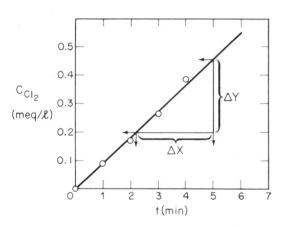

Figure 5-10

Before proceeding to log-log or semilog graphs, we shall review the measurement of slopes on linear plots of equations of the form

$$y = mx + b \qquad \text{(AE-1)}$$

In working through this example on linear plotting techniques, we shall refer to Figure 5-10. The method below is presented in the hope that it will facilitate learning the technique of measuring slopes on semilog paper. First, measure Δx and Δy with a ruler. For this particular example, $\Delta x = 1.093$ in. and $\Delta y = 1.031$ in. Next, determine the linear scale for each variable; [for C, .254 (meq/l)/in.; for t, 2.54 min/in.]:

$$\Delta C = 1.031 \text{ in.} \times .254 \text{ meq}/l/\text{in.} = .262 \text{ meq}/l$$

$$\Delta t = 1.093 \text{ in.} \times 2.54 \text{ min/in.} = 2.77 \text{ min}$$

$$\text{Slope} = k = \frac{\Delta C}{\Delta t} = \frac{.262}{2.77} = .0945 \text{ meq}/l/\text{min}$$

E2.–THE USE OF LOG-LOG PLOTS IN REACTION ORDER ANALYSIS

Log-log graph paper is normally used when the dependent variable, for instance y, is proportional to the independent variable x raised to some power m:

$$y = bx^m \qquad \text{(AE-2)}$$

In many engineering situations it is necessary to determine experimentally the power m to which the independent variable is raised by making measurements on x and y. One such ex-

477

ample is the following relationship between the reaction rate $-r_A$ and the concentration C_A given by

$$-r_A = kC_A^\alpha \qquad (AE\text{-}3)$$

where the specific reaction rate constant k and the reaction order α are to be determined.

$$\boxed{\text{Let } y = -r_A \qquad m = \alpha, \qquad x = C_A, \qquad b = k}$$

If we take the log of both sides of Equation (AE-2), we get

$$\log y = \log b + m \log x \qquad (AE\text{-}4)$$

Let

$$u = \log y$$
$$v = \log x$$
$$B = \log b$$

Then Equation (AE-4) becomes

$$u = B + mv \qquad (AE\text{-}5)$$

Observe that Equation (AE-2) has now been rearranged so that a plot of u vs. v will be linear with slope m.

Example. Determine α and k, assuming that Equation (AE-3) can be used to fit the following data (see Tables E–1 and E–2):

TABLE E–1

x (C_A, gmole/l)	1	2	3	4
y ($-r_A$, gmole/l/hr)	3	12	27	48

TABLE E–2

$\log x$	0	.303	.478	.606
$\log y$	.478	1.080	1.410	1.672

Nomenclature note: $\log \equiv \log_{10}$; $\ln \equiv \log_e$.

Plot the data given in either Table E–1 or E–2 on the graph in Frame (AE-1) to determine the parameters m and b.

(AE-1)

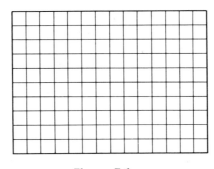

Figure E-1

(AE-1)

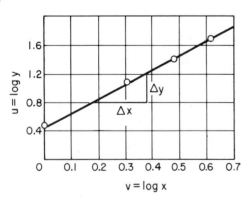

Figure E-2

Slope $= \dfrac{1.2 - .8}{.38 - .18} = 2 = m$

$v = 0, \quad u_0 = B = .478$

$B = \log b, \quad b = 10^{.478} = 3$

$y = 3x^2$

$-r_A = 3C_A^2$ gmole/l/hr

$k = 3$ l/gmole/hr

One may circumvent the calculations of the logs in Table E–1 by plotting the variables x and y directly on log-log paper. In the log-log diagram shown in Frame AE-2, plot the variables x and y, taking the appropriate values from either Table E–1 or E–2.

(AE-2)

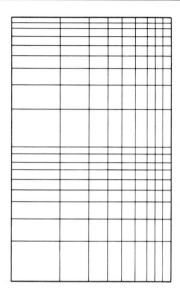

Figure E-3

To measure the slope directly (i.e., to measure Δy and Δx, and then calculate $\Delta y/\Delta x$) with a ruler or similar instrument, be sure that the length of each cycle is the same on each axis. Determine the constants n and b from the previous plot by measuring the slope directly from the data you plotted above.

(AE-3)

E3.–USE OF SEMILOG PLOTS IN RATE DATA ANALYSIS

Semilog graph paper is used when one is dealing with either exponential growth or decay, such as

$$y = be^{mx} \tag{AE-6}$$

For the first-order elementary reaction

$$A \longrightarrow \text{products}$$

which is carried out at constant volume, the rate of disappearance of A is given by

$$\frac{-dC_A}{dt} = kC_A \tag{AE-7}$$

when $t = 0$; $C_A = C_{A0}$, where the units of C_A are gmoles/l; and t is expressed in minutes. What are the units of k?

(AE-2)

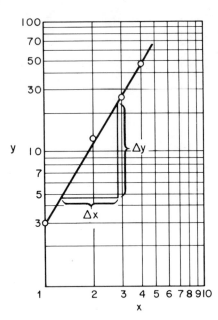

Figure E-4

The length of one cycle on the x axis is 4.35 cm, and the length of one cycle on the y axis is 3.6 cm; consequently you cannot take the ratio of the measured lengths (i.e., $\Delta y/\Delta x$) directly to calculate the slope. Choosing the values from Table E-1 we obtain the plot shown above.

(AE-3)

$$\left.\begin{array}{l}\Delta y = \dfrac{2.55 \text{ cm}}{3.6 \text{ cm/cycle}} \\[6pt] \Delta x = \dfrac{1.55 \text{ cm}}{4.35 \text{ cm/cycle}}\end{array}\right\} \quad \dfrac{\Delta y}{\Delta x} = \dfrac{2.55}{1.55} \times \dfrac{4.35}{3.60} = 1.99 \cong 2.0$$

$y = bx^{2.0}$ when $x = 1$, $y = 3$

$3 = b(1)^2, \quad b = 3$

$y = 3(x)^2$

$-r_A = 3C_A^2 \text{ gmole}/l/\text{hr}$

482

(AE-4)

(AE-5)

Integrating the rate equation, we obtain

$$\ln\left(\frac{C_A}{C_{A0}}\right) = -kt \qquad \text{(AE-8)}$$

We wish to determine the specific reaction rate constant, k; consequently a plot of the $\ln C_A$ vs. t should produce a straight line whose slope is $-k$. We may eliminate the calculation of the logs of each concentration data point by plotting our data directly on semilog graph paper.

Plot the points in Table E-3 on the semilog graph shown below.

TABLE E-3

t, min	0	2	4	8	14
C, gmole/l	2.0	1.64	1.38	.95	.60

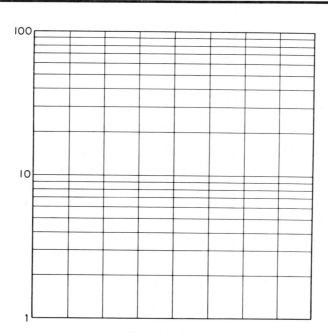

Figure E-5

483

(AE-4)

$$k = -\frac{1}{C_A}\frac{dC_A}{dt}$$

$$[k] = \frac{[C_A]}{[C_A][t]} \quad \text{where [ ] means dimensions of}$$

$$k = \frac{1}{[t]} = \text{time}^{-1} \quad (\text{e.g., min}^{-1})$$

(AE-5)

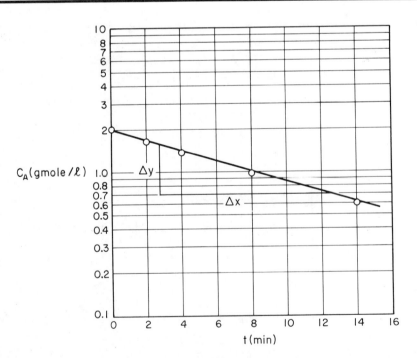

Figure E-6

Algebraic Method. Draw the best straight line through your data points. Choose two points on this line, t_1 and t_2, and the corresponding concentrations at these times:

$$\ln \frac{C_{A1}}{C_{A0}} = -kt_1, \quad \ln \frac{C_{A2}}{C_{A0}} = -kt_2$$

$$\ln C_{A2} - \ln C_{A1} = -k(t_2 - t_1), \quad k = -\left(\frac{\ln C_{A2} - \ln C_{A1}}{t_2 - t_1}\right) \quad \text{(AE-9)}$$

From the previous semilog plot: When $t = 8$, C_A is 1.05; when $t = 12$, C_A is .750:

$$k = -\left(\frac{.75 - 1.05}{12 - 8}\right) = +.075 \quad \text{(AE-10)}$$

What are the units of k in Equations (AE-8)–(AE-10)?

(AE-6) _____

_____

Graphical Technique. In the previous example we had

$$\ln \frac{C_A}{C_{A0}} = -kt$$

Converting to the base 10 we get

$$\frac{\ln (C_A/C_{A0})}{2.3} = \log \frac{C_A}{C_{A0}} = \frac{-kt}{2.3} \quad \text{(AE-11)}$$

The slope of a plot of log C_A vs. time should be a straight line with slope $-k/2.3$. Referring to Figure (E-5), we draw a right triangle with the acute angles located at points $C_A = 1.6$, $t = 2.8$ and $C_A = .7$, $t = 12.8$. Next the distances Δx and Δy are measured with a ruler. These measured lengths in y and x are 1.35 and 4.65 cm, respectively:

$$\Delta y = -1.35 \text{ cm} \times \frac{1 \text{ cycle}}{3.9 \text{ cm}} = -.35 \text{ cycle}$$

$$\Delta x = 4.65 \text{ cm} \times \frac{14 \text{ min}}{6.7 \text{ cm}} = 9.7 \text{ min}$$

$$\text{Slope} = \frac{-.35}{9.7} = -.0361$$

$$k = -2.3 \text{ slope} = -2.3 (-.0361) \text{ min}^{-1}$$
$$= .083 \text{ min}^{-1}$$

A modification of the algebraic method is possible by drawing a line on semilog paper so that the dependent variable changes by a factor of 10. Equation (E-9) can be rewritten in the form

$$k = -\frac{\ln C_{A2}/C_{A1}}{t_2 - t_1} = \frac{\ln C_{A1}/C_{A2}}{t_2 - t_1}$$

$$= \frac{2.3 \log C_{A1}/C_{A2}}{t_2 - t_1} \quad \text{(AE-12)}$$

Choose the points (C_{A1}, t_1) and (C_{A2}, t_2) so that $C_{A2} = .1 C_{A1}$:

$$k = \frac{2.3 \log 10}{t_2 - t_1} = \frac{2.3}{t_2 - t_1} \quad \text{(AE-13)}$$

This modification will be referred to as the *decade method*.

(AE-6) The units of k in Equation (AE-10) are gmole/l/min, and these units are inconsistent with the units of k in the original equation, which are sec^{-1}. The correct method of determining k is given below:

$$k = -\frac{\ln C_{A2} - \ln C_{A1}}{t_2 - t_1} = \frac{\ln(C_{A1}/C_{A2})}{t_2 - t_1} \tag{AE-6-1}$$

When $t = 8$, $C_A = 1.05$; when $t = 12$, $C_A = .75$. *Equation (AE-10) is incorrect* because we took the differences in the concentrations rather than the differences in the natural logarithms of the concentration. The correct value of k is

$$k = \frac{\ln(1.05/.75)}{12 - 8} = \frac{.336}{4}$$

$$= .084 \text{ min}^{-1}$$

F

GUIDED DESIGNS

F1.–Solid Waste Disposal to Produce Crude Oil

The November 17, 1969 issue of *Chemical and Engineering News* (p. 43) discusses a process that could be a significant step in the direction of improving waste disposal. However, this process has been carried out only with laboratory-bench-type equipment. Based on the information given in this article, scale up this process so that it could be used to dispose of the garbage of the city of Ann Arbor (100,000 population). Although you should consider scale-up of as many pieces of equipment as possible, primary attention should be focused on reactor scale-up.

In your reactor design you may want to consider the following points, after reading the *News* article:

1. What is the chemical formula of the oil?

2. From the information given in the article, calculate the pounds of product (oil + water) per pounds of garbage. Then calculate the pounds of oil per pound of organic feed.

3. Write a balanced reaction between carbon monoxide and cellulose (cellulose monomer unit $C_6H_{10}O_5$). From this reaction determine the pounds of oil per pound of organic feed. How does this compare with your answer to part 2?

4. How much carbon monoxide would this require? What about regeneration?

5. In the laboratory experiment, the investigators let the reaction proceed for 20 min. What would be the effect of "holding time"?

6. What phases are present in the reaction of garbage and carbon monoxide? Would you expect surface area to be one of the variables?

7. List any inconsistencies or misprints you may find in the article. For example, in the second column, 1:4 should be 1.4.

8. Could a coat of oil form over the garbage particles during reaction to inhibit further reaction?

9. Additional considerations: particle size, mixing effects, catalysts, garbage purity, reactor size.

Another major section that should be included in your report is the type of experimental program you should set up. What additional information do you need to complete the scale-up process? *How do you propose to obtain this information?* Suggest specific experiments, if possible.

F2.–Automobile Catalytic Afterburners[†]

For the purpose of evaluating a proposed catalytic reactor for the removal of nitric oxide from automobile exhaust, you are asked to study the article by A. Stodehnia-Korrani and K. Nobe, *I.E.C. Process Design Development* 9 (1970), 455. See Figure F-2.

Later we shall attempt to size the reactor, but first let us look at the economics of ethylene consumption. Specifically, estimate the cost of ethylene per gallon of gasoline consumed, needed to completely remove all traces of nitric oxide from the exhaust of a typical American passenger vehicle. As a first step, read the article mentioned above and decide on the assumptions and data that may be required.

1. Derive an expression for the total weight of catalyst for a packed bed reactor in terms of the reaction rate defined by Equation (2) in the above article and expressed as a function of conversion; to do this you will need to proceed to part 2.

2. Express the reaction rate as given by Equation (3) in the article in terms of the fractional conversion X of NO, the initial mole fraction of NO, and the stoichiometric ratio of ethylene to NO in the feed. (Point out any inconsistencies in the nomenclature used in the paper.)

3. Size a catalytic reactor to reduce the concentration of nitrogen oxide from a typical value of 1500 ppm to 5 ppm of NO (approximate California emission standards proposed for 1980) in a 50. standard cu ft per min automobile exhaust. You may assume that all nitrogen oxides are present as NO and that the copper-silica catalyst used in the above-cited work is selective to NO reduction by ethylene in the presence of O_2. Prepare a plot of the total weight W of the required catalyst vs. the feed ratio θ, i.e., ethylene to NO, for two different reactor temperatures T, 100°C and 400°C. You may use data from the work cited. About four to five points of W vs. θ at each T should suffice, in the range $.1 \leq \theta \leq 10$ (moles of ethylene fed per mole of NO fed).

4. The following data can be used to determine the optimum size of a catalytic reactor for use in $(NO)_x$ automobile emission control.
   a. Ethylene cost: $.05 cents/lb.
   b. Catalyst cost: $800/cu ft, based on apparent catalyst density. Assume a catalyst lifetime of 1 yr or 12,000 miles, whichever occurs first.
   c. Ethylene tank, ethylene injection system, and catalytic converter housing: $200.00 installation cost. Assume a lifetime of 3 yr or 36,000 miles, whichever occurs first. These figures can be regarded as cost to the consumer.

Additional references:

1. G. C. Hass et al., "Influence of Vehicle Operating Variables on Exhaust Emissions," *J. Air Pollut. Control Assoc.* 17, No. 6, (1967), 384.

2. W. Limville et al., "Evaluation of Methods for Controlling Vehicular Exhaust," *J. Air Pollut. Control Assoc.* 10, No. 1 (1960), 21.

3. B. H. Eggelston and R. W. Hurn, *U.S. Bur. Mines Rep. of Inv.*, No. 7390, May 1970.

[†]Guided design suggested by M. L. Cadwell and J. D. Goddard.

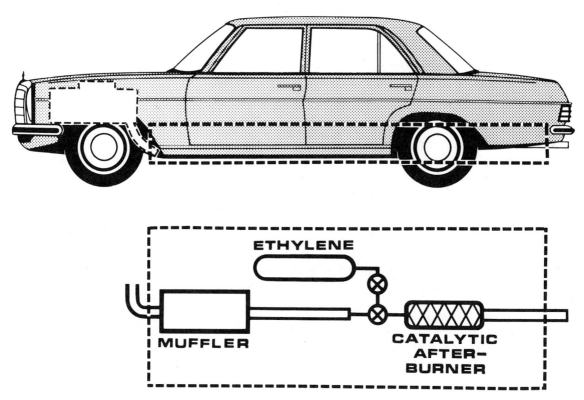

Figure F-2

F3.–Hollow Fiber Artificial Kidney[†]

The use of *enzymes as catalysts* in the engineering sense has received increased attention in recent years (c.f., the review of Carbonell and Kostin[‡]). In particular, as we have seen on p. 267. Levine and LaCourse[§] have executed the preliminary design of an artificial kidney using microencapsulated urease. As an alternative to the microencapsulated enzyme reactor, Rony[||] has reviewed some of the equations useful for the design of reactors employing enzymes immobilized within hollow fibers.

These reactors are similar in nature to the standard shell and tube heat exchanger except that the tubes are membranes. The membranes are permeable to the nitrogenous waste products in the bloodstream but not to the larger enzyme molecules. Thus, the circulation of blood through the shell side of the reactor leads to the selective removal of waste products by the enzyme solution inside the tubes.

Using the data provided by Levine and LaCourse and by Rony, calculate the size of the hollow fiber artificial kidney required to give the same conversion and useful lifetime as that indicated for the microcapsule reactor. Are there any advantages provided by the hollow fiber reactor? Finally, can the efficiency of the hollow fiber reactor be improved by placing the enzyme on the shell side and passing the blood through the tubes?

[†]Guided design suggested by D. J. Fink.
[‡]R. G. Carbonell and M. D. Kostin, "Enzyme Kinetics and Engineering," *A.I.Ch.E.J.* 18, No. 1 (1972), 1.
[§]S. N. Levine and W. C. LaCourse, "Materials and Design Consideration for a Compact Artificial Kidney," *J. Biomed. Mater. Res.* 1 (1967), 275.
[||]P. R. Rony, "Multiphase Catalysis II. Hollow Fiber Catalysts," *Biotech. Bioeng.* 13, No. 3 (1971), 431.

F.4 Experimental Study of Fermentation Kinetics[†]

Fermentation is usually defined as the *chemical reactions catalyzed by enzyme systems which, in turn, are produced during the growth of microorganisms*. The chemical reactions that occur during the fermentation process are many and complex but they may be summarized by an overall equation:

$$C_{12}H_{22}O_{11} + H_2O \longrightarrow 4C_2H_5OH + 4CO_2$$

As the concentration of alcohol increases, yeasts die because the sugar is being used up or because the alcohol concentration exceeds 12% (by volume), at which time the fermentation process will stop. (Usually 1–3 weeks are needed for the fermentation to reach completion.)

From one of the references given at the end of this section, choose a *recipe* for a juice or fruit you would like to ferment. Bring enough ingredients to prepare 1 gal of *must* (raw juice).

1. Mix the ingredients in a 1-gal container and determine the sugar concentration with a hydrometer. Tables are available that give the relation between specific gravity and sugar concentration (see p. 44 of Berry). If possible, also measure the absorbancy of the solution with a colorimeter.

2. Place an air lock and a 10-in. hypodermic syringe in a rubber stopper which is placed on the bottle. The syringe will be used to withdraw small samples that will be read in a colorimeter to follow the yeast concentration in an approximate manner. The rate of reaction can be followed by measuring the CO_2 evolution with a wet test meter or by water displacement. Knowing the initial sugar concentration and the total amount of CO_2 evolved up to time t, one can determine the sugar concentration at time t. Take special care to make your system airtight.

In a chemical reaction catalyzed by enzymes, when E, S, W, E·S, and P represent enzyme, substrate, water, enzyme-substrate complex, and products respectively, the rate equation reduces, under appropriate assumptions, to the form called the Michaelis-Menten equation (see Chapter 7):

$$-r_s = \frac{k'(S)(E_t)}{(S) + K_m} \tag{AF-1}$$

where $-r_s$ = rate of disappearance of S
(E_t) = total concentration of enzyme (yeast)
K_m = Michaelis constant

As a first approximation, assume that during the first stages of yeast cell growth the total yeast concentration is proportional to the absorbancy A measured in the colorimeter.

$$-r_s = \frac{(\alpha k')(S)A}{(S) + K_m} = \frac{k(S)A}{(S) + K_m} \tag{AF-2}$$

Rearranging,

$$\frac{A}{-r_s} = \frac{1}{k} + \frac{K_m}{k}\left(\frac{1}{S}\right)$$

After processing the data to obtain $-r_s$ as a function of S, determine whether this analysis is applicable to your experiment. What modifications of the analysis (if any) might be necessary to treat your data? (Hint: Rework your data assuming ethanol acts as a competitive inhibitor.) A schematic diagram of the experimental setup is shown in Figure F-4.

[†]Guided design suggested by Thomas Owens and Ming Kung Li.

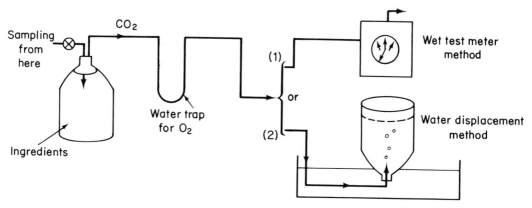

Figure F-4

REFERENCES:

1. C. J. J. Berry, *First Steps in Winemaking*, 4th ed., Standard Press, Andover, Harts C., England.

2. L. B. Church, "The Chemistry of Winemaking," *J. Chem. Ed. 49*, No. 3 (March 1972), 174.

3. N. Blakebrough, *Biochemical and Biological Engineering Science*, Vol. 1, Academic Press, New York, 1967 (to be referred to for basic knowledge of bacteria growth and fermentation kinetics).

4. J. B. Anderson, "Biological Reactions: Kinetics of Yeast Growth," *Chem. Engr. Ed. 6* (1972), 134.

Index

A

Absorption, 462
Activited complex (*see* Activate intermediates)
Activated molecule, 245
Activate intermediates, 245
Activation energy, 75, 367
   apparent, 442, 463
Active-site, 187
Adiabatic reaction:
   batch reactor, 349, 358
   CSTR example, 327
   SO_2 oxidation example, 335
   tubular (plug flow) example, 329
Adsorption:
   chemisorption, 185
   physical, 185

Adsorption—(*cont.*)
   rate limiting (controlling), 199
   rate of, 191
Air pollution, 33, 117
Alkylation-dealkylation catalyst, 227
Analysis of rate data, 133
Apparent activation energy, 442, 463
Arrhenius equation, 75, 153, 325, 327, 376
Artificial kidney, 267, 490
Automobile exhaust 33, 117

B

Backmix reactor (*see* Continuous stirred tank reactor)
Baker's yeast, 303
Basis of calculation, 35

Batch reactors:
 adiabatic operation, 349, 358
 advantages, 25
 design equation, 17, 47
 differential and integral form of design equation, 47
 example problems, 87, 109, 113, 121, 137, 141, 147, 149, 383
 non-isothermal operation, 349
 variable pressure (constant volume), 19, 109, 113, 121, 137, 273
 variable volume (constant pressure), 19, 179
Bench scale analysis:
 reactors (*see* Reactors, laboratory)
Bench scale experiments:
 chemical analysis, 173
 data analysis, 133ff
Bimolecular, 89
Blood, 178, 267, 490
Boundary layer, 443
By-products (*see* Simultaneous reactions)

C

California Professional Engineers Chemical:
 Engineering Examination (PECEE) xvi, 5, 129, 130, 180, 236, 352, 402, 461
Carberry reactor, 163, 165, 181
Cascade of CSTR(s), 59, 70
Catalyst:
 classification of, 227
 definition, 186
 poisoning, 225, 238, 461
 pore diffusion, 427, 431, 459
 rate limiting step (*see* Rate limiting step)
 regeneration, 225, 285
 single pellets, 427, 447
 steps in a catalytic reaction, 189
 surface area, 187, 429, 435
Catalyst pellet:
 constriction, 431
 modeling, 189, 427
 porosity, 431
 tortuosity, 431
Catalytic afterburner, 488
Catalytic reaction steps, 189
Catalytic surface:
 definition, 187
Chain reactions, 261, 263, 265, 299, 300
Chain transfer (*see* Propagation)
Chemical identity, 3
Chemical species definition, 2
Chemisorption (*see* Adsorption)
Closed system, 309, 349 (*see also* Batch reactors)
Comparison of plug flow and CSTRs:
 reactor volume, 53, 56
 selectivity, 365, 368, 371
Competitive inhibition of enzyme reactions, 227
Compressibility, 81, 83
Computer program for SO_2 oxidation, 345

Concentration:
 as a function of conversion, 80, 82, 85, 95, 98
 batch, 39, 95
 flow, 43, 95
 multiple reactions, 379, 380
 of total active site, 187
 of vacant sites, 187
Concentration profile down a plug flow reactor, 115
Concentration profile (internal), 437
Consecutive reactions (*see* Series reactions)
Constriction in catalyst pores, 431
Continuous flow reactors, 21
Continuous stirred tank reactor (CSTR):
 adiabatic operation, 327
 advantages, 27
 design equation, 21, 22, 48
 example problems, 111, 125, 327
 non-isothermal operation, 307ff
 perfectly mixed assumption, 11, 21
Controlling mechanism (*see* Rate limiting step)
Conversion:
 batch, 37
 flow, 41
 multiple reactions, 377
Conversion factors, 472
Crickets, 99
Cumene decomposition, 195
Cyclization, 187, 227

D

Deductive reasoning, xv
Dehydration, 231
Dehydrogenation, 229
Delta, δ (change in total number of moles), 82
Desorption, 197, 207
Determination of reaction orders, 75, 113
Dichloroethane decomposition example, 107
Differential mass balance, 15, 417, 420
Differential method of data analysis, 135, 468
Differential reactor, 157, 209
Diffusion:
 bulk, 415
 coefficient:
  order of magnitude, 425
  prediction of binary gas diffusivities, 475
  temperature and pressure dependence, 423
 definition, 411
 effective diffusivity, 431
 Fick's first law, 413
 fundamentals, 411ff
 Knudsen, 417, 423, 431
 pore (*see* Catalyst, pore diffusion)
 through a stagnant film, 419
 with reaction, 421
Diffusivities, 423, 475
Digital computer solution, 345
Dihydrofuran decomposition example, 113
Disguised kinetics (*see* Falsified kinetics)
Dual site mechanism, 195

493

E

Eadie plot, 283, 284, 285
Effectiveness factors:
    definition of, 427
    internal, 427, 437, 449, 452
    overall, 427, 447, 449, 452, 453
    Thiele modulus, 435
Elementary reaction, 77
Endothermic reaction, 77
Energy balance:
    derivation, 307
    steady state design equation, 325
    unsteady state design equation, 349, 350
Enthalpy:
    change in, 311, 312
    definition, 309
    of formation, 311
Enzymes:
    coenzyme, 289
    cofactor, 289
    definition, 265
    example problems, 271, 273, 291, 293, 296
    holoenzyme, 289
    inhibition:
        competitive, 277, 283
        non-competitive, 279, 283
        uncompetitive, 277, 283
    Michaelis constant, 269
    Michaelis Menten rate law, 269
    pH dependence of rate law, 275
    regeneration, 285
Epidemiology, 301
Epsilon, ϵ (volume change with reaction), 83
Equal Molar Counter Diffusion (EMCD), 415
Equilibrium:
    concentration, 120
    conversion, 120
Equilibrium constant:
    adsorbtion, 191, 198
    calculation of, 201, 474
    desorption, 199
    surface reaction, 200
    temperature dependence, 339, 351, 474
Excess reactant, 135
Exothermic reaction, 77
Experimental:
    design, 167
    interpretation of rate data, 167
    methods of analysis for chemical species, 123
    planning, 171

F

Falsified kinetics, 463
Fermentation, 490
Fick's first law, 413
Film theory, 443, 462
Fireflies, 99
First order reaction, 79, 107, 113
Fixed bed (see Packed bed reactor)
Flow reactors (see also Continuous stirred tank reactor, Fluidized bed, Packed bed, and Tubular reactor):

Flow reactors—(cont.)
    advantages, 27
    concentration profile in a plug flow reactor, 115
    CSTRs design equation, 48
    fixed bed design equation, 219
    fluidized bed design equation, 160, 219
    in series, 55ff, 70
    tubular (plug) flow design equation, 49
Fluidized bed:
    advantages, 31
    design equation, 106, 222
    example problem, 221
Flux:
    heat, 415
    molar, 411
Foam reactor, 32
Forced convection mass transfer, 417, 433
Fourier's law, 415
Free radical, 261, 263, 265
Frequency factor, 75

G

Garbage disposal, 487
Gas-solid catalytic reactors (see Fluidized and Packed bed reactors)
Gas-solid reactions (see Heterogeneous reactions)
Gas constant R, 472
Gas hourly space velocity (GHSV), 65, 69
Generalized reaction, 35, 375
General mole balance equation, 15, 17
Gradientless, 159, 165
Graphical differentiation, 468
Graphical integration, 465
Graphical techniques of solution, 53, 54, 145, 333, 465, 468, 479

H

Half life:
    definition, 165
    method of analysis, 165, 166, 181
Halogenation catalyst, 233
Heat capacity:
    delta C_p, $\Delta \tilde{C}_p$, 321
    mean, between T and T_o, $\tilde{C}_{p_i}$, 317
    mean, between T and T_R, $\hat{C}_{p_i}$, 321
    temperature dependence, 318
Heat of adsorption, 185
Heat of reaction:
    at reference temperature, 321
    definition, 77, 315, 316
    endothermic, 77
    exothermic, 77
    temperature dependence, 321
Heat transfer, 415, 445
Heat transfer coefficient, 314, 445
Henry's law, 290
Heterogeneous reactions, 183ff
Holding time (see Space time)

Hollow fiber artificial kidney, 489
Hydration catalyst, 231
Hydrogenation catalyst, 229
Hydrogen bromide reaction mechanism, 263

I

Inductive reasoning, xv
Industrial reactors, 25
Inhibitors:
    enzymes (*see* Enzymes, inhibition)
    heterogeneous, 199, 203
Initial rate, 155
Initiation, 255, 263, 265
Instrumental methods of chemical analysis, 173
Integral method analysis, 141
Integration:
    graphical, 465
    method of analysis, 141
    table of reactor design integrals, 468
Intermediates (*see also* Activate intermediates), 245, 253, 267, 289
Interpretation of rate data, 133ff, 167
Interstage cooling, 465
Irreversible reaction, 77
Isomerization, 10, 229

K

Kinetic energy, 309
Kinetic rate expression (*see* Rate law)

L

Langmuir isotherm, 194
Least squares (*see* Multiple regression)
Limiting reactant, 52
Line-Weaverburk plot, 272, 273, 283
Liquid hourly space velocity (LHSV), 65, 69
Liquid-solid reaction, 157
Los Angeles Basin, 33

M

Mass transfer (*see also* Diffusion):
    boundary layer, 443
    correlations, 445, 447, 448
    resistance:
        external, 433, 441
        gas phase, 427, 433
        internal (catalyst), 427, 441
        liquid phase, 462
Mass transfer coefficient, 443, 455
Material balance, 15
Maximum reaction velocity, 275
Measurement of slopes:
    linear paper, 147, 479
    log-log paper, 481
    semi-log paper, 481

Mechanism of reaction:
    heterogeneous, 189, 197, 213, 215
    homogeneous, 253, 261
Method of excess, 135
Methods of chemical analysis:
    electrochemical, 175
    interphase separation, 175
    nuclear, 175
    optical, 173
    physical, 173
Methods of plotting rate data, 141, 145, 217
Michaelis:
    pH function, 275
    constant, 269
Michaelis-Menten kinetics, 269
Modeling:
    catalytic afterburner, 488
    CSTR, 21
    diffusion in catalyst pellets, 427
    diffusion with reaction, 417
    heterogeneous reactions, 189, 195
    hollow fiber artificial kidney, 490
    homogeneous non-elementary reactions, 261
    multiple reactions, 383
    of the Los Angeles Basin, 33
    tubular reactor, 23
Mojave Desert, 33
Mole balances, 14, 45
Molecular diffusion, J_A, 411
Molecularity, 89
Monomolecular layer, 185, 194
Multiple operating states (*see* Multiplicity of the steady state)
Multiple reactions:
    conversion, 377
    dealkylation of mesitylene, 385
    heat effects, 405ff
    optimization of, 363ff, 391, 397
    parallel, 361
    reactor design, 383, 393, 397
    selectivity, 183, 375, 396
    selectivity parameter, 365
    series, 361
    stoichiometric table, 375ff
    yield, 183, 375, 396
Multiple regression, 181
Multiplicity of the steady state, 72, 302, 355

N

Nonelementary reactions, 93, 243
Nonisothermal reactor operation, 307ff
n^{th} order reaction, 75

O

Optimization:
    inlet temperature, 353
    in reactor staging, 59
    of conversion in multiple reactions, 391, 397
    of selectivity parameter, 365
Order of reaction, 75

Overall effectiveness factor, 427, 447, 449, 452, 453
Oxidation catalyst, 231

P

Packed bed reactor:
   adiabatic operation, 335
   advantages, 31
   design equation, 219
   examples, 221, 235
   Houdry reactor, 225
   interstage cooling of, 354
   SO_2 oxidation example, 335
Packed catalyst bed, 189, 221, 451
Parallel reactions, 361
Partial pressure, 93, 105, 191
Phase change, 70, 97
   during reaction, 95
   enthalpy, 312
Plug flow (*see* Tubular reactor)
Poisoning of catalysts (*see* Catalyst, poisoning)
Pore diffusion (*see* Catalyst, pore diffusion)
Pores, 189, 429, 459, 461
Porosity, 431
Potential energy, 309
Preditor-prey relationships, 303
Pressure variations as a method of analysis, 103, 137
Propagation, 255, 263
Pseudo-rate constant, 135, 136
Pseudo-steady state approximation, 249

R

Rate:
   of absorption, 462
   of adsorption, 191
   of desorption, 207
   of diffusion (*see* Diffusion)
   of disappearance, 9, 13
   of formation, 9, 13
   of mass transfer (*see* Mass transfer)
   of reaction definition:
      heterogeneous, 9, 11, 77
      homogeneous, 9, 13, 77
   of surface reaction, 205
Rate constant (*see* Specific reaction rate)
Rate data, 133ff, 209
Rate law:
   analysis of, 133ff
Rate law determination:
   from experimental observation, 133
   from the literature, 75
   from theory, 183
Rate limiting step:
   adsorption, 199, 214
   desorption, 207
   diffusion:
      external, 443
      internal, 427
   surface reaction, 205, 216

Reaction:
   catalysts, 185, 227
   classification, 79
   elementary, 77
   endothermic, 77
   exothermic, 77
   expression (*see* Rate law)
   irreversible, 77
   reversible, 91
   schemes, 55, 371
Reaction front, 347
Reaction mechanism:
   enzyme, 267, 287, 289, 293
   free radical, 261
   heterogeneous, 189, 197
   table of steps in homogeneous reaction, 253, 261
Reaction order, 75
Reactor design (*see* various types of reactors)
Reactors:
   advantages and disadvantages of reactor types, 25, 27, 31
   batch (*see* Batch reactor)
   Carberry reactor, 163, 165, 181
   continuous stirred tank reactor (CSTR) (*see* Continuous stirred tank reactor)
   differential, 157, 209
   fixed bed (*see* Packed bed reactor)
   fluidized bed (*see* Fluidized bed)
   laboratory:
      Carberry, 163, 165, 181
      differential, 157, 209
   non-isothermal operation, 307ff
   packed bed (*see* Packed bed reactor)
   semibatch, 25, 131
   spinning basket (*see* Carberry reactor)
   staging, 55
   tubular flow (plug flow) (*see* Tubular reactor)
Reference state:
   Gibbs free energy, 339, 474
   heat of reaction, 323, 329
Regeneration of catalyst, 29, 225, 285
Relative rates of reaction for the abstract reaction, 46
Residence times, 21
Resistance to mass transfer, 409, 427, 443
Reversible reaction, 91

S

Santa Ana wind, 33
Saponification, 49
Schmidt number, 447
Searching for a mechanism:
   heterogeneous reaction, 209
   homogeneous reaction, 253, 261
Selection of reaction schemes, 55, 57, 59, 371
Selectivity:
   definition of, 183, 375
   optimization of, 391, 397
Selectivity parameter:
   definition, 365
   maximization of, 365, 367, 369
Semibatch reactors, 25, 131

Separation, 363
Series reactions, 361
Shaft work, 309
Sherwood number, 445, 447
Side product, 363
Simultaneous reactions, 361ff
Single particle:
    effectiveness factor (internal), 440
    external mass transfer resistance, 447
    internal diffusion, 427
Size comparison of flow reactors, 54
Slopes (*see* Measurement of slopes)
Smog, 117
Solid-gas catalysis (*see* Heterogeneous reactions)
Space time, 65, 69
Space velocity, 65, 69
Specific reaction rate, 73, 75, 325, 327
Specific volume, 309
Steps in a catalytic reaction, 189
Stirred tank (*see* Continuous stirred tank reactor)
Stoichiometric table:
    multiple reactions:
        batch, 379
        flow, 384, 388
    single reactions:
        batch, 39, 52, 80, 82
        flow, 44, 98
Successive reactions (*see* Series reactions)
Surface:
    accessibility, 427
    concentration, 187
Surface area of catalysts, 187, 429, 435
Surface reaction:
    dual site mechanism, 193
    rate limiting, 205, 216
    single site mechanism, 195
Surface reaction controls, 205
Surroundings, 309
System volume, 13

T

Table:
    of conversion factors, 472
    of diffusivity relationship, 423
    of integrals, 467
    of methods of analysis, 173
    of steps in modeling of multiple reactions, 383
    of steps in searching for a mechanism, 253, 261
    of stoichiometric relationships, 39, 44, 52, 379, 384, 388
    of types of boundary conditions, 423
Temperature dependence:
    of equilibrium constant, 339, 474
    of heat capacity, 318, 323
    of heat of reaction, 323
    of specific reaction rate:
        from collision theory, 75, 325, 327
        from transition state theory, 75
Temperature effects:
    in batch reactors, 349, 350
    in continuous stirred tank reactors, 327
    in packed bed reactor, 335

Temperature effects—(*cont.*)
    in plug flow reactors, 329
Temperature profile in SO_2 oxidation, 346
Termination step, 255, 263, 265
Thermodynamics, 307, 474
Thiele Modulus, 435, 440, 441
Tortuosity factor, 431, 432
Total pressure data analysis, 103, 107, 113
Transition state theory, 75
Tubular reactor (plug flow):
    adiabatic operation, 329
    advantages, 27
    design equation, 23, 49, 124
    example problems, 115, 121, 329
    non-isothermal operation, 307ff

U

Ultrasonically induced chemical reaction, 145
Undesired product, 363
Unimolecular reaction, 89
Unsteady state energy balance, 349

V

Vacant sites, 187
Variable pressure batch reactor (*see* Batch reactor)
Variables affecting rate of reaction, 73
Variable volume batch reactor (*see* Batch reactor)
Viscous dissipation, 309, 326
Volume, 39
Volume change due to reaction:
    derivation, 81
    multiple reactions, 379, 380
    single reactions, 83
Volumetric flow rate, 43

W

Weight of catalyst:
    design equations, 160, 219
    examples, 222, 224
Work, 309

X

Xylene (meta), 385

Y

Yeast:
    Baker's, 303
    wine, 490
Yield definition, 183, 375, 396

Z

Zero order reaction, 127, 147, 149